Condensed
Matter Physics

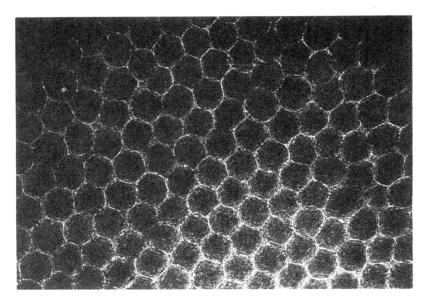

A dimple electron lattice on liquid ⁴He [courtesy of P. Leiderer]. This photograph shows the surface deformation about 6 s after an electric field was increased above a critical field. The temperature is 3.5 K.

Condensed Matter Physics

A. ISIHARA

Dover Publications, Inc.
Mineola, New York

Bibliographical Note

This Dover edition, first published in 2007, is an unabridged republication of the work originally published by Oxford University Press, New York and Oxford, in 1991.

Library of Congress Cataloging-in-Publication Data

Isihara, A. (Akira)
    Condensed matter physics / A. Isihara.
        p. cm.
    Originally published: New York : Oxford University Press, 1991.
    Includes bibliographical references and index.
    ISBN-13: 978-0-486-45877-9
    ISBN-10: 0-486-45877-6
        1. Condensed matter. I. Title.

QC173.454.I85 2007
530.4'1—dc22

                                                        2007003421

Manufactured in the United States of America
Dover Publications, Inc., 31 East 2nd Street, Mineola, N.Y. 11501

PREFACE

The term "condensed matter physics" has become more widely used in recent years than the traditional term "solid state physics." This change has taken place because quite often phenomena in apparently different condensed systems stem from the same physics, and many solid state phenomena can be studied most effectively in relation to phenomena in some other form of condensed matter. The universality of scaling laws at critical temperatures is a typical example.

In spite of this trend, there is virtually no book on condensed matter physics that can be used both as an introduction to research and also as a graduate text. Consequently, it is often difficult to get an overview on condensed matter physics from very many, often voluminous, specialized books. The rapid progress and new discoveries in this field enhance this difficulty. The present book has been written for this reason.

It is not an easy task to cover numerous systems and subjects in condensed matter in a handy volume. In fact, this author's main effort was in "condensing" widely spread systems and subjects in this tremendously rich and rapidly growing field.

A dozen subjects are presented in this book in separate chapters. Each chapter is designed to be directly accessible to the reader without going through the previous chapters. The emphasis is not on detailed theoretical derivations and techniques but rather a quick approach to a variety of systems and phenomena. The relatively small size of this book will hopefully enable the reader to get a general appreciation of the vast areas of condensed matter physics.

The author is grateful to the colleagues and organizations who gave permission to reproduce figures. The latter include the American Physical Society, the Royal Society of London, the Physical Society of Japan, the Japan Society of Applied Physics, Les Editions de Physique, IOP Publishing Ltd., the National Institute of Standards and Technology, International Business Machines Corporation, North-Holland Publishing Company, Plenum Publishing Company, Pergamon Press, Inc., and Springer-Verlag. He is especially indebted to Jeffrey Robbins, Harold Underwood, Atsusi and Hikaru for valuable suggestions and correcting of errors.

Buffalo, New York   A. I.
January, 1991

CONTENTS

Condensed
Matter Physics

1

SIMPLE LIQUIDS

One could well argue that it is more difficult to treat theoretically a liquid than a gas with a random distribution of molecules or a crystal with a regular structure. Nevertheless, its description is a good place to begin discussing condensed matter physics since the liquid state is no stranger to us and yet has strong molecular correlations. Theoretical techniques developed for a liquid are often applicable to other condensed matter systems.

1.1. Pair distribution function

The liquid state is similar to the solid state in having a free surface, low compressibility, and high density. In fact, its density is only about 5% lower than that of a solid, although there are exceptional cases such as water. However, its fluidity and molecular order are unique. Molecules in liquids can change position continuously, fill out a vessel with a formation of a meniscus, and do not have the long-range order of solids but only short-range order. Both liquids and gases are called *fluids* but these two are clearly distinguishable except near the critical point.

Since the molecular distribution of a liquid differs from that of a gas or a crystal, its study is important. This study can be performed effectively using two experimental methods: X-ray or electromagnetic wave diffraction and neutron scattering. The former method probes a static distribution that is quite different from those of gases and crystals. The latter provides information on the dynamic correlation of molecules, which is also distinctive. We shall discuss the former in this section and the latter in the next section.

X-ray scattering from a liquid depends not only on the molecular distribution but also on the molecular species. Therefore, we consider only a simple liquid (without any complex molecular structure), homogeneous and in equilibrium [1].

X-ray studies probe the *radial distribution function* $g(r)$, which is connected with the *pair distribution function* $\rho_2(r)$ such that

$$\rho_2(r) = n^2 g(r). \tag{1.1}$$

The pair distribution function $\rho_2(r)$ represents the probability of finding two molecules at a relative distance r. For homogeneous and isotropic systems, this probability depends only on the magnitude of r and approaches n^2 as $r \to \infty$, where n is the number density. Hence, the radial distribution function $g(r)$ has the asymptotic property:

$$g(r) \to 1, \qquad (r \to \infty).$$

Its deviation from this asymptotic value represents the molecular correlations. For a given distance r, $ng(r)4\pi r^2\, dr$ is the number of molecules in the spherical shell of width dr at r.

The pair distribution function satisfies the normalization:

$$V \int \rho_2(r)\, d\mathbf{r} = \langle N(N-1) \rangle, \qquad (1.2)$$

where V is the volume and N is the number of molecules in the liquid. The average $\langle \cdots \rangle$ is taken if this total number is fluctuating. In such cases, $\rho_2(r)$ is defined in a *grand ensemble* where $\langle \cdots \rangle$ represents a grand-ensemble average.

From Eqs. (1.1) and (1.2), we learn that the radial distribution function satisfies

$$\frac{1}{V} \int [g(r) - 1]\, d\mathbf{r} = \frac{\langle N^2 \rangle - \langle N \rangle^2 - \langle N \rangle}{\langle N \rangle^2}. \qquad (1.3)$$

For large $\langle N \rangle$, the right-hand side represents the square average fluctuations of the total number N about its average $\langle N \rangle$. One can show statistically [2] that

$$\frac{\langle N^2 \rangle - \langle N \rangle^2}{\langle N \rangle} = -\frac{\langle N \rangle kT}{V^2} \left(\frac{\partial V}{\partial p} \right)_T,$$

$$= I(0), \qquad (1.4)$$

where $I(0)$ is proportional to the intensity of scattered radiation at zero scattering angle and p is the pressure. Note that the quantity

$$\kappa_T = -\frac{1}{V} \left(\frac{\partial V}{\partial p} \right)_T$$

$$= (nkT)^{-1} I(0) \qquad (1.5)$$

is the *isothermal compressibility*. Light scattering will be discussed shortly.

Generally, $g(r)$ stays 0 near the origin at $r = 0$ due to the strong short-range repulsion of molecules, reaches a maximum at a point where nearest-neighbor molecules gather, and then shows weaker maxima and

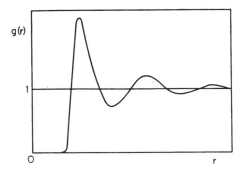

FIG. 1.1. Radial distribution function of a simple liquid.

gradually approaches its asymptotic value of 1. Its schematic curve is illustrated in Fig. 1.1.

The *structure factor* $S(q)$ is a quantity that is related to $g(r)$ through the Fourier transform:

$$S(q) = 1 + n \int [g(r) - 1] e^{i\mathbf{q} \cdot \mathbf{r}} \, d\mathbf{r}$$

$$= 1 + n \int [g(r) - 1] \frac{\sin qr}{qr} \, d\mathbf{r}. \tag{1.6}$$

As a function of q, the graph of $S(q)$ resembles that of $g(r)$ in that it is small for small q, approaches 1 for $q \to \infty$, and wiggles in between. However, $S(q)$ for small q corresponds to $g(r)$ at large r. In particular,

$$S(0) = nkT\kappa_T$$

$$= I(0). \tag{1.7}$$

The structure factor $S(q)$ can be expressed in a series in q^2 starting from the above value with the coefficients given by moments of $[g(r) - 1]$ if the sine function in the integrand of Eq. (1.6) is expanded. The pair distribution function or the structure factor can be determined theoretically on the basis of certain approximations or numerical calculations. Since the determination is one of the major theoretical tasks in the theory of liquids and statistical mechanics, many techniques have been developed, and its detailed discussion must be sought elsewhere, although some theoretical methods are outlined in Appendix for convenience's sake. In what follows, we present an approximate but very useful approach.

Ornstein and Zernike [3] assumed that the *total correlation function*

$$h(r) = g(r) - 1 \tag{1.8}$$

consists of a *direct correlation function* $C(r)$ and an *indirect correlation function* via intermediates molecules in such a way that

$$h(r) = C(r) + n \int C(r_{13})h(r_{32})\, dr_3. \tag{1.9}$$

Fourier transformation of this equation yields

$$S(q) = \frac{1}{1 - n\tilde{C}(q)}, \tag{1.10}$$

where $\tilde{C}(q)$ is the Fourier transform of the direct correlation function.

We assume that $\tilde{C}(q)$ can be expanded in a Taylor series. If $C(r)$ is spherically symmetric, its Fourier transform can be expressed to order q^2 such that

$$\tilde{C}(\mathbf{q}) = C_0\left[1 - \frac{q^2}{6}\langle r^2 \rangle\right], \tag{1.11}$$

where

$$C_0 = \int C(r)\, d\mathbf{r};$$
$$\langle r^2 \rangle C_0 = \frac{1}{6}\int C(r)r^2\, dr. \tag{1.12}$$

The notation $\langle \cdots \rangle$ here represents an average. Using Eq. (1.10) for $q = 0$ we arrive at

$$S(q) = \frac{S(0)}{1 + \xi^2 q^2}, \tag{1.13}$$

where

$$\xi^2 = \frac{nC_0}{6}S(0)\langle r^2 \rangle. \tag{1.14}$$

From Eq. (1.13) we find that the Ornstein and Zernike equation yields

$$h(r) = nkT\kappa_T \frac{e^{-r/\xi}}{r} - \delta(r). \tag{1.15}$$

Hence, the molecular correlation is short-ranged as long as ξ stays finite. This short-rangeness of molecular correlation has been found experimentally to be the case except near a critical temperature. Let us examine this point from a somewhat different angle by expanding the free energy density $F_1(\mathbf{r})$ about its equilibrium value $\langle F_1 \rangle$. We adopt the Landau expansion:

$$F_1(r) - \langle F_1 \rangle = \tfrac{1}{2}a[n(r) - n]^2 + \tfrac{1}{2}b(\nabla n)^2 + \cdots, \qquad (1.16)$$

where a and b are expansion coefficients. The constant a should vanish at the critical point because

$$a = \left(\frac{\partial^2 F_1}{\partial n^2}\right)_T = \frac{1}{n}\left(\frac{\partial p}{\partial n}\right)_T. \qquad (1.17)$$

The pair distribution function is related to the correlation function of local density fluctuations. The density–density correlation function is given by

$$\langle [n(\mathbf{r}) - n][n(\mathbf{r}') - n] \rangle = n\delta(\mathbf{r} - \mathbf{r}') + \rho_2(|\mathbf{r} - \mathbf{r}|) - n^2, \qquad (1.18)$$

where $n(\mathbf{r})$ represents the density at $\mathbf{r}$. One can express the density fluctuation $n(\mathbf{r}) - n$ in terms of its Fourier transform $n_\mathbf{q}$:

$$n(\mathbf{r}) - n = \frac{1}{(2\pi)^3}\int n_\mathbf{q} e^{i\mathbf{q}\cdot\mathbf{r}}\, d\mathbf{q}. \qquad (1.19)$$

The Fourier transform $n_\mathbf{q}$ must satisfy

$$n_\mathbf{q}^* = n_{-\mathbf{q}},$$

because $n(\mathbf{r}) - n$ is real. From Eqs. (1.18) and (1.19) we learn that

$$\langle |n_\mathbf{q}|^2 \rangle = nV + V\int e^{i\mathbf{q}\cdot\mathbf{r}}[\rho_2(r) - n^2]\, d\mathbf{r}. \qquad (1.20)$$

The free energy of the entire liquid with volume V is then expressed as

$$F - \langle F_1 \rangle V = \frac{1}{2V}\sum_q (a + bq^2)|n_\mathbf{q}|^2. \qquad (1.21)$$

Therefore, we adopt the following as the probability of having $|n_\mathbf{q}|^2$:

$$\frac{\exp[-(a + bq^2)|n_\mathbf{q}|^2/2VkT]}{\int \exp[-(a + bq^2)|n_\mathbf{q}|^2/2VkT]\, dn_\mathbf{q}},$$

to arrive at the average:

$$\langle |n_{\mathbf{q}}|^2 \rangle = \frac{VkT}{a + bq^2}. \tag{1.22}$$

Using this result in Eq. (1.20) we obtain for $b \neq 0$ an expression that is equivalent to Eq. (1.15):

$$\rho_2(r) - n^2 + n\delta(r) = \frac{kT}{4\pi br} \exp[-(a/b)^{1/2}r]. \tag{1.23}$$

In accordance with Eq. (1.17), $a = 0$ at the critical temperature so that the pair distribution function becomes long-ranged:

$$\rho_2(r) - n^2 + n\delta(r) = \frac{kT_c}{4\pi br}. \tag{1.24}$$

Near the critical temperature T_c the parameter a may be assumed to vary such that

$$a = a_0(T - T_c). \tag{1.25}$$

We find then

$$\kappa_T = \frac{1}{n^2 a_0(T - T_c)}. \tag{1.26}$$

That is, the isothermal compressibility diverges. As we shall discuss later, the exponent of this divergence in real liquids is slightly larger than 1. Correspondingly, the correlation function is frequently expressed as [4]

$$g(r) - 1 \sim \frac{e^{-r/\xi}}{r^{1+\eta}}, \tag{1.27}$$

where η is a small correction. This divergence of κ_T is due to the long-range molecular correlation that also causes $I(0)$ to diverge. Strong light scattering at T_c, i.e., *critical opalescence*, has long been observed.

The structure factor is determined by the intensity of scattered radiation in accordance with Eq. (1.13). That is, the Ornstein–Zernike theory predicts that $1/S(q)$ is expected to be proportional to q^2.

The structure factor can be determined by light or X-ray scattering. This depends on the interference of scattered waves from intra- and interatomic electrons in liquids. The former is given by the so-called *atomic structure factor*, which is denoted by f. The latter is determined by the pair distribution function. The combination of these two factors yields the following expression for the scattered intensity:

$$J(s) = NI_0[1 + I(s)], \tag{1.28}$$

where N is the total number of molecules, I_0 is the intensity of incident radiation and $I(s)$ is given by

$$I(s) = n \int [g(r) - 1]e^{is \cdot r} dr$$
$$= S(q) - 1. \tag{1.29}$$

Here

$$s = k - k_0 = q \tag{1.30}$$

represents the difference between the wave vectors k_0 and k of the incident and scattered radiation respectively. If no frequency change takes place upon scattering, the magnitude of s can be expressed as

$$s = \frac{4\pi \sin(\theta/2)}{\lambda}, \tag{1.31}$$

where λ is the wavelength of the radiation. In short, $J(s)$ is determined by $S(q)$ and

$$g(r) = 1 + \frac{1}{2\pi^2 nr} \int_0^\infty s[I(s)] \sin(rs) \, ds. \tag{1.32}$$

Hence, light scattering can directly determine the correlation function. Equations (1.27) and (1.29) show that the reciprocal scattered intensity is proportional to q^2 for large q as in the Ornstein–Zernike theory but deviations take place for small q. The solid line in Fig. 1.2 shows the q^2 proportionality. This proportionality has been observed [5]. The dotted lines extend the linear variation to $q = 0$ where the ordinate is a constant which is proportional to $kTn\kappa_T$. The dashed curves represent the deviation that depends on the exponent η. This is a small parameter close to 0.05.

Once the pair distribution function is determined, thermodynamic functions can be generated. For instance, the equation of state of a classical system is given exactly by [1, 2]

$$\frac{p}{nkT} = 1 - \frac{n}{6kT} \int \frac{d\phi(r)}{dr} g(r) \, dr, \tag{1.33}$$

where $\phi(r)$ is a molecular potential. This equation is called the *virial equation of state*. From Eqs. (1.5) and (1.6) we find another exact relation:

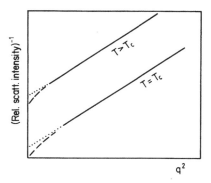

FIG. 1.2. Inverse relative scattered intensity near a critical point T_c. The dotted lines extrapolate the linear variation and the dashed lines represent actual behaviors.

$$n \int [g(r) - 1] \, d\mathbf{r} = kTn\kappa_T - 1. \tag{1.34}$$

Equation (1.29) shows that the left-hand side of the above equation is equal to the intensity $I(0)$ in the incident direction in agreement with Eq. (1.7). Thus, the q^2 plot in Fig. 1.2 should yield the constant on the right-hand side. At the same time, we learn that the higher the temperature the higher the value at $q = 0$ in Fig. 1.2.

It is often convenient to introduce a coupling constant λ to the interacting part H_1 of the Hamiltonian so that the strength of the interaction can be varied. The total Hamiltonian is then of the following form:

$$H = H_0 + \lambda H_1;$$
$$H_1 = \sum_{i<j} \phi(r_{ij}). \tag{1.35}$$

One can show that the equation of state is given by

$$p = p_0 - \frac{1}{2V} \int_0^1 d\lambda \int \phi(r) \rho_2(r, \lambda) \, d\mathbf{r}, \tag{1.36}$$

where the pair distribution function is now dependent on λ and p_0 is the pressure corresponding to H_0.

The radial distribution function of a liquid depends on the nature of the molecules, temperature, and density. Generally, its first peak is reduced when the temperature is increased. However, liquid helium shows the opposite temperature dependence below the λ point.

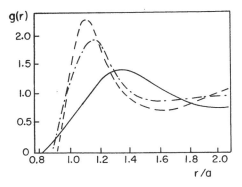

FIG. 1.3. Radial distribution function of three liquids. Solid curve, liquid ^{4}He at 1.94 K; dashed curve, Ne at 35.05 K; dashed-dotted curve, Ar at 163 K. (From Raveché and Mountain [6])

The quantity $4\pi r^2 ng(r)$ yields the number of molecules in the spherical shell of unit width at a distance r from the origin. Approximating the first peak by a Gaussian distribution, it has been found that the number of "nearest-neighbor atoms" in liquid argon is around 10.6 at 84.4 K, indicating that the liquid has a local structure that is somewhere between those of the fcc and bcc lattices.

There are quantum effects on the first peak of $g(r)$. Figure 1.3 compares the radial distribution functions of three liquids [6]. The solid, dashed, and dashed-dotted curves represent respectively liquid ^{4}He at 1.94 K, $n = 0.0245$ Å^{-3}; Ne at 35.05 K, $n = 0.0317$ Å^{-3}; and Ar at 163 K, $n = 0.08$ Å^{-3}. The abscissa is a reduced distance r/a, where $a = 2.556$ Å for He, 2.786 Å for Ne, and 3.405 Å for Ar. Note that the first peak of liquid helium is lower and appears at larger r in comparison with the cases of argon and neon.

1.2. Dynamic structure factor

The effectiveness of probing condensed matter by X-ray or neutron scattering depends on the energy and momentum that are transferred into the system. With energies of order 10 keV, X-ray and Mössbauer γ-ray experiments probe only spatial correlations. Neutrons can have energies in the range 1–100 meV and wavelengths in the range 1–10 Å. The wavelength λ in angstroms of a neutron and its energy ε in electronvolts are related to each other by

$$\lambda = \frac{0.286}{\varepsilon^{1/2}}.$$

On the other hand, the correlation time of a liquid is of the order of 10^{-13} s and the correlation length is of the order of 10^{-8} cm. Therefore, it is possible to investigate both spatial and temporal correlations because neutrons can spend time comparable to the correlation time over the correlation length in a liquid. Neutron scattering provides information concerning diffusion and vibration of molecules, but since the thermal energy kT is $0.86T \times 10^{-4}$ eV, the observation of energy gain of more than 0.1 eV is difficult. For such a case, energy-loss techniques have been developed.

The space-time correlation function introduced by van Hove [7] can be used effectively to describe neutron scattering. This function measures the correlation of the local density at time 0 with that at time t through the definition

$$G(\mathbf{r}, t) = \frac{1}{n} \langle n(0, 0) n(\mathbf{r}, t) \rangle, \tag{2.1}$$

where n is the average number density,

$$n(\mathbf{r}, t) = \sum_i \delta(\mathbf{r} - \mathbf{r}_i(t)) \tag{2.2}$$

is the local number density at time t, and $\langle \cdots \rangle$ is a statistical average.

The same function can be defined in a somewhat more dynamical way by following the motions of molecules. Let us denote the position of the ith molecule at time 0 by $\mathbf{r}_i(0)$. Let $\mathbf{r}$ be the point where the correlation is measured, and denote by $\mathbf{r}'$ the sum

$$\mathbf{r}_i(0) + \mathbf{r} = \mathbf{r}'.$$

A molecule j may be at this point at time t:

$$\mathbf{r}_j(t) = \mathbf{r}'.$$

By the time t, the ith molecule may have moved out of the position $r_i(0)$, and can even be at $\mathbf{r}'$. In this case $j = i$, and $G_s(\mathbf{r}, t)$ yields the probability of finding a particle at position $\mathbf{r}$ after time t when it was at the origin at time 0. Or, if $j \neq i$, another molecule j can be at $\mathbf{r}'$.

The corresponding $G(\mathbf{r}, t)$ can be introduced such that it represents the probability of finding any particle at $\mathbf{r}$ at time t when a particle is at origin at time 0. According to van Hove, the space-time correlation function is defined by [7]

$$G(\mathbf{r}, t) = \frac{1}{N} \left\langle \sum_{ij} \int d\mathbf{r}' \, \delta[\mathbf{r} + \mathbf{r}_i(0) - \mathbf{r}'] \, \delta[\mathbf{r}' - \mathbf{r}_j(t)] \right\rangle, \tag{2.3}$$

where

$$\mathbf{r}_j(t) = e^{iHt/\hbar} \mathbf{r}_j e^{-iHt/\hbar}$$

is the position operator of the jth molecule in the Heisenberg representation, .

H is the Hamiltonian of the system without the neutrons, and $\langle \cdots \rangle$ represents a statistical average. Note that the molecules i and j are not necessarily different. If they are the same, the notation $G_s(\mathbf{r}, t)$ is used to represent a self-correlation function. Hence,

$$G_s(\mathbf{r}, t) = \frac{1}{N} \sum_i \left\langle \int d\mathbf{r}' \, \delta[\mathbf{r} + \mathbf{r}_i(0) - \mathbf{r}'] \, \delta[\mathbf{r}' - \mathbf{r}_i(t)] \right\rangle. \tag{2.4}$$

Especially,

$$G_s(\mathbf{r}, 0) = \delta(\mathbf{r}). \tag{2.5}$$

As can be shown easily, the correlation function satisfies

$$G(\mathbf{r}, t) = G(-\mathbf{r}, -t). \tag{2.6}$$

For instant correlations in homogeneous liquids, the relation

$$G(\mathbf{r}, 0) = \delta(\mathbf{r}) + ng(\mathbf{r}) \tag{2.7}$$

holds, where $g(\mathbf{r})$ is the radial distribution function, which depends only on the distance r. This instantaneous correlation function can be probed by X-ray scattering.

If there is no thermal motion and if $n(\mathbf{r}, t) = n$, $G(\mathbf{r}, t)$ is reduced to n. This is the limit that is expected in the long-time limit:

$$G(\mathbf{r}, \infty) = n. \tag{2.8}$$

Note that in the same limit,

$$G_s(\mathbf{r}, \infty) = 0. \tag{2.9}$$

The Fourier transform of the deviation of $G(\mathbf{r}, t)$ from its limiting value n is called the *dynamic structure factor*. This is defined by

$$S(q, \omega) = \frac{1}{2\pi} \int_{-\infty}^{\infty} \int e^{i(\mathbf{q} \cdot \mathbf{r} - \omega t)} [G(\mathbf{r}, t) - n] \, d\mathbf{r} \, dt. \tag{2.10}$$

The Fourier transform $S_s(\mathbf{q}, \omega)$ of $G_s(\mathbf{r}, t)$ can also be defined. These two Fourier transforms are related to the differential cross-sections for coherent and incoherent scattering respectively such that

$$\frac{d^2\sigma^{\text{coh}}}{d\omega \, d\varepsilon} = \frac{\bar{b}^2 Nk}{\hbar k_0} S(q, \omega). \tag{2.11}$$

$$\frac{d^2\sigma^{\text{inc}}}{d\omega \, d\varepsilon} = \frac{(\overline{b^2} - \bar{b}^2)Nk}{\hbar k_0} S_s(q, \omega). \tag{2.12}$$

Here, b is the strength of the Fermi pseudopotential V_i of a given nucleus i:

$$V_i = \frac{2\pi\hbar^2}{m} b_i \delta(\mathbf{r} - \mathbf{r}_i(t)).$$ (2.13)

Through such a potential, the nucleus scatters an incident neutron beam with momentum $\hbar k_0$ into momentum $\hbar k$ such that $\mathbf{q} = \mathbf{k}_0 - \mathbf{k}$. $\bar{b}$ and $\overline{b^2}$ are the averages of b_i and b_i^2 over all the nuclei. The differential cross-section for coherent scattering by N molecules is then given by Eq. (2.11), and that for incoherent scattering cross-section is given by Eq. (2.12).

The dynamic structure factor is reduced to the static structure factor upon integration over ω:

$$\int S(q, \omega) \, d\omega = S(q).$$ (2.14)

In addition, it satisfies the sum rule:

$$\int \hbar\omega S(q, \omega) \, d\omega = \frac{(\hbar q)^2}{2m},$$ (2.15)

and is related to the energy-loss cross-section $S(-q, \omega)$ by

$$S(q, \omega) = e^{\hbar\omega/kT} S(-q, -\omega).$$ (2.16)

This relation can be understood by taking the ratio of the cross-section $\sigma(\mathbf{k}_0 \to \mathbf{k})$ and its inverse $\sigma(\mathbf{k} \to \mathbf{k}_0)$ in consideration of the detailed balancing condition. Note in this respect that

$$\hbar\omega = \frac{(\hbar k_0)^2}{2m} - \frac{(\hbar k)^2}{2m}.$$

Equation (2.15) represents the first moment of the dynamic structure factor. There are of course many other moments.

Equation (2.16) generates a relation:

$$\frac{S(q, \omega) - S(q, -\omega)}{2} = \tanh\left(\frac{\hbar\omega}{kT}\right) \frac{S(q, \omega) + S(q, -\omega)}{2}.$$ (2.17)

That is, the real and imaginary parts of the correlation function are related to each other through [8]

$$i \int_{-\infty}^{\infty} e^{-i\omega t} \operatorname{Im} G(\mathbf{r}, t) \, dt = \tanh\left(\frac{\hbar\omega}{2kT}\right) \int_{-\infty}^{\infty} e^{-i\omega t} \operatorname{Re} G(\mathbf{r}, t) \, dt.$$ (2.18)

Neutron or laser light scattering has been applied to classical and quantum liquids, polymers, magnetic, and many other condensed systems [9]. Some of the applications will be addressed in later chapters. We shall discuss in the remainder of this section the especially important role the dynamic structure factor plays for critical light scattering.

When condensed matter is considered as a continuum in the so-called hydrodynamical regime, the equation of continuity, the equation of motion, and the energy conservation equation relate the density, pressure, and temperature fluctuations with each other. The fluctuating hydrodynamical modes represented by these macroscopic quantities are nonlinearly coupled with each other. In particular, the Fourier transform $n_q(\omega)$ of the density fluctuations in time and space, which determines $S(q, \omega)$, is obtained by solving these equations by linearization, decoupling, and some other approximations. Its standard solution is of the form [10]:

$$n_q(\omega) = n_q(0) \frac{z^2 + (a + b)q^2 z + abq^4 + c^2(1 - \gamma^{-1})q^2}{z^3 + (a + b)q^2 z^2 + (c^2 q^2 + abq^4)z + ac^2 q^4 \gamma^{-1}}. \quad (2.19)$$

Here,

$$z = -i\omega; \qquad a = \lambda/(nc_v); \qquad b = (\tfrac{4}{3}\eta + \zeta)/(nm). \quad (2.20)$$

c is the sound velocity at zero frequency; $\gamma = c_p/c_v$ is the specific heat ratio; b is called the longitudinal kinetic viscosity, which depends on the shear and bulk viscosities η and ζ; n is the equilibrium density; m is the mass; and λ is the thermal conductivity.

The roots of the denominator of Eq. (2.19) can be used to identify the processes for the decay of density fluctuation. If the last term with q^4 is neglected, the three roots are given approximately by

$$\omega = -iD_T q^2; \qquad \pm cqi\Gamma q^2, \quad (2.21)$$

where

$$D_T = a/\gamma; \\ \Gamma = [a(1 - \gamma^{-1}) + b]/2. \quad (2.22)$$

The quantity D_T is called the *thermal diffusivity* and Γ is the attenuation coefficient of sound. Equation (2.19) can now be simplified as a sum of three terms:

$$n_q(\omega) = n_q(0)\left[\frac{1 - \gamma^{-1}}{z + D_T q^2} + \frac{1}{2\gamma}\left(\frac{1}{z + \Gamma q^2 - icq} + \frac{1}{z + \Gamma q^2 + icq}\right)\right]. \quad (2.23)$$

One can show that the first and second terms are related to the decay of entropy fluctuations and that of pressure respectively [10]. The Rayleigh line corresponds to the nonpropagating entropy fluctuations at constant pressure and the Brillouin lines to sound.

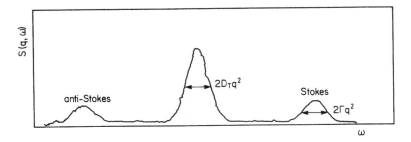

FIG. 1.4.   Scattered intensity in the hydrodynamical regime. Rayleigh line, center and Brillouin lines, both sides.

The inverse Fourier transform of Eq. (2.23) yields

$$n_q(t) = n_q(0)[(1 - \gamma^{-1}) \exp(-D_T q^2 t) + \gamma^{-1} \cos(cqt) \exp(-\Gamma q^2 t)]. \quad (2.24)$$

On the other hand, $S(q, \omega)$ is given by the Fourier transform:

$$S(q, \omega) = \frac{1}{2\pi N} \int_{-\infty}^{\infty} e^{i\omega t} \langle n_q(t) n_{-q} \rangle \, dt. \quad (2.25)$$

The above form of $n_q(t)$ yields

$$S(q, \omega) = \frac{1}{2\pi} S(q) \left((1 - \gamma^{-1}) \frac{2D_T q^2}{\omega^2 + D_T^2 q^4} + \frac{\gamma^{-1}\Gamma q^2}{(\omega + cq)^2 + \Gamma^2 q^4} \right.$$
$$\left. + \frac{\gamma^{-1}\Gamma q^2}{(\omega - cq)^2 + \Gamma^2 q^4} \right). \quad (2.26)$$

Accordingly, the spectrum consists of three Lorentzian lines, the *Rayleigh line* at $\omega = 0$ and two *Brillouin lines* at $\omega = \pm cq$, which are also called *Stokes* (+) and *anti-Stokes* (−) lines respectively. The spectrum is illustrated schematically in Fig. 1.4.

The Rayleigh line has strength $(1 - \gamma^{-1})$ and half-width

$$\Delta \omega_R = \frac{\lambda}{mnC_p} q^2 = D_T q^2. \quad (2.27)$$

This width is around 10^7 rad/s for laser light. When the two Brillouin lines are combined the strength is γ^{-1}. The half-width is given by

$$\Delta \omega_B = \Gamma q^2, \quad (2.28)$$

which is typically 10^9 rad/s. The Brillouin shift

$$\omega_B = cq \qquad (2.29)$$

is around 10^{11}–10^{12} rad/s.

The total intensities of the Rayleigh line and each of the Brillouin lines are given by

$$I_R = (1 - \gamma^{-1})S(q); \qquad (2.30)$$

$$I_B = S(q)/(2\gamma). \qquad (2.31)$$

Hence, the total scattered intensity is

$$I_R + 2I_B = S(q). \qquad (2.32)$$

The intensity ratio of the Rayleigh and Brillouin lines is called the *Landau–Placzek ratio*. This is given by

$$\frac{I_R}{2I_B} = 1 - \gamma^{-1}. \qquad (2.33)$$

Thus, dynamical scattering provides a wealth of information; the total scattered intensity yields the isothermal compressibility, the intensity ratio of Rayleigh and Brillouin lines, the specific heat ratio, $\Delta\omega_R$ the thermal diffusivity, ω_B the sound velocity, and $\Delta\omega_B$ the sound attenuation.

We have been concerned so far with the hydrodynamical regime. This regime is specified by the requirement that the correlation length ξ is small in comparison with q^{-1}:

$$q\xi \ll 1.$$

However, as the critical temperature is approached the correlation length diverges, causing deviations from this condition. In fact, the critical region is characterized by the opposite condition:

$$q\xi \gg 1.$$

The transport coefficients depend on wave number $\mathbf{q}$ and frequency ω in this regime due to the spatial and temporal correlations between the molecules, which are reflected into intense coupling between fluctuating quantities. Thus, deviations from the above Rayleigh linewidth $\Delta\omega_R$ take place as $q\xi$ increases from its hydrodynamical limit. These deviations have been studied theoretically and experimentally and important progress in the critical regime has been made in several stages [11–13] since the late 1960s. As a result, *dynamical scaling theory* has been developed.

A phenomenological approach to the critical regime can be made by using the following power-law variations of the relevant quantities in the above hydrodynamical results:

$$\xi \sim \varepsilon^{-\nu}; \quad C_p \sim \varepsilon^{-\gamma}; \quad \lambda \sim \varepsilon^{-\psi}; \quad D_T \sim \varepsilon^{\gamma-a}; \quad c \sim \varepsilon^{-\alpha/2}. \quad (2.34)$$

The approximate values of the exponents are $a = 0.58$, $\nu = 0.67$, $\psi = 0.5$ and $\gamma = 1.25$, where $\varepsilon = T/T_c - 1$.

The Rayleigh line is predominantly determined by the heat diffusion mode. According to the critical variations in Eq. (2.34), the Rayleigh linewidth approaches zero as

$$\Delta\omega_R \sim \varepsilon^{\gamma-\psi}. \quad (2.35)$$

The exponent $\gamma - \psi \sim \nu$ has been found to be around 0.6 for CO_2, Xe, and SF_6 [12] as theoretically expected. The corresponding decrease in the thermal diffusivity toward the critical point is called the *critical slowing down* of fluctuations.

The Brillouin shift is determined by the sound velocity c at zero frequency. Since this velocity approaches zero as $\varepsilon^{\alpha/2}$ the Brillouin shift is expected to approach zero as

$$\omega_B \sim \varepsilon^{\alpha/2}. \quad (2.36)$$

If Γ/c_v is dominant, the Brillouin linewidth diverges as

$$\Delta\omega_B \sim \varepsilon^{\alpha-\psi}. \quad (2.37)$$

Thus, the Brillouin lines approach the Rayleigh line and predominantly determine the linewidth. Note that from these variations critical exponents can be determined. Note that $D_T \sim \xi^{-(\gamma-a)/\nu}$ is ξ-dependent and that γ is not the specific heat ratio, although the customary notation has been used.

According to dynamical scaling, the half-width $\Delta\omega_R$ is a homogeneous function of q and inverse correlation length ξ^{-1}. If the order of homogeneity is s, then

$$\Delta\omega_R = \phi(q, \xi^{-1}) = q^s\phi\left(1, \frac{1}{q\xi}\right). \quad (2.38)$$

In particular, in the limit $q\xi \to \infty$,

$$\Delta\psi_R \sim q^s. \quad (2.39)$$

Experiment shows that the linewidth is proportional to q^3 so that the degree of homogeneity s can be 3. From Eqs. (2.27) and (2.34) we find that indeed

$$s = 2 + (\gamma - a)/\nu$$

$$\sim 3. \tag{2.40}$$

A more specific expression for $\Delta\omega_R$ is given by

$$\Delta\omega_R = \frac{kT}{6\pi\xi^3\eta} K(q\xi), \tag{2.41}$$

According to Kawasaki [11], the function is given in the first approximation by

$$K(x) = \tfrac{3}{4}[1 + x^2 + (x^3 - x^{-1})\tan^{-1} x]. \tag{2.42}$$

From the behaviors of $K(x)$ for small and large x, we find

$$\Delta\omega_R \rightarrow \begin{cases} \dfrac{kT}{6\pi\eta\xi} q^2[1 + \tfrac{3}{5}(q\xi)^2 + \cdots], & (q\xi \rightarrow 0); \tag{2.43a} \\[2ex] \dfrac{kT}{16\eta} q^3. & (q\xi \rightarrow \infty). \tag{2.43b} \end{cases}$$

We can express Eq. (2.43a) in the form of Eq. (2.27):

$$\Delta\omega_R = D_T^* q^2 \tag{2.44}$$

with an effective thermal diffusivity given by

$$D_T^* = \frac{kT}{6\pi\eta\xi^*}, \tag{2.45}$$

where

$$\xi^* = \xi[1 + \tfrac{3}{5}(q\xi)^2 + \cdots]^{-1}. \tag{2.46}$$

The quantity $6\pi\eta\xi^*$ in the denominator of D_T^* can be interpreted as an effective Stokes friction of a sphere of radius ξ^*. Note that the effective diffusivity D_T^* is q-dependent as a result of strong couplings between fluctuations of relevant quantities in the critical region. These couplings occur because the Fourier transforms of the macroscopic variables such as velocity, energy, and pressure in the equations of motion, continuity, and conservation are all dependent on ξ. These variables determine the propagation modes of acoustic waves, viscous flows, and heat diffusion. The couplings between these modes are enhanced through their ξ dependences as the critical point is approached.

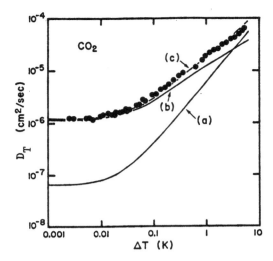

FIG. 1.5. Thermal diffusivity of CO_2 as a function of $\Delta T = T - T_c$. Curve a, theoretical background contribution; curve b, the contribution from the critical part of the linewidth based on Kawasaki's function $K(x)$ of Eq. (2.42); curve c, the sum of the critical and background contributions. The circles are the data. (From Swinney and Henry [12])

Figure 1.5 illustrates D_T as a function of $\Delta T = T - T_c$ for carbon dioxide on the critical isochore given by Swinney and Henry [12]. The theoretical curves represent: (a) the background contribution, (b) the contribution of the critical part of the linewidth; (c) the sum of the critical and background contributions. The black circles represent the measured values at scattered angle 90°. The large background contribution, which is not included in the above formula, causes ambiguity in analyzing the theoretical result.

We note in Eq. (2.44) that

$$\tau^{-1} = \frac{D_T^*}{\zeta^2} \tag{2.47}$$

is a characteristic frequency, its inverse being the relaxation time τ for thermal diffusion. Acoustic modes with frequently larger than τ^{-1} are expected to show dispersion since they are no longer effectively coupled with the heat-diffusion mode. Such a coupling is important for the Brillouin components, as can be seen from Eqs. (2.22). As the critical point is approached and the characteristic frequency drops to zero, the number of these modes increases and the sound velocity is no longer constant but decreases. Hence, the study of the sound velocity near T_c becomes an important subject. Kawasaki's result in Eq. (2.42) is based on the replacement of shear viscosity by a frequency- and q-independent constant. Such an approximation is not

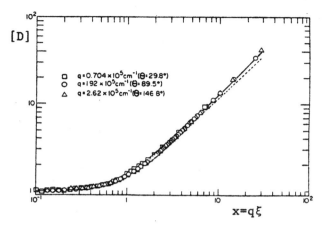

FIG. 1.6.   Reduced diffusion coefficient $[D]$ as a function of the scaling variable $x = q\xi$. Dashed and solid curves represent respectively Eq. (2.42) and $K^*(x)$ of Eq. (2.48).

valid for $q\xi \gg 1$. Hence, theoretical improvements have been attempted [13]. For instance, Burstyn et al. [13] have introduced a new function:

$$K^*(x) = K(x)[S(x)]^{z_\eta} \tag{2.48}$$

where $z_\eta \sim 0.06$ is a universal critical exponent and $S(x)$ behaves such that

$$S(x) \to \begin{cases} a_0 \\ a_\infty x \end{cases} \tag{2.49}$$

where a_0 and a_∞ are constants. They also proposed a reasonable procedure to eliminate the background contribution that is not included in Eqs. (2.41) or (2.48). Representing this background contribution by $\bar{D}$, they determined the dimensionless diffusion coefficient

$$[D] = \frac{6\pi\eta\xi}{kT} [D - \bar{D}] \tag{2.50}$$

as a function of $x = q\xi$, as shown in Fig. 1.6. The dashed curve represents Kawasaki's result $K(x)$, which shows deviations for large x. The solid curve is their approximate result, while the symbols are the data for three scattering angles as indicated. $[D]$ represents $K(x)$ or $K^*(x)$.

1.3.   Theory of condensation

The condensation of a gas into a liquid is one of the oldest known and most familiar phase transitions. In fact, the important discovery of the existence of a critical temperature was made by Andrews in 1869. Therefore, it is understandable that considerable effort has been made toward its theoretical description. This effort has undoubtedly contributed to the developments of statistical mechanics and many-body theory. However, no completely satisfactory theory of condensation has been constructed as yet. Nevertheless, this phenomenon involves some basic features of interacting particles and serves as a challenging classical subject for condensed matter physics.

The liquid state is realized not only by condensation but also by melting. These transitions into a liquid are generally *first-order* because the first-order thermodynamic derivatives of the Gibbs free energy are discontinuous. It follows the path of minimum values of the free energy, but a metastable phase with a higher free energy may still occur. A supercooled liquid, which we observe as in the form of icy rain, is a typical example. Its stability contrasts with that of a glass, which is also in a metastable state.

1.3.1.   *Yang–Lee theory*

Yang and Lee [14] developed in 1952 a general theory of condensation for a system of molecules with a hard core. This theory starts with the recognition that only a finite number of hard-sphere molecules can be put in a given volume V. Then, the grand partition function $\Xi(T, V, z)$ must be a finite polynomial:

$$\Xi(T, V, z) = \sum_{N=0}^{M} z^N Q_N, \tag{3.1}$$

where Q_N is the configurational partition function of N molecules:

$$Q_N = \frac{1}{N!} \int e^{-\beta\Phi} \, d\mathbf{r}^N. \tag{3.2}$$

and M is the maximum number of molecules that can be put in the volume.

The grand partition function can be rewritten such that

$$\Xi(T, V, z) = \prod_{k=1}^{M} \left(1 - \frac{z}{z_k}\right). \tag{3.3}$$

The roots z_k of the grand partition function must be complex because all the Q_N are positive. This means that no divergence is expected for

$$\frac{pV}{kT} = \ln \Xi,$$

when the fugacity z is changed in accordance with density variations.

However, this situation may change in the *thermodynamic limit* in which

$$M \to \infty, \qquad V \to \infty, \qquad M/V = n = \text{constant}.$$

In order to find what would happen in this important limit, we note that a term with z_k must be accompanied by its complex conjugate with z_k^* because Ξ and z are real. Writing $z_k = \exp(i\theta_k)$ and combining a term with its complex conjugate we obtain

$$\ln\left(1 - \frac{z}{z_k}\right) + \ln\left(1 - \frac{z}{z_k^*}\right) = 2 \ln \frac{[z^2 - 2zr_k \cos \theta_k + r_k^2]^{1/2}}{r_k}$$

$$= -V[\phi_k(z) - \phi_k(0)], \qquad (3.4)$$

where for convenience a new function

$$\phi_k(z) = -\frac{1}{V} \ln[z^2 - 2zr_k \cos \theta_k + r_k^2] \qquad (3.5)$$

has been introduced.

The equation of state is given by summing the terms of the form of Eq. (3.4). Since we have combined two terms, such a sum runs from $k = 1$ to $M/2$, but we can stretch the sum to M again by introducing a factor $1/2$. Hence,

$$\frac{p}{kT} = \frac{1}{V} \ln \Xi$$

$$= -\frac{1}{2} \sum_{k=1}^{M} [\phi_k(z) - \phi_k(0)]. \qquad (3.6)$$

In grand ensemble theory, the fugacity z is determined from the number density condition:

$$n = \frac{1}{V} \frac{\partial \ln \Xi}{\partial \ln z} = \frac{z}{2} \sum_k E_k(z). \qquad (3.7)$$

The functional relation between the functions $\phi_k(z)$ and $F_k(z)$ suggests that the former can be interpreted as the electrostatic potential and the latter as the electric field at point z. Figure 1.7a illustrates the electric field E at point Q, which is located on the real axis at a distance z from the origin. The field E is due to a line of charges on $P'P$ that is perpendicular to the z plane as illustrated in Fig. 1.7b. The electrostatic potential at Q is given by

$$\phi_k = -2\sigma \ln[z^2 - 2zr_k \cos \theta_k + r_k^2]^{1/2},$$

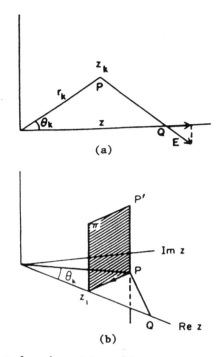

FIG. 1.7. Convergence of complex roots to a point z_1.

where σ is a linear charge density. Thus, an electrostatic analogue has been established for the relation between p and n.

As the volume and number of molecules of the system increase, the number of roots of the grand partition function will also increase. When the temperature or the density is changed, these roots will move about in the complex z plane.

Let us consider increasing z from the origin following density changes. At first, no zero is expected to appear on the real z axis, and the system is gaseous. In fact, for small z the grand partition function can be expanded in ascending powers of z, resulting in a gaseous equation of state.

As $N \to \infty$ and $V \to \infty$, some of the zeros may "condense" onto the real axis. If a complex root approaches a point z_1 on the real axis, its complex conjugate will also approach the same point. Many other roots may also converge upon the same point. Let z_1 be the smallest of such points. As z increases from the origin, z_1 represents the first fugacity at which a phase transition takes place. Let us imagine that a plane π of a continuous charge distribution is formed as in Fig. 1.7b as the line of charges PP' moves towards z_1. According to electrostatics, the potential, i.e., p/kT, is continuous across

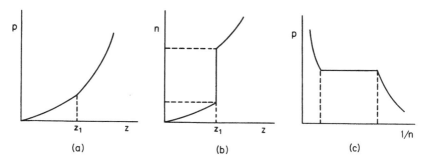

Fig. 1.8. Schematic variations of pressure and density as functions of z.

such a plane, but the electric field will be discontinuous by the amount

$$E(z_1 +) - E(z_1 -) = 4\pi\sigma.$$

That is, the density will jump, resulting in a first-order phase transition. The point z_1 may be interpreted to correspond to the condensation point. The critical temperature is then the point above which no root appears on the real axis.

The above consideration of the variations of p and n as functions of z is illustrated in Fig. 1.8. The pressure p in (a) is continuous at z_1 but the density jumps as in (b). The p–v curve is shown schematically in (c). Corresponding to the crystallization of liquids, there must be at least one more point on the real z axis where roots of the grand partition function converge.

The above interpretation of condensation is general. A difficult open question is to prove that such a convergence of zeros actually occurs. Moreover, it is desirable to find how many real roots appear on the real axis in the case of simple liquids. While these crucial points await future theoretical developments, at least the case of a two-dimensional lattice gas has been investigated on the basis of Onsager's exact solution of the Ising lattice. It has been shown that the roots are distributed on a unit circle at the origin. Hence, there is a real and positive z at which a phase transition takes place. No other roots appear on the real axis.

1.3.2. *van der Waals equation*

In 1873, a few years after the discovery of the critical temperature in CO_2, van der Waals showed that the gas and liquid phases can be represented by a single equation of state:

$$p = \frac{kT}{v - b} - \frac{a}{v^2}. \tag{3.8}$$

Note that the first term on the right-hand side diverges at $v = 1/n = b$ due to the hard core of molecules. This hard-sphere pressure is reduced by the second term, which comes from the attractive interaction. If the attractive potential is represented by the van der Waals potential

$$\phi_v(r) = -\frac{\phi_0}{r^6},$$

its influence at an average distance $\langle r \rangle = n^{-1/3}$ can be proportional to

$$\phi_v(\langle r \rangle) = -\phi_0 n^2 = -\frac{\phi_0}{v^2}. \tag{3.9}$$

The second term represents this type of the contribution from attractive forces.

The critical temperature is determined from the conditions:

$$\frac{dp}{dv} = 0; \quad \frac{d^2 p}{dv^2} = 0 \quad (T = T_c). \tag{3.10}$$

The critical point is specified by

$$v_c = 3b; \quad p_c = \frac{a}{27b^2}; \quad kT_c = \frac{8a}{27b}. \tag{3.11}$$

One of the important features of the van der Waals equation of state is that it satisfies the *law of corresponding states*. That is, it can be expressed in a universal form in terms of the reduced variables $p^* = p/p_c$, $v^* = v/v_c$ and $t^* = T/T_c$:

$$p^* = \frac{8T^*}{3v^* - 1} - \frac{3}{v^{*2}}. \tag{3.12}$$

Because of this universality, the equation of state played a historically important role in the practical realization of the liquefaction of gases. As a result van der Waals was awarded a Nobel prize in 1910.

Below T_c the van der Waals equation shows a wiggle. That is, it contains a thermodynamically unstable portion. For equilibrium this portion has to be removed from the isotherm in accordance with a *Maxwell construction*; by drawing a horizontal line such that the two areas between the wiggling van der Waals isotherm and the horizontal line are equal to each other.

Near the critical temperature, the van der Waals equation follows

$$\Delta n \sim (\Delta T)^\beta, \quad (\beta = \tfrac{1}{2}), \tag{3.13}$$

where Δn is the difference in the densities of the liquid and gas phases and $\Delta T = T_c - T$. Equation (3.13) represents a typical power relation that is followed also by some other classical theories, including the Weiss theory of ferromagnetism. In contrast, actual systems are characterized by

$$\beta = \tfrac{1}{3} \tag{3.14}$$

We shall discuss such critical power behaviors later.

Van der Waals introduced his equation of state phenomenologically. In 1963, Kac, Uhlenbeck, and Hemmer [15] derived a van der Waals equation exactly for a one-dimensional hard-sphere system with an exponential attractive force in the long-range and weak limit called the *van der Waals limit*. Their theory is mathematically involved. A simpler approach can be made [15] based on a linear potential model. A generalization of the KUH theory to arbitrary dimensions has also been made [16, 17].

Let us consider a three-dimensional system with a short-range repulsive potential and a long-range attractive potential:

$$\Phi(\mathbf{r}^N) = \Phi_R(\mathbf{r}^N) + \Phi_A(\mathbf{r}^N)$$
$$= \sum_{i<j} \{\phi_R(r_{ij}) + \phi_A(r_{ij})\}, \tag{3.15}$$

where the indices R and A correspond respectively to the repulsive and attractive parts. The attractive potential is assumed to have a form

$$\phi_A(r) = \gamma^\nu w(\gamma r), \tag{3.16}$$

where γ is a strength parameter and ν is the dimension of a given system. When γ is reduced, this potential decreases but the integral

$$\lim_{\gamma \to 0} \int \phi_A(r)\, dr^\nu = \alpha \tag{3.17}$$

is assumed to stay constant. The limit $\gamma \to 0$ under this condition represents the van der Waals limit. The constant α is simply the integral of $w(x)$ in the ν-dimensional space.

While the case of arbitrary ν can be treated, it is convenient to discuss the case of $\nu = 3$. We remark first that several potentials fall into the above category. For instance, the van der Waals attractive potential $\phi_v(r)$ can be written as

$$\phi_v(r) = -\mu r_0^{-3} \left(\frac{r_0}{r}\right)^6. \tag{3.18}$$

For this potential, we consider the limiting process:

$$\varepsilon \to 0; \qquad r_0 \to 0,$$

in which

$$\varepsilon r_0^3 = \mu$$

is constant. In this case, the constant α is given by

$$\alpha = -\frac{4\pi}{3}\mu. \tag{3.19}$$

Corresponding to the two parts of the total interaction potential, we write down the configurational partition function Q as follows [16]:

$$\begin{aligned} Q &= \frac{1}{N!}\int e^{-\beta\Phi}\, d\mathbf{r}^N \\ &= Q_R\langle e^{-\beta\Phi_A}\rangle, \end{aligned} \tag{3.20}$$

where

$$Q_R = \frac{1}{N!}\int e^{-\beta\Phi_R}\, d\mathbf{r}^N \tag{3.21}$$

is the configurational partition function of a system with only the repulsive potential ϕ_R. This system will be called the *reference system*, $\langle \cdots \rangle$ represents an average in this reference system:

$$\langle \cdots \rangle = \frac{\int(\cdots)e^{-\beta\Phi_R}\, d\mathbf{r}^N}{\int e^{-\beta\Phi_R}\, d\mathbf{r}^N}. \tag{3.22}$$

The logarithm of the partition function can be expressed as

$$\ln Q = \ln Q_R + \langle e^{-\beta\Phi_A} - 1\rangle_c, \tag{3.23}$$

where $\langle \cdots \rangle_c$ represents a cumulant average obtained by expanding the Boltzmann factor in powers of β and by replacing each term by its corresponding cumulant average. For instance, the first three cumulants are defined by

$$\begin{aligned} \langle\Phi_A\rangle_c &= \langle\Phi_A\rangle, \\ \langle\Phi_A^2\rangle_c &= \langle(\Phi_A - \langle\Phi_A\rangle)^2\rangle, \\ \langle\Phi_A^3\rangle_c &= \langle(\Phi_A - \langle\Phi_A\rangle)^3\rangle. \end{aligned} \tag{3.24}$$

Since the total potential is a sum of pair potentials, the statistical averages of its given power may break into a product of several subaverages.

For instance,

$$\langle \phi_{ij}\phi_{kl} \rangle = \langle \phi_{ij} \rangle \langle \phi_{kl} \rangle$$
$$\langle \phi_{ij}\phi_{jk} \rangle = \langle \phi_{ij} \rangle \langle \phi_{jk} \rangle.$$

As can easily be seen, the corresponding cumulant averages all vanish (*cumulant theorem*) [2]. In terms of graphs, the cumulant averages can be represented by irreducible-type graphs since all the reducible-types give no contribution.

Any cumulant average that emerges from the second term of Eq. (3.23) can be analyzed systematically in terms of the power of the attractive potentials and the number of particles. For the given form of the attractive potential, such an average can be expressed by a dimensionless variable $x = \gamma r$. One immediately realizes that unless the power matches the number of particles such that the average is completely expressed in terms of x, the average will vanish in the process $\gamma \to 0$.

The only remaining average is

$$\langle \Phi_A \rangle_c = {}_N C_2 \int e^{-\beta \Phi_R} \phi_A(r)\, d\mathbf{r}^N \bigg/ \int e^{-\beta \Phi_R}\, d\mathbf{r}$$

$$= -\tfrac{1}{2} V \int \phi_A(r) \rho_2^R(r)\, d\mathbf{r}, \tag{3.25}$$

where $\rho_2^R(r)$ is the pair distribution function of the reference system. The potential form given by Eq. (3.16) can be introduced into the right-hand side. For a single phase region where the pair distribution function is analytic, we find

$$\langle \Phi_A \rangle_c = -\tfrac{1}{2} V \lim_{\gamma \to 0} \int w(x) \rho_2^R(x/\gamma)\, d\mathbf{x}$$

$$= -\tfrac{1}{2} N n \int w(x)\, d\mathbf{x}; \qquad (\gamma \to 0), \tag{3.26}$$

where the integration should be performed outside of the hard-sphere diameter. Hence, if we define

$$\alpha = \int w(x)\, d\mathbf{x}, \tag{3.27}$$

the equation of state can be expressed as

$$p = p_R + \frac{n^2}{2}\alpha. \tag{3.28}$$

The pressure p_R of a reference system can be obtained explicitly and exactly for one dimension. It is called *Tonks' equation* and is given by

$$p = \frac{kT}{v - a}, \tag{3.29}$$

where v is the length per particle and where we have omitted the index R from p because there is no attractive force. A similar form may be adopted for two and three dimensions. The equation of state is then of the van der Waals form.

We might remark that if in general the Hamiltonian of a system is split into two parts:

$$H = H_0 + H_1, \tag{3.30}$$

the free energy can also be split into two parts:

$$F = F_0 + F_1, \tag{3.31}$$

where F_0 belongs to H_0, and F_1 represents the effects of H_1. One can then prove that F_1 has upper and lower bounds as follows [18]:

$$N\alpha_u \geqq F_1 \geqq N\alpha_l. \tag{3.32}$$

The upper and lower bounds are

$$\alpha_u = \frac{n}{2} \int \phi_a(r) g_0(r) \, d\mathbf{r},$$
$$\alpha_l = \frac{n}{2} \int \phi_a(r) g(r) \, d\mathbf{r}. \tag{3.33}$$

Here $g(r)$ is the radial distribution function of the system with the Hamiltonian H, and $g_0(r)$ is that of the system with H_0.

One can choose H_0 as the Hamiltonian of the reference hard-sphere system and H_1 as $\Phi_a(r)$ of Eq. (3.15). The van der Waals equation in the form of Eq. (3.28) can be derived in the van der Waals limit $\gamma \to 0$ under the assumption that $g(r)$ is analytic.

1.4. Scaling and renormalization

In a *second-order phase transition*, certain characteristic physical properties of a system disappear completely at a point called the *critical temperature*. The disappearance of the local order of a liquid at its critical temperature, of ferromagnetism at a Curie point, and of superfluidity in liquid ^{4}He at the λ-point are such examples. Near these critical temperatures, experiments have revealed universality expressed by the same power laws as hold among thermodynamically corresponding quantities in different phase transitions. Analyses of such power laws have caused considerable progress in the study of phase transitions. Moreover, a very important *renormalization group theory* has been developed.

1.4.1. Critical exponents

The isotherm in gas–liquid condensation and the magnetization curve of a magnet resemble each other if the magnetic field is plotted in the latter case as a function of the intensity of magnetization. If Δ denotes deviations from critical values, the two correspond to each other such that

$$\Delta M \rightleftarrows \Delta n;$$

$$\Delta H \rightleftarrows \Delta p.$$

Moreover, such corresponding quantities follow the same power laws. For example, near the Curie temperature of a ferromagnet, the spontaneous magnetization M vanishes following

$$M(T) = A(T_c - T)^\beta \tag{4.1}$$

where β is again $1/3$ as in Eq. (3.14). Experimental critical indices are listed in Table 1.1 for gas–liquid and magnetic systems.

Table 1.1. Critical power laws and indices

	$c_V \sim |\Delta T|^{-\alpha}$	$n_g - n_l \sim (\Delta T)^\beta$	$\kappa \sim (\Delta T)^{-\gamma}$	$\Delta p \sim |\Delta n|^\delta$
Ar		0.34 ± 0.02	1.20 ± 0.05	4.2 $(+0.6, -0.3)$
Xe	0.08 ± 0.02	0.35 ± 0.015	1.203 ± 0.02	4.4
^{4}He	0.127	0.3534 ± 0.0028	1.17 ± 0.0005	
^{3}He	$\alpha < 0.3$	0.361 ± 0.001	1.15 ± 0.03	3.5 ± 0.1
H_2		0.375 ± 0.015		
CO_2		0.34 ± 0.015	1.20 ± 0.02	4.2
		$M \sim |\Delta T|^\beta$	$\chi \sim |\Delta T|^{-\gamma}$	$H \sim M^\delta$
Fe		0.33 ± 0.015	1.333 ± 0.015	4.22 ± 0.1
Co			1.21 ± 0.04	
Ni		0.365 ± 0.02	1.35 ± 0.02	
CrO_2			1.63 ± 0.02	
MnF_2		0.335 ± 0.005		

The specific heat exponent is close to zero so that a logarithmic law can be used. Below T_c, the index α' is often used. Below the λ-point of ^{4}He, the specific heat exponent $\alpha' = 0.159$ in comparison with $\alpha = 0.127$ above this point. For an antiferromagnet such as MnF_2, the sublattice magnetization has the same exponent $\beta = 1/3$. In Table 1.1, χ is the susceptibility, which corresponds to the isothermal compressibility κ_T of fluids.

In general, thermodynamic quantities are derived from the free energy. Hence, the critical exponents may be related to each other. For instance, thermodynamically we expect

$$\delta = \frac{\gamma}{\beta} + 1. \tag{4.2}$$

In the case of the van der Waals equation of state, $\beta = 1/2$, $\gamma = 1$, and therefore $\delta = 3$. Thus, its critical indices deviate from the experimental values. Actually, they follow classical mean-field theory.

1.4.2. Scaled equation of state

In thermodynamics, the Gibbs free energy $G(T, p)$ is the appropriate thermodynamic function when pressure and temperature are chosen as independent variables. Being the Gibbs free energy per particle, the chemical potential μ can also be used. In particular, near a critical point, quantities such as

$$\Delta\mu = \mu - \mu_c; \qquad \Delta n = n - n_c$$

are relevant.

Along the isotherm and near T_c of a fluid, the functional form of $\Delta\mu$ can be guessed from the experimental power laws. Vincentini-Missoni et al. [19] proposed a universal scaled form

$$\Delta\mu = |\varepsilon|^{\beta+\gamma} = \frac{p_c}{n_c} g\left(\frac{\Delta n}{|\varepsilon|^\beta}\right), \tag{4.3}$$

where the function g is made dimensionless by the prefactor p_c/n_c. Widom [19] proposed a slightly different form:

$$\Delta\mu = \frac{p_c}{n_c}\left(\frac{\Delta n}{n_c}\right)\Psi(\varepsilon, |\Delta n/n_c|^{1/\beta}), \tag{4.4}$$

where Ψ is a homogeneous function of degree γ in its two variables and

$$\varepsilon = \frac{T}{T_c} - 1.$$

From the homogeneity with respect to τ, one finds

$$\Delta\mu = \frac{p_c}{n_c}\left(\frac{\Delta n}{n_c}\right)|\varepsilon|^{\gamma}\psi(\pm 1, |\Delta n/n_c|^{1/\beta}/|\varepsilon|). \tag{4.5}$$

This is effectively the same as Eq. (4.3) except now the two branches of Ψ, above $(+)$ and below $(-)$ T_c, are indicated. Moreover, homogeneity in the second variable leads to

$$\Delta\mu = \frac{p_c}{n_c}\frac{\Delta n}{n_c}\left|\frac{\Delta n}{n_c}\right|^{\gamma/\beta}\Psi(\varepsilon/|\Delta n/n_c|^{1/\beta}, 1). \tag{4.6}$$

According to Eq. (4.2), the ratio $\gamma/\beta = \delta - 1$. Hence, Eq. (4.6) can be rewritten in the form of Eq. (4.3).

These equations are designed to reproduce the correct critical power laws. For instance, since $\Delta\mu$ is proportional to Δp, we find

$$\Delta p \sim \Delta n^{\delta}. \tag{4.7}$$

δ is slightly larger than 4 as in Table 1.1. The variation of $|\Delta\mu|$ of argon as a function of $\Delta\rho = (n - n_c)/n_c$ is compared with that of Δp in Fig. 1.9(a) and (b). The solid and dashed curves correspond respectively to the high- and low-density sides of the isotherm respectively. Note that the dashed curves appear below the solid curves in (b), even though the solid curves look similar to those in (a). The curves in (c) show Δp as a function of the volume difference Δv. The isothermal compressibility diverges at T_c in accordance with

$$\kappa_T = \left[n^2\frac{\partial\Delta\mu}{\partial\Delta n}\right]^{-1} \sim \varepsilon^{-\gamma} \qquad (T > T_c; n = n_c). \tag{4.8}$$

For xenon, γ is around 1.2 as in Table 1.1.

Similar scaled expressions can be obtained near the Curie temperature of a magnetic system. These scaled equations are under certain symmetry assumptions. For instance, in Eq. (4.3), the difference in the exponent γ above and below T_c is not explicitly introduced. More generally, the scaling laws assume certain symmetries in the thermodynamic variables. In the magnetic case, the spontaneous magnetization is symmetric about the line $M = 0$, whereas the field is antisymmetric in M. These symmetry properties are present in the Ising model and in the corresponding lattice gas. Indeed, the lattice gas has a coexistence curve which is symmetric in the density. On the other hand, its chemical potential is antisymmetric with respect to the critical isochore. For real gases, it appears that the use of a set of variables $(\Delta\mu, \Delta\rho)$ is slightly better than the equivalent choice $(\Delta p, \Delta v)$. In Fig. 1.9 the overlap region of the former set in (a) is much larger than that for the latter in (c).

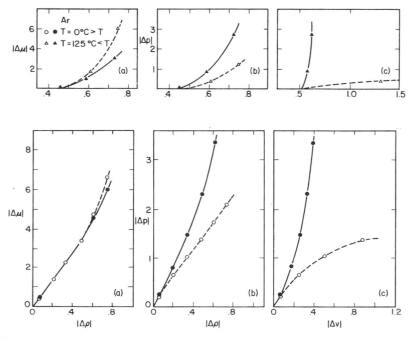

Fig. 1.9. The critical behaviors of argon: (a) $\Delta\mu$ against $|\Delta\rho|$; (b) $|\Delta p|$ against $|\Delta\rho|$; (c) $|\Delta p|$ against $|\Delta v|$, v being molar volume. The solid (dashed) curves and closed (open) symbols correspond to the high (low) density sides. (From Vincentini-Missoni et al. [19])

1.4.3. *Renormalization group theory*

The concept of renormalization of physical quantities has been known for many years. An effective charge or an effective mass of particles is a consequence of renormalization. Such a concept stems from the belief that a correct result should not depend on a particular order in interaction and is free from the divergences that inadequate perturbational series bring in.

A phase transition can be the result of increases in the role played by molecular interaction. In such a case, a perturbational approach may not be useful. Instead, theory could be based on universality underlying various critical phenomena independently of interactions.

In the 1960s, Kadanoff [20] introduced the concept of rescaling the Ising statistics in terms of block spins. Although somewhat incompatible with the block spin concept, nearest-neighboring interactions were assumed, and the scaling laws were derived by adjusting the degree of freedom in order to accommodate these block spins.

Subsequently in 1971 Wilson [21] introduced renormalization group theory. In this theory it is recognized that a partition function constructed for blocks of spins in an Ising lattice should be of the same form as that of the site spins, if the former is summed over all the internal states of the blocks. The process of forming block spins can be repeated to larger and larger blocks, generating a *renormalization group*.

The concept of renormalization is general, but is is most convenient to discuss the case of an Ising lattice. Therefore, let us consider an Ising spin lattice for which the partition function is expressed as

$$\exp[-Nf(K)] = \sum_{\{s_i\}} \exp(H\{s_i\}), \qquad (4.9)$$

where N is the total number of spins with spin variables $\{s_i\}$, $f(K)$ is the free energy per spin (divided by kT), which depends on the interaction parameter K, and H is the spin–spin interaction (divided by $-kT$). The spins $\{s_i\}$ are called *site spins*.

Equation (4.9) is exact if the interaction Hamiltonian involves no approximation. Generally, the interaction can be expressed as

$$H = \sum_{\alpha} K_{\alpha} s_{\alpha}, \qquad (4.10)$$

$$s_{\alpha} = \prod_{i \in \alpha} s_i, \qquad (4.11)$$

where α represents all the subsystems of the set of spins $\{s_i\}$. For the Ising problem, α represents a set consisting of all the nearest-neighboring spin pairs. Note that the free energy $f(K)$ is a function of the set $\{K_{\alpha}\}$ of interaction parameters.

The site spins in a block about a site spin form a *block spin*. If the total number of block spins is N', the ratio N/N' is determined by the ratio of the size l of each block to the original cell size. If d is the dimension of the lattice,

$$\frac{N}{N'} = l^d. \qquad (4.12)$$

Each block spin can be characterized by the largest and the smallest values of the total spins within the block. Since there are several spins in each block, it is appropriate to introduce "internal spin states" to be denoted by σ_I. The total degree of freedom of each block remains the same as that of the original site spins within the block. If each site spin takes on values ± 1, the total spin in a block takes on values $d, d - 1, \ldots, -d$, where d is the largest and $-d$ is the smallest. We can choose new units in which $d = 1$. We define

$$S'_I = \text{sign}\left(\sum_{i \in I} s_i\right). \qquad (4.13)$$

S'_i, representing a block spin, takes on the same values ± 1 as the original site spins. Corresponding to each value of S'_I, there are internal spin states that will be represented by σ_I. For instance, in the case of a square lattice, block spins may consist of 9 site spins. In this case, there are 2^8 internal spin states.

The partition function can be constructed based on block spins in the form:

$$\exp[-N'f'(K')] = \sum_{\{S'_I\}} \sum_{\{\sigma_I\}} \exp[H(\{S'_I\}, \{\sigma_I\}). \qquad (4.14)$$

This expression is exact when all the "internal" and "external" spin states, $\{S'_i\}$ and $\{\sigma_i\}$, are taken into consideration in evaluating the partition function. In Eq. (4.14), if a partial sum over $\{\sigma_I\}$ is performed, the partition function is given by

$$\exp[-N'f'(K')] = \sum_{\{S'_I\}} \exp[H'(\{S'_I\})], \qquad (4.15)$$

where the new Hamiltonian H' is of the form

$$H' = \sum_\alpha K'_\alpha S'_\alpha. \qquad (4.16)$$

Here,

$$S'_\alpha = \prod_{I \in \alpha} S'_I. \qquad (4.17)$$

Here, α covers all the blocks. The free energy f' is now a function of the set of interactions $\{K'_\alpha\}$.

Let us now assume that the functions $f(K)$ and $f'(K')$ have the same singular properties because the same critical behavior may be expected from both functions in the thermodynamic limit. That is, the two functions are assumed to differ from each other only in the regular parts. Hence, one can write

$$N'f'(K') = Nf(K) + \text{regular function}. \qquad (4.18)$$

From Eq. (4.12), this equation is

$$l^{-d}f'(K') = f(K) + \text{regular function}. \qquad (4.19)$$

These two expressions show that the two interaction constants K' and K are related to each other. This relation is expressed as

$$K' = K'(\{K\}), \qquad (4.20)$$

which denotes that the set $\{K\}$ has been renormalized into $\{K'\}$. In this sense, Eq. (4.20) is a *renormalization equation*. The interaction parameters K and K' are functions of temperature T. The particular point at which

$$K' = K^* = K'(\{K^*\}) \tag{4.21}$$

is called *a fixed point*. At this point, renormalization processes do not cause any further change. A critical temperature is expected to be a fixed point.

Let us assume that K' is a regular function of K and introduce an infinitesimal transformation matrix with elements given by

$$T_{\alpha\beta} = \left(\frac{\partial K'_\alpha}{\partial K_\beta}\right)_{K^*}. \tag{4.22}$$

The derivatives are defined at a fixed point K^*. The eigenvalue problem associated with this matrix is

$$\sum_\alpha T_{\alpha\beta}\phi_\alpha^{(i)} = \lambda_i\phi_\beta^{(i)}. \tag{4.23}$$

In order to discuss trajectories near a fixed point, two vectors u_i and u'_i are introduced such that

$$u_i = \sum_\alpha \phi_\alpha^{(i)}(K_\alpha - K_\alpha^*),$$
$$u'_i = \sum_\beta \phi_\beta^{(i)}(K_\beta - K_\beta^*). \tag{4.24}$$

By definition,

$$u_{i'} = \lambda_i u_i. \tag{4.25}$$

The eigenvalue problem given by Eq. (4.23) is now represented by this relation between the two vectors near K^*. On the other hand, Eqs. (4.24) represent transformations of parameters ΔK_α and $\Delta K'_\beta$ into u_i and u'_i. Hence the critical behavior should be describable in terms of this new set. The free energy will then be given by

$$f(u_1, u_2, \ldots) = l^{-d}f(\lambda_1 u_1, \lambda_2 u_2, \ldots). \tag{4.26}$$

This is a homogeneous equation. In particular, along the direction in which only u_1 varies,

$$f(u_1, 0, \ldots) = l^{-d}f(\lambda_1 u_1, 0, \ldots). \tag{4.27}$$

Hence, if a power law

$$f(u_1, 0, \ldots, 0) = au_1^{\alpha_1} \tag{4.28}$$

is assumed,

$$Au_1^{\alpha_1} = l^{-d}A(\lambda_1 u_1)^{\alpha_1},$$

where A is an unknown constant. This relation means

$$\alpha_1 = \frac{\ln l_d}{\ln \lambda_1}. \tag{4.29}$$

The critical index α_1 in the chosen direction is determined from this equation.

In order to show more explicitly what is involved in these expressions, let us make use of the critical variations of the magnetization and specific heat. These quantities follow

$$\frac{\partial f}{\partial H} = M \sim H^{1/\delta}, \tag{4.30}$$

$$\frac{\partial^2 f}{\partial T^2} = c \sim \left(1 - \frac{T}{T_c}\right)^{-\alpha'} \quad (T < T_c). \tag{4.31}$$

The former equation suggests

$$f \sim H^{1/\delta + 1}, \tag{4.32}$$

and the latter

$$f \sim \left(1 - \frac{T}{T_c}\right)^{2 - \alpha'}. \tag{4.33}$$

If variables u_T and u_H are considered to correspond respectively to T and H, α_1 in Eq. (4.29) is expected to depend on T or H. If the critical indices α and δ are used, and with an appropriate suffix T or H, Eqs. (4.28) and (4.33) yield

$$\alpha' = 2 - \alpha_T. \tag{4.34}$$

Similarly, from Eqs. (4.28) and (4.32) we find

$$\delta^{-1} = \alpha_H - 1. \tag{4.35}$$

It is known that α is close to 0 and δ for an Ising lattice is 15. If the size of a block is $l = 3$, we arrive at

$$\lambda_T = 3, \qquad \lambda_H = 9^{15/16} = 7.845. \tag{4.36}$$

1.5.   Condensed matter under extreme pressure

Crystallization and melting are inverse transitions. Yet, due to the difference in the original states, the latter has been approached theoretically more than

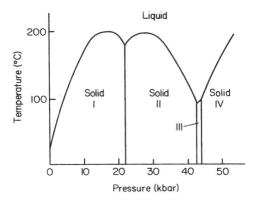

FIG. 1.10. Phase diagram of caesium. (After Jayaraman et al. [22])

the former. Melting transition is first order because symmetry changes are involved.

1.5.1. Existence of a maximum melting temperature

In general, melting points of most substances increase with pressure. However, not only does the rate of this increase tend to slow down with increasing pressure, but also there are cases with decreasing melting points. For instance, ice, Sb, Bi, Ga, and Ge belong to this category.

In the case of cerium, the melting point is minimum at 33 kbar and 662°C. Beyond this pressure, the melting point increases with pressure. Hence, the phase equilibrium curve is concave.

The melting point of caesium shows a complex behavior as shown in Fig. 1.10 [22]. First, it increases with pressure, reaches a maximum, and then decreases to a cusp-type minimum where the crystal changes its structure from bcc to fcc. Beyond this point the melting point increases again, and after a second maximum, starts decreasing. That is, up to around 40 kbar the phase equilibrium curve consists of two convex curves meeting with each other at the cusp minimum. Beyond 40 kbar, there is a transition into another fcc structure in the crystal phase, where the melting point shows again a cusp-type minimum in correspondence to a transition from solid II to solid III. The characteristic electronic structure of caesium in crystal II is 6s, while it is 5d in III. With a further increase of pressure, another phase transition takes place in the crystal phase where the melting point shows a kink.

A maximum melting point has been observed in Ba, K, and Rb at around 18 kbar, 80 kbar, and 40 kbar respectively. Other examples include Te, Eu, C, Bi, Pb, Bi_2Te_3, KNO_3, $NaClO_3$. It is interesting to observe that the lighter potassium requires a higher pressure than heavier metals to reach the maximum melting point. The existence of a maximum melting point

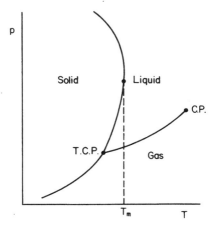

FIG. 1.11. Maximum melting point.

suggests that these systems are liquids under extreme pressure. A schematic phase diagram for such a case is shown in Fig. 1.11. Here, three important points are indicated: the maximum melting point T_m, the critical point, and the tricritical point. The gas and liquid phases meet at a critical point, but a similar critical point does not seem to exist between the liquid and solid phases. The behavior of the melting curve at absolute zero is unknown. The curve may either terminate at a finite pressure or increase indefinitely.

In contrast to the case of condensation of a gas into a liquid, melting is due essentially to the hard core of molecules, which depends on the electronic structure. In the case of caesium, the outer shell is formed by the 6s orbital while the inner 5d shell is vacant. At a very high pressure the electron in the 6s shell will experience strong Coulomb repulsion from the neighboring 6s electrons so that it makes a transition into the 5d shell.

The energy difference between the s and d shells of alkali atoms decreases with the atomic number. For instance, the energies of the 4s and 3d orbitals of potassium are -0.319 Rydberg and -0.123 Rydberg respectively, while the energies of 6s and 5d orbitals of a single caesium atom are -0.286 and -0.154 Rydberg respectively. As pressure increases, the energy difference between the two orbitals is expected to decrease, and finally a transition to the d state takes place. In correspondence with this change, Cs III has a smaller volume than Cs II, and the former is favored at a high pressure. A similar situation exists in the case of alkaline earth atoms. Pressure-induced transitions may involve changes such as 4s → 3d, 5s → 4d, 6s → 5d, and 6s → 4f.

Reentrance into a liquid state at a sufficiently high pressure is consistent with the view that the core of the earth consists of molten iron, nickel, silicon, and other atoms. Note that, as in the case of caesium, electronic transitions can take place at very high pressure. In fact, it is expected that all solids become metallic due to ionization under sufficient compression.

As the size of an astronomical body increases, further transitions are expected. After burning its hydrogen fuel, a star can start collapsing due to its own gravitational force. A star of the size of the sun is expected to become a *white dwarf star*, while a much larger star will further collapse into a *neutron star*. The outer crust of a neutron star is believed to consist of fully ionized nuclei and a highly degenerate relativistic electron plasma. The inner crust might contain a highly degenerate neutron liquid that is superfluid, relativistic electrons, and increasingly neutron-rich nuclei. This layer might be followed by superfluid neutrons, superfluid protons, and normal electron liquid. Finally, the core may consist of hadrons. All these changes take place in about 10 km radius, and the density changes from 10^7 g/cm^3 at the outer crust, to 10^{12} g/cm^3 for the inner crust, and finally 10^{14} g/cm^3 at the core.

1.5.2. *Two-dimensional melting*

Under constant pressure melting takes place upon heating. Although no exact theory of melting exists, an effective theory of two-dimensional melting has been given by Kosterlitz and Thouless [23] based on a consideration of topological order. Two-dimensional systems do not have true long-range order of the type which is observed in three-dimensional systems, but phase transitions do exist, as is clear from the case of an Ising lattice. Kosterlitz and Thouless treated melting in two dimensions based on translational and rotational singularities called *dislocations* and *disclinations* respectively. In two dimensions, these singularities can be specified by a point, rather than a line as in three dimensions, allowing simple treatments.

In a continuum description, the elastic energy of a 2D crystal is given by

$$H_0 = \frac{1}{2} \int d\mathbf{r}[2\mu u_{ij}u_{ij} + \lambda u_{ii}^2], \qquad (5.1)$$

where μ and λ are Lamé's constants and the strain tensor u_{ij} is given by

$$u_{ij} = \frac{1}{2}\left(\frac{\partial u_j}{\partial r_i} + \frac{\partial u_i}{\partial r_j}\right), \qquad (5.2)$$

where $u_i(\mathbf{r})$ is the displacement in the i direction at position $\mathbf{r}$. The displacement vector $\mathbf{u}(\mathbf{r})$ may be decomposed into the phonon and dislocation parts:

$$\mathbf{u}(\mathbf{r}) = \mathbf{u}_1(\mathbf{r}) + \mathbf{u}_0(\mathbf{r}), \qquad (5.3)$$

such that

$$\oint \mathbf{u}_1(\mathbf{r})\, dl = 0; \qquad \oint \mathbf{u}_0(\mathbf{r})\, dl = \mathbf{b}. \qquad (5.4)$$

Here, the integration is performed along a closed contour and **b** is a Burgers' vector if the contour surrounds only one dislocation. Otherwise, it is the total Burgers' vector of the dislocations within the contour. If a closed contour circulates around a disclination in a square lattice, a unit vector rotates by a multiple of $\pi/2$ and **u** changes by an amount that is proportional to the radius of the contour.

In two dimensions, an isolated disclination has an energy proportional to the area of a system and is large. Hence, these singularities may be neglected. In contrast, the energy of an isolated dislocation with Burgers' vector b grows logarithmically with the area L^2:

$$E_{id} = \pi J \ln(L/a_c) + O(1), \tag{5.5}$$

where a_c is the size of the dislocation core and

$$J = \frac{1}{2\pi^2} \frac{\mu(\mu + \lambda)}{(2\mu + \lambda)} b^2. \tag{5.6}$$

On the other hand, the probability of finding isolated dislocations is proportional to b^2/L^2. Hence, the entropy is proportional to the logarithm of the size of the system, and the free energy of an isolated dislocation can be expressed as

$$F_{dis} = (\pi J - 2kT) \ln(L/a_c). \tag{5.7}$$

The melting temperature is then approximately given by

$$T_m = \frac{\pi J}{2k}, \tag{5.8}$$

where the free energy vanishes.

Although isolated dislocations can in principle appear above T_m, they may not be stable because their energy is logarithmically large for a large system. However, the strain energy of a pair of dislocations with equal but oppositely directed Burgers' vectors depends on their distance r and is not so large. It is given approximately by

$$E_{pd} = 2\pi J \left(\ln \frac{r}{a_c} - \tfrac{1}{2} \cos 2\theta \right) + E_0 \qquad (r \gg a_c), \tag{5.9}$$

where θ is the angle between **b** and **r** and E_0 is the energy required to create two dislocations one lattice spacing apart. Therefore, isolated dislocation will form pairs unless the temperature is sufficiently high.

Formula (5.8) gives a melting point if J is given. It has been successfully applied to the melting of a *Wigner crystal* of 2D electrons. However, in general the elastic constants in the constant J are temperature-dependent. Therefore, a renormalization process must be taken in order to determine the melting point.

The above theory makes use of the logarithmic dependence of the energy and entropy, and hence is restricted to two dimensions. Three-dimensional melting can also be dependent on topological singularities that are created thermally. However, since their positions are specified by a line rather than a point, the determination of the melting point is more difficult than for the 2D case.

The Kosterlitz and Thouless theory predicts a continuous phase transition due to dislocation unbinding. If disclination singularities are included, there can be another transition into an intermediate phase [24], although for small core energy of dislocations a first-order melting due to grain-boundary unbinding may take place [25]. We shall discuss this intermediate phase in Section 11.5.

1.5.3. Computer simulations

Melting can be studied by evaluating the equation of state numerically. Since there are many studies and because melting is considered to depend primarily on the hard core of molecules, let us comment only on the case of a hard-sphere system. Such a system has been shown to be in a solid phase and "melts" into a fluid state. In this case there is no clear distinction between the gas and liquid phases.

Several Monte Carlo methods have been developed to evaluate thermodynamic properties of a hard-sphere gas. We remark only one approach to the equation of state through the radial distribution function.

According to the *virial equation of state* the pressure p of a 3D fluid is given by

$$\frac{p}{nkT} = 1 - \frac{2\pi n}{3kT} \int_0^\infty \frac{d\phi}{dr} g(r) r^3 \, dr. \tag{5.10}$$

The abrupt change in the potential function $\phi(r)$ at a hard-sphere diameter a may be expressed by

$$\frac{d\phi}{dr} = -kT\delta(r - a).$$

Hence the equation of state of a hard-sphere system is given by

$$\frac{p}{nkT} = 1 + bng(a). \tag{5.11}$$

Here, b is the van der Waals co-volume which is one-half of the volume excluded by a single molecule. For three dimensions it is

$$b = \frac{2\pi}{3} a^3.$$

According to Eq. (5.11) the pressure p is determined by the radial distribution function at the hard-sphere diameter. One can then make use of the radial distribution function obtained numerically to evaluate the equation of state. Although the value of $g(a)$ is not generally available, the 1D case is exactly given by

$$g(a) = \frac{1}{1 - bn} \tag{5.12}$$

with b given by $a/2$.

For a hard-sphere system the entropy per particle at melting in excess of the ideal gas value at the same temperature and volume is given by

$$s_e = s - s_0$$
$$= -4.86k.$$

On the other hand, the packing function defined by

$$\eta = \frac{\pi d^3}{6v},$$

where d is the diameter of a hard-sphere molecule and v is the volumes per particle, takes on the value

$$\eta = 0.494.$$

The volume exclusion due to the hard core of molecules is expected to be strongest in one dimension, weaker in two dimensions, and still weaker in three dimensions. This observation suggests that for three dimensions,

$$g(a) \sim \frac{1}{(n_c - n)^t}, \tag{5.13}$$

where $0 < t < 1$. An analysis [26] of numerical results indicates that

$$t = 0.76 \pm 0.02. \tag{5.14}$$

In molecular dynamics, the motion of molecules is simulated. If the ith molecule at r_i changes its velocity by Δv_{ij} in time t upon collision with the molecule j at r_j, the equation of state may be evaluated numerically based on

$$\frac{p}{nkT} = 1 + \frac{1}{N\langle v^2 \rangle} W, \tag{5.15}$$

where

$$W = \frac{1}{t} \sum_i (r_i - r_j) \cdot \delta v_{ij}, \tag{5.16}$$

and the temperature is determined by the square average velocity:

$$\langle v^2 \rangle = \frac{3kT}{m}. \tag{5.17}$$

.A numerical simulation [27] with 879 hard-disk particles shows a transition in two dimensions: pure solid and pure fluid phases are observed for $A/A_0 < 1.26$ and $A/A_0 > 1.33$, respectively, where A is the area and A_0 is the area at close packing, and in between, a van der Waals type loop. The density change associated with this phase transition is found to be approximately 4%. Another numerical study [28] indicates that the fcc phase is stable near close packing of a hard-sphere solid. Also, a Monte Carlo simulation [29] of dislocation vector systems in two dimensions reveals two possible types of phase transition. These two depend on the core energy of dislocations. For a large core energy the melting is continuous and due to dislocation unbinding, in agreement with the Kosterlitz and Thouless theory. On the other hand, if the core energy is small, the melting is caused by the nucleation of grain-boundary loops and the transition is first order.

REFERENCES

1. Several books on simple liquids are available. See for instance: J. P. Hansen and I. R. McDonald, *Theory of Simple Liquids* (Academic Press, New York, 1976). H. N. V. Temperley, J. S. Rowlinson, and G. S. Rushbrooke (eds.), *Physics of Simple Liquids* (Wiley, New York, 1968). C. A. Croxton (ed.), *Progress in Liquid Physics* (Wiley, New York, 1978). N. H. March and M. P. Tosi, *Atomic Dynamics in Liquids* (McMillan, New York, 1977). J. S. Rowlinson and F. K. Swinton, *Liquids and Liquid Mixtures* (Butterworth, London, 1982).
2. A. Isihara, *Statistical Physics* (Academic Press, New York, 1971).
3. L. S. Ornstein and F. Zernike, *Proc. Amst. Acad. Sci.* 17, 793 (1914).
4. M. E. Fisher, *Rep. Prog. Phys.* 30, 615 (1967).
5. See for instance: M. S. Green (ed.), *Critical Phenomena* (Academic Press, New York, 1971).
6. Harold J. Raveché and Raymond D. Mountain, *Phys. Rev. A* 9, 435 (1974).

7. L. Van Hove, *Phys. Rev.* **95**, 249 (1954); *Physica* **24**, 404 (1958).
8. P. Schofield, *Phys. Rev. Lett.* **4**, 239 (1960).
9. Y. Yamada and J. D. Axe (eds.), *Neutron Scattering in Condensed Matter* (North Holland, Amsterdam, 1983). Stephen W. Lovesey, *Theory of Neutron Scattering from Condensed Matter* (Oxford University Press, 1985). P. A. Egelstaff (ed.), *Thernal Neutron Scattering* (Academic Press, New York, 1965).
10. See the articles by H. Z. Cummins and also by D. McIntyre and J. V. Sengers in *Physics of Simple Liquids* in Ref. 1. H. Z. Cummins and A. P. Levanyuk (eds.), *Light Scattering Near Phase Transitions* (North-Holland, Amsterdam, 1983). L. P. Kadanoff and P. C. Martin, *Ann. Phys. (N.Y.)* **24**, 419 (1963). Raymond D. Mountain, *Rev. Mod. Phys.* **38**, 205 (1966). H. Eugene Stanley, *Introduction to Phase Transition and Critical Phenomena* (Oxford University Press, New York, 1987).
11. M. Fixman, *J. Chem. Phys.* **36**, 310 (1962). W. Botch and M. Fixman, *J. Chem. Phys.* **42**, 199 (1965). K. Kawasaki, *Phys. Rev.* **150**, 291 (1966); *Phys. Lett. A* **30**, 325 (1969); *Phys. Rev. A* **1**, 1750 (1970). Leo P. Kadanoff and Jack Swift, *Phys. Rev.* **166**, 89 (1968). B. I. Halperin and P. C. Hohenberg, *Phys. Rev.* **177**, 952 (1969). P. C. Hohenberg and B. I. Halperin, *Rev. Mod. Phys.* **49**, 1 (1977). R. A. Ferrell, *Phys. Rev. Lett.* **24**, 1169 (1970).
12. J. V. Sengers and P. H. Keyes, *Phys. Rev. Lett.* **26**, 70 (1971). T. K. Lim, H. L. Swinney, K. H. Langley, and T. A. Kachnowski, *Phys. Rev. Lett.* **27**, 1776 (1971).
H. L. Swinney and D. L. Henry, *Phys. Rev. A* **8**, 2586 (1973). R. Mohr and K. H. Langley, *J. de Phys.* Suppl. **33**, C1–81 (1972). G. T. Feke, G. A. Hawkins, J. B. Benedek, K. H. Langley, and P. B. Elterman, *Opt. Commun.* **7**, 13 (1973).
13. K. Kawasaki and S. M. Lo, *Phys. Rev. Lett.* **29**, 48 (1972). S. M. Lo and K. Kawasaki, *Phys. Rev. A* **5**, 421 (1972), *A* **8**, 2176 (1973). R. Perl and R. A. Ferrell, *Phys. Rev. Lett.* **29**, 51 (1972); *Phys. Rev. A* **6**, 2358 (1972). H. L. Swinney and B. E. Saleh, *Phys. Rev. A* **7**, 747 (1973). H. C. Burstyn, J. V. Sengers, J. K. Bhattacharjee, and R. A. Ferrell, *Phys. Rev. A* **28**, 1567 (1983).
14. C. N. Yang and T. D. Lee, *Phys. Rev.* **87**, 404, 410 (1952).
15. M. Kac, G. E. Uhlenbeck, and P. C. Hemmer, *J. Math. Phys.* **4**, 229 (1963). M. Wadati and A. Isihara, *Physica* **58**, 212 (1972). A. Isihara, *Physica* **64**, 497 (1973).
16. A. Isihara, *Physica* **71**, 597 (1974).
17. J. L. Lebowitz and O. Penrose, *J. Math. Phys.* **7**, 98 (1966). A. Isihara, *Physica* **71**, 75 (1974).
18. A. Isihara, *J. Phys. A* **1**, 539 (1968). A. Isihara, *Physica* **71**, 75 (1974).
19. M. Vicentini-Missoni, J. M. H. Levelt Sengers, and M. S. Green, *J. Res. Nat. Bureau of Stand.* **73A**, 563 (1969). B. Widom, *J. Chem. Phys.* **43**, 3898 (1965). M. Fisher, *Rev. Mod. Phys.* **46**, 597 (1974).
20. L. P. Kadanoff, *Rev. Mod. Phys.* **39**, 395 (1967). B. Widom, *J. Chem. Phys.* **43**, 3898 (1965).
21. K. G. Wilson, *Phys. Rev. B* **4**, 3184 (1971).
22. A. Jayaraman, R. C. Newton, and J. M. McDonough, *Phys. Rev.* **159**, 527 (1967).
23. J. M. Kosterlitz and D. J. Thouless, *J. Phys. C* **6**, 1181 (1973). D. J. Thouless, *J. Phys. C* **11**, L189 (1978). R. Morf, *Phys. Rev. Lett.* **43**, 931 (1979).
24. D. R. Nelson and B. I. Halperin, *Phys. Rev. B* **19**, 2457 (1979). A. P. Young, *Phys. Rev. B* **18**, 1855 (1979).

25. S. T. Chui, *Phys. Rev. Lett.*, **48**, 933 (1982).
26. Y. Song, R. M. Stratt, and E. A. Mason, *J. Chem. Phys.* **88**(2), 1126 (1988).
27. B. J. Alder and T. E. Wainwright, *Phys. Rev.* **127**, 359 (1962).
28. K. W. Kratky, *Chem. Phys.* **57**, 167 (1981).
29. Y. Saito, *Phys. Rev. B* **26**, 6239 (1982).

2

APPROACH TO ELECTRON SYSTEMS

At low temperatures and high densities, electrons are gaseous and mobile, as in the case of metals. As their density is increased, their Coulombic correlations become stronger relative to the kinetic energy so that they enter a liquid or crystal phase. In reality, electrons are confined in a finite volume only in the presence of opposite charges, which are usually positive ions. Hence, their properties and descriptions depend on the status of the latter. In what follows, we shall introduce some basic approaches that do not require explicit treatments of the host ions.

2.1. Fermi liquid theory

When electron correlations are strong, the term *electron liquid* or more widely *Fermi liquid* is used in place of an electron gas or a Fermi gas, even though the distinction between "gas" and "liquid" may not be clearly defined. The name Fermi liquid is convenient and commonly used because it can include liquid ^{3}He and some heavy nuclei. However, with only long-range Coulomb forces, the difference between an electron gas and an electron liquid is subtle in comparison with their difference from an electron solid known as a *Wigner crystal*.

Experimentally, the specific heat of most metals and liquid ^{3}He is linear with temperature, although some deviations may take place due to interactions or phase transitions. The linear variation is in accord with Sommerfeld's theory of an ideal Fermi gas if an effective mass is used. In generalization of the ideal gas theory, Landau tried to describe fermion systems in terms of quasiparticles that are more or less free. His basic idea is that if the interaction is switched on adiabatically, the energy states may still be specified in terms of the free-particle states. However, the states depend on many other particles in the system, and therefore can be a functional of the distribution function of quasiparticles.

In the absence of excitation above the Fermi energy the total energy E of a fermion system is given by the ground-state energy E_g. By labeling the excited states by momentum $\mathbf{p}$ as in the ideal gas, E is expressed as

$$E = E_g + \sum_{\mathbf{p}} \varepsilon(p) n_p, \tag{1.1}$$

where n_p is the distribution function of quasiparticles in the excited state with momentum **p**. The entropy of the system is

$$S = -k \sum_p [n_p \ln n_p + (1 - n_p) \ln (1 - n_p)]. \qquad (1.2)$$

In equilibrium, the distribution function is determined by temperature and is determined by the condition that the free energy is minimum. It is

$$n_p(\varepsilon) = \frac{1}{e^{[\varepsilon(p) - \mu]/kT} + 1}. \qquad (1.3)$$

Note that the energy $\varepsilon(p)$ here may depend on temperature, although for the linear specific heat this dependency does not exist. In order to see how such temperature dependency enters, let us consider the spin susceptibility. The quasiparticle energy is changed by the amount due to spin–magnetic field coupling. This change may be expressed as

$$\delta\varepsilon(p, \sigma) = -\mu_B^* \boldsymbol{\sigma} \cdot \mathbf{H}, \qquad (1.4)$$

where μ^* is an effective Bohr magneton. This energy is considered to consist of two terms. One is the direct coupling energy given by

$$\delta_1\varepsilon(p, \sigma) = -\mu_B \boldsymbol{\sigma} \cdot \mathbf{H}, \qquad (1.5)$$

where μ_B is the Bohr magneton of the ideal case. The other is the energy due to changes in the quasiparticle distribution and is assumed to be given by

$$\delta_2\varepsilon(p, \sigma) = \sum_{p's'} f(\mathbf{p}\sigma, \mathbf{p}'\sigma') \, \delta n(\mathbf{p}', \sigma'). \qquad (1.6)$$

The function f has yet to be specified in consideration of the magnetic interaction between two quasiparticles. For weak magnetic fields, we assume that it is of the following form:

$$f(\mathbf{p}\sigma, \mathbf{p}'\sigma') = f_0(\mathbf{p}, \mathbf{p}') + (\boldsymbol{\sigma} \cdot \boldsymbol{\sigma}')f'(\mathbf{p}, \mathbf{p}'). \qquad (1.7)$$

Using these expressions in Eq. (1.4), we arrive at a self-consistent equation to determine μ^*:

$$\mu_B^* = \mu_B + 2\mu_B^* \sum_{p'} f(\mathbf{p}, \mathbf{p}') \frac{\delta n_p(\varepsilon')}{\delta\varepsilon'}. \qquad (1.8)$$

On the right-hand side, we have made use of the equilibrium distribution function.

Near absolute zero, the derivative may be replaced by $-\delta(\varepsilon_F - \varepsilon')$, where ε_F is the Fermi energy (chemical potential). Also, the magnitude of $\mathbf{p}$ and $\mathbf{p}'$ can be replaced by p_F, the Fermi momentum. As a result, one can use

$$f'(\mathbf{p}, \mathbf{p}') = \phi(p_F, \theta), \tag{1.9}$$

where θ is the angle between $\mathbf{p}$ and $\mathbf{p}'$. Thus, Eq. (1.8) leads us to

$$\mu_B^* = \frac{\mu_B}{1 + s} \tag{1.10}$$

with

$$s = g(\varepsilon_F)\langle\phi(p_F, \theta)\rangle. \tag{1.11}$$

Here, $g(\varepsilon_F)$ is the density of states at the Fermi surface in which the mass should be renormalized in consideration of electron–phonon and electron–electron interactions. $\langle\phi\rangle$ is the average of ϕ on the Fermi surface.

The magnetic susceptibility is given per unit volume by

$$\chi_p = \frac{\chi_p^0}{1 + s}, \tag{1.12}$$

where

$$\chi_p^0 = \mu_B^2 g(\varepsilon_F). \tag{1.13}$$

One can show easily that the density of states $g(\varepsilon_F)$ also characterizes the specific heat. Equation (1.12) then shows that the susceptibility is related to the specific heat and that it deviates from the Pauli susceptibility χ_p^0 by the denominator factor $(1 + s)$, where s is due to an average spin–spin interaction. If s is negative, the susceptibility is enhanced. In reality, the ratio χ_p/χ_p^0 is 1.75 and 1.58 in sodium and potassium respectively.

Note that the above theory is not restricted to electrons. In fact, it has been applied to liquid ^{3}He successfully. Note also that similar considerations can be given for nonequilibrium cases, resulting in viscosity and other transport coefficients.

2.2. Dielectric function and plasmons

A *one-component plasma (OCP) model* provides a basic description of electrons in metals. In this model, the electrons are considered in a background of positive ions forming a cloud of neutralizing charges. Being neutral, the response of this plasma to an external electric field is characterized by a *dielectric function*. In general, this is a function of position in the system and time. Therefore, its Fourier transform $\varepsilon(\mathbf{q}, \omega)$ is a function of wavenumber $\mathbf{q}$ and frequency ω. This function relates the dielectric displacement vector $\mathbf{D}(\mathbf{q}, \omega)$ to the electric field $\mathbf{E}(\mathbf{q}, \omega)$ such that

$$\mathbf{D}(\mathbf{q}, \omega) = \varepsilon(\mathbf{q}, \omega)\mathbf{E}(\mathbf{q}, \omega). \tag{2.1}$$

According to electrostatics, the dielectric displacement is determined through Poisson's equation by an externally introduced test charge density, while the electric vector depends not only on the test charge but also on the charge fluctuations induced by the test charge. That is,

$$i\mathbf{q} \cdot \mathbf{D}(\mathbf{q}, \omega) = 4\pi\rho_t(\mathbf{q}, \omega),$$

$$i\mathbf{q} \cdot \mathbf{E}(q, \omega) = 4\pi[\rho_t(q, \omega) + \langle\rho(q, \omega)\rangle].$$

The average density $\langle\rho\rangle$ may be represented by the free particle polarization function given by

$$\chi(q, \omega) = \frac{2}{(2\pi)^3} \int \frac{f(\varepsilon_{\mathbf{k}+\mathbf{q}}) - f(\varepsilon_k)}{\varepsilon_{\mathbf{k}+\mathbf{q}} - \varepsilon_k + \omega + i0} \, d\mathbf{k}, \tag{2.2}$$

where ε_k is the kinetic energy, the factor 2 is due to spin, and the imaginary notation $i0$ in the denominator means that it is brought to zero after integration. The dielectric function is then given by

$$\varepsilon(q, \omega) = \frac{1}{1 + u(q)\chi(q, \omega)}, \tag{2.3}$$

where

$$u(q) = \frac{4\pi e^2}{q^2} \tag{2.4}$$

is the Fourier transform of the Coulomb potential. The dielectric function thus obtained represents a result based on the *random-phase approximation* (RPA).

In Eq. (2.3), $u(q)$ is the bare Coulomb potential. In a self-consistent approach this potential is replaced by $u(q)/\varepsilon$. In this case the dielectric function is given by

$$\varepsilon(q, \omega) = 1 - u(q)\chi(q, \omega). \tag{2.5}$$

The polarization function corresponds to three dimensions. The static polarization function, $\chi(q, 0)$ is called *Lindhard's function*. The polarization function depends on dimension. For one, two, and three dimensions, its analytic expression is

$$\chi(q, 0) = \begin{cases} \dfrac{k_F}{2\pi^2}\left[1 - \dfrac{1}{4}\left(s - \dfrac{4}{s}\right)\ln\left|\dfrac{s^2 + 2s}{s^2 - 2s}\right|\right], & \text{(3D)} \\[3mm] \dfrac{1}{2\pi}\begin{cases} 1, & s \leq 2 \\ 1 - (1 - 4s^{-2})^{1/2}, & s \geq 2 \end{cases}, & \text{(2D)} \\[3mm] \dfrac{1}{2\pi q}\ln\left|\dfrac{s^2 + 2s}{s^2 - 2s}\right|, & \text{(1D)} \end{cases} \tag{2.6}$$

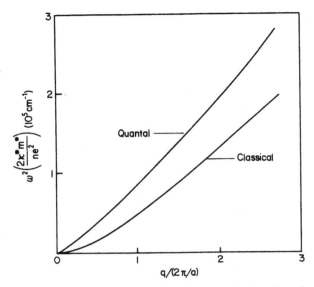

FIG. 2.1. Classical and quantal plasmon dispersion curves for two dimensions.

Figure 2.1 illustrates these expressions in units of $\chi(0, 0)$. The abscissa is a reduced variable q/k_F, k_F being the Fermi wavenumber. The 1D curve with a divergence at $q = 2k_F$ shows relatively the strongest response to the test charge. In the 2D case, the derivative of $\chi(q, 0)$ is singular at the same point, and the 3D curve has an inflection point. In these two cases, the response is strongest for $q = 0$. In all the three cases, $\chi(q, 0)$ approaches 0 as $q \to \infty$.

The dielectric function is imaginary because of the small imaginary number in the integrand. Note that

$$\frac{1}{x + i0} = P\frac{1}{x} - i\pi\delta(x), \qquad (2.7)$$

where P stands for taking the principal value and the symbol $i0$ is a small imaginary part which is brought to zero after it has been used.

The real part of the dielectric function is related to screening. In general, the real part is large for small ω, indicating that screening in the static case is effective. The imaginary part is related to the Coulomb energy. In terms of the density operator ρ_q defined by

$$\rho_q = \sum_j \exp(-i\mathbf{q} \cdot \mathbf{r}_j), \qquad (2.8)$$

the Hamiltonian of an OCP can be expressed as

$$H = H_0 + H_c$$
$$= \frac{1}{2m} \sum_i p_i^2 + \sum_{q \neq 0} \frac{2\pi e^2}{q^2} (\rho_q^* \rho_q - N). \tag{2.9}$$

On the other hand, by using Eq. (2.7) one can show that

$$\text{Im}\left(\frac{1}{\varepsilon(q, \omega)}\right) = u(q) \sum_n |\langle n \langle \rho_q | 0 \rangle|^2 \{\delta(\omega + \omega_{n0}) - \delta(\omega - \omega_{n0})\}, \tag{2.10}$$

where $\omega_{n0} = \varepsilon_n - \varepsilon_0$ is the energy difference between the state n and the ground state. Hence, the expectation value of the interaction Hamiltonian H_c in the ground state is obtained from the imaginary part of the dielectric function in accordance with

$$E_{\text{int}}(e^2) = -\sum_q \left(\frac{1}{2\pi} \int d\omega \, \text{Im}\left(\frac{1}{\varepsilon(q, \omega)}\right) + \frac{2\pi n e^2}{q^2}\right). \tag{2.11}$$

Here, the interaction energy is considered as a function of the interaction parameter e^2. The ground-state energy E_g is given by

$$E_g = \left\langle 0 \left| \sum_i p_i^2/2m \right| 0 \right\rangle + E_{\text{int}}(e^2)$$
$$= \frac{3}{10} \frac{N p_F^2}{m} + \int_0^{e^2} \frac{de^2}{e^2} E_{\text{int}}(e^2). \tag{2.12}$$

The second equation is obtained by differentiating the first equation with respect to the interaction parameter and then integrating the result.

In terms of the exact state $|\psi(r, t)\rangle$ in the presence of a test charge, the average of the density fluctuations induced in the system at time t by the test charge is given by

$$\langle \rho(q, t) \rangle = \langle \psi(r, t) | \rho_q | \psi(r, t) \rangle.$$

This average is expected to follow the time change of the test charge. If the test charge oscillates as $\exp(-i\omega t)$, we can introduce

$$\langle \rho(r, t) \rangle = \langle \rho(q, \omega) \rangle e^{-i\omega t} \tag{2.13}$$

with

$$\langle \rho(q, \omega) \rangle = \langle \psi(r, t) | \rho_q e^{i\omega t} | \psi(r, t) \rangle. \tag{2.14}$$

On the other hand, the dynamic structure factor is given by

$$S(q, \omega) = \sum_{ij} \frac{1}{2\pi} \int dt \langle \exp[-i\mathbf{q}\cdot\mathbf{r}_i(0)] \exp[i\mathbf{q}\cdot\mathbf{r}_j(t)] \rangle \qquad (2.15)$$

$$= \frac{1}{2\pi} \int dt e^{-i\omega t} \langle \rho_q(t)\rho_q(0)^* \rangle. \qquad (2.16)$$

Hence, the dielectric function and the dynamic structure factor are related with each other through

$$Im\left(\frac{1}{\varepsilon(q, \omega)}\right) = u(q)[S(q, -\omega) - S(q, \omega)]. \qquad (2.17)$$

The dynamic structure factor measures the energy transfer from a probe, i.e., the test charge in the present case, to the system. It is natural that this dissipative process is related to the imaginary part of the dielectric function.

The RPA dielectric function has been used frequently, but it does not provide a good approximation for large q. Therefore, the dielectric function given by Eq. (2.3) is modified such that

$$\varepsilon(q, \omega) = 1 - \frac{u(q)\chi}{1 + u(q)G(q, \omega)\chi} \qquad (2.18)$$

$G(q, \omega)$ is a correction function. This function is unknown, but it is expected to approach 1 for $q \to 0$ because the RPA is generally good in this limit.

The dielectric function can vanish at a *plasmon frequency* that depends on q. The equation

$$\varepsilon(q, \omega_p) = 0$$

provides a plasmon dispersion relation. For three dimensions, the plasmon in the long-wavelength limit is well defined and is given by

$$\omega_p = \left(\frac{4\pi n e^2}{m}\right)^{1/2}. \qquad (2.19)$$

For electrons with density $n \sim 10^{23}\,\mathrm{cm}^{-3}$ and mass $m \sim 10^{-27}\,\mathrm{g}$, this frequency is approximately given by

$$\omega_p \sim 10^{16}\,\mathrm{s}^{-1}. \qquad (2.20)$$

In the $q \to 0$ limit, the dielectric function varies as

$$\varepsilon(q, \omega) = 1 - \frac{\omega_p^2}{\omega^2}. \qquad (2.21)$$

For small q and low temperatures, the plasmon dispersion relation is quadratic:

$$\omega = \omega_p \left(1 + \frac{6}{5} \frac{v_F q^2}{\omega_p^2} \right), \qquad (2.22)$$

where $v_F = \hbar k_F / m$ is the Fermi velocity.

For two dimensions, the plasmon dispersion relation at absolute zero is given in the long-wavelength limit by

$$\omega = \left(\frac{2\pi n e^2}{m} q \right)^{1/2} (1 + \tfrac{3}{8} a_0 q). \qquad (2.23)$$

Hence, a 2D plasmon is not defined in this limit. Here a_0 is the Bohr radius.

For inversion layer electrons an effective Bohr radius must be used. It is given by

$$a_0^* = \frac{\hbar^2 \kappa^*}{m^* e^2} \frac{1}{g_v}, \qquad (2.24)$$

where κ^* is an effective dielectric constant, and g_v is the valley degeneracy. For [100] silicon inversion layers, $g_v = 2$. The effective mass m^* in this case is $0.19 m_0$, m_0 being the bare mass. For a given system, the quantum correction tends to increase the plasmon frequency such that the smaller the effective mass and the lower the valley degeneracy the larger the correction.

The dispersion relation given by Eq. (2.23) is linear in the limit $q \to 0$, and increases faster for a small but finite q due to a quantum effect as indicated by the appearance of the Bohr radius. For a 2D classical electron gas between Si and SiO_2 in a metal–oxide–semiconductor field-effect transistor, MOSFET, the dispersion relation for small q is given by

$$\omega = \left(\frac{n e^2}{2\kappa^* m^*} q \right)^{1/2}. \qquad (2.25)$$

If the dielectric constants of semiconductor and oxide are κ_s and κ_o respectively, and if d is the thickness of the oxide layer, κ^* is given by

$$\kappa^* = \tfrac{1}{2}[\kappa_s + \kappa_{ox} \coth(qd)]. \qquad (2.26)$$

For a silicon MOSFET, κ_s and κ_{os} correspond respectively to the Si and SiO_2 layers. For a GaAs/GaAlAs heterostructure, κ_s is the dielectric constant of GaAs layer and κ_o corresponds to that of the GaAlAs layer. The thickness d is the sum of all these layers. The dispersion curve vanishes in the long-wavelength limit. This is a 2D characteristic that has been confirmed by experiment.

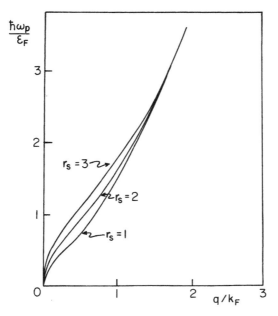

FIG. 2.2.   Coulomb effects on quantal 2D plasmon dispersion.

Figure 2.1 illustrates this 2D characteristic for the classical case given by Eq. (2.25) with Eq. (2.26) and the quantal case of Eq. (2.23). The parameters are $\kappa_s = 11.4$, $\kappa_{ox} = 3.8$, $d = 0.244$ µm, and $a = 5$ mm. The quantal case is above the classical case.

Figure 2.2 shows Coulomb effects on quantal 2D plasmon dispersion. The curvature for small q increases with r_s. The ordinate is the ratio of the plasmon energy $\hbar\omega_p$ and the Fermi energy.

For silicon inversion layer electrons it is appropriate also to add a finite thickness correction. This correction has been calculated for an exponentially decreasing wavefunction. The result shows that a finite thickness due to this form of the wavefunction reduces the plasmon frequency. Therefore, this effect and the quantum effect act oppositely. As a consequence, the classical results given by Eqs. (2.25) and (2.26) without these corrections are better) than expected. Figure 2.3 illustrates the 2D plasmon dispersion relation of GaAs/GaAlAs heterostructures obtained by Batke et al. based on far-infrared transmission spectroscopy[1]. The white and black circles represent the data points of two samples. The curve illustrates Eq. (2.25) with both finite thickness and quantum corrections. The electron density is 6.7×10^{11} cm^{-2}, the dielectric constants are $\kappa_s = 12.8$ and $\kappa_o = 11$ for GaAs and AlGaAs layers respectively and $d = 800$ Å.

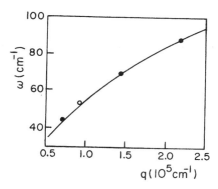

FIG. 2.3. Two-dimensional plasmon dispersion in GaAs/AlGaAs heterostructures. Solid and open circles are data; solid curve is Eq. (2.17) with all necessary corrections. (After Batke et al. [1])

2.3. Long-distance impurity effect in metals

Charged impurities in metals cause a long-distance effect at low temperatures, as observed in the Knight shift of nuclear magnetic resonance. Such a long-distance effect can be studied theoreticlly by placing a test charge at the origin. This charge will have correlations with all the electrons in the system. In order to investigate how such correlations due to Coulomb forces vary with distance, let us investigate the limiting behavior $r \to \infty$ of the radial distribution function for a pair of electrons. One of the electrons in the distribution function can be considered as a test charge placed at the origin. [2]

The correlated part of the radial distribution function is given by

$$G(r) = n^{-2}[\rho_2(r) - n^2 + I_2(r)], \tag{3.1}$$

where $\rho_2(r)$ is the pair distribution function and $I_2(r)$ is the ideal gas contribution. For long distances, the chain diagram approximation can be effectively adopted for the pair distribution function. The corresponding formula is given by Eq. (A.17) in the Appendix, but for long distances the exchange-chain contribution in that formula may be neglected.

The ideal gas term $I_2(r)$ is given by

$$I_2(r) = \left(\frac{1}{(2\pi)^3} \int f(p) e^{i\mathbf{p} \cdot \mathbf{r}} \, d\mathbf{p} \right)^2, \tag{3.2}$$

where $f(p)$ is the Fermi distribution function. For absolute zero, the evaluation of the right-hand side integral is easy. We arrive at

$$I_2(r) = \left(\frac{1}{2\pi^2} \frac{\sin k_F r - k_F r \cos k_F r}{r^3} \right)^2. \tag{3.3}$$

This contribution is long-ranged. However, we are interested in the asymptotic form of $G(r)$ due to Coulomb interaction..

Near absolute zero, it is convenient to use a continuous variable y defined by

$$y = \frac{2\pi j}{\beta \varepsilon_F},$$

where ε_F is the Fermi energy and $\beta = 1/kT$. Also, let us introduce reduced variables defined by

$$x = rk_F, \qquad s = \frac{q}{k_F}.$$

The chain-diagram formula is then expressed as

$$\begin{aligned}
G(r) &= -\frac{3}{2\pi^2}\left(\frac{3}{2\pi}\right)^{1/3} r_s H(x) \\
&= -0.1188 r_s H(x).
\end{aligned} \tag{3.4}$$

$$H(x) = \frac{1}{x}\int_0^\infty dy \int_0^\infty ds \, \frac{sF^2(s, y)\sin sx}{s^2 + s_0^2 F(s, y)} \tag{3.5}$$

where

$$\begin{aligned}
F(s, y) = 1 - \frac{1}{8}\left(s - \frac{4}{s} - \frac{y^2}{s^3}\right) \ln\left|\frac{(s^2 + 2s)^2 + y^2}{(s^2 - 2s)^2 + y^2}\right| \\
- \frac{y}{2s}\left[\tan^{-1}\left(\frac{s^2 + 2s}{y}\right) - \tan^{-1}\left(\frac{s^2 - 2s}{y}\right)\right]. \tag{3.6}
\end{aligned}$$

The constant s_0 is given by

$$s_0^2 = \frac{2}{\pi a_0 k_F} = 0.331\,722 r_s, \tag{3.7}$$

where r_s is a dimensionless parameter which is used frequently for electron systems. It is this radius of a sphere per electron measured in units of the Bohr radius:

$$\frac{4\pi}{3}(r_s a_0)^3 = \frac{1}{n}.$$

That is,

$$r_s = \left(\frac{3}{4\pi n}\right)^{1/3}\frac{me^2}{\hbar^2}. \tag{3.8}$$

Note that the Rydberg unit of energy, which is also used for electron systems, is given by

$$R = \frac{me^4}{\hbar^2 r_s^2}. \tag{3.9}$$

This energy equals 13.61 eV. That is, the parameter r_s can be used as a density parameter or as an energy parameter. Note that when the Coulomb potential e^2/r is expressed in the Rydberg unit, it is proportional to r_s. Therefore, r_s is an effective interaction parameter.

Let us now investigate analytically the asymptotic behavior of $H(x)$ based on the Lighthill theorem. This theorem states that the asymptotic behavior of a Fourier transform is determined by the singularity of the function to be transformed. We note that the function $F(s, 0)$ is singular at $s = 2$, which corresponds to $q = 2k_F$. Hence, the relevant Fourier transform is given by

$$H_0(x) = \frac{1}{x} \int_0^\infty \frac{sF^2(s, 0)}{s^2 + s_0^2 F(s, 0)} \sin sx \, dx, \tag{3.10}$$

where

$$F(s, 0) = 1 + \phi(s);$$

$$\phi(s) = \frac{1}{s}\left(1 - \frac{s^2}{4}\right) \ln\left|\frac{s + 2}{s - 2}\right|. \tag{3.11}$$

Since the absolute magnitude of $\phi(s)$ does not exceed 1, it is possible to obtain the asymptotic expression of $H_0(x)$ for small r_s or high density by expanding the denominator. The dominant contributions are given by

$$H_0(x) = \frac{1}{x}(J_0 + J_1 + J_2). \tag{3.12}$$

Here, the Fourier transforms of the terms with $\phi(s)^0$, $\phi(s)^1$, and $\phi(s)^2$ are represented respectively by J_0, J_1, and J_2. They are

$$J_0 = \frac{\pi}{2}\exp(-s_0 x).$$

$$J_1 = \frac{8 + s_0^2}{(4 + s_0^2)^2} \frac{\pi}{2x^3}(2x\cos 2x - \sin 2x). \tag{3.13}$$

$$J_2 = -\frac{32\pi}{(4 + s_0^2)^3} \frac{\sin 2x}{x^3}(\gamma - \tfrac{3}{2} + \ln|x| - \ln 4).$$

In J_2, γ is the Euler constant. Note that J_1 and J_2 are long-ranged as they decrease only algebraically.

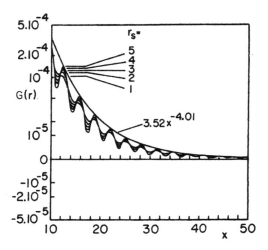

FIG. 2.4. Radial distribution function of a 3D electron gas at absolute zero, showing long-distance correlations. (From Isihara and Ten Seldam [2])

A more correct result has been obtained by a numerical integration [2]. It has been found that the radial distribution function reaches the first peak at around $x = 3$. This peak is higher and occurs at a smaller x as r_s increases, reflecting increasing electron correlations toward low densities. After several wiggles, the function decreases more slowly toward zero with characteristic oscillations. Figure 2.4 illustrates the asymptotic result for the correlated part $G(r)$ of the radial distribution function for five values of r_s. The solid curve is a smoothed-out curve that is roughly proportional to x^{-4}. Note that the period of these oscillations is independent of the interaction parameter r_s. It is determined by the Fermi wavevector. That is, these oscillations are essentially due to the sharpness of the Fermi distribution function.

2.4. Bloch electrons and magnetic Bloch electrons

In an OCP model of electrons in metals the ions are smeared out to form a cloud of neutralizing background charges, and the electrons are essentially in free space. The model allows treating of electron–electron correlations but a more realistic model should include the lattice structure of the ions. We discuss such a model in this and the next sections.

A single electron may be considered in a periodic potential due to the ions. Let $V(r)$ be an ion potential that is periodic with the periodicity of the lattice and the Schrödinger equation of the electron be

$$\left(\frac{1}{2m} p^2 + V(r)\right)\psi(r) = \varepsilon\psi(r). \tag{4.1}$$

According to the Bloch theorem, its wavefunction is given in the following form without a normalization constant:

$$\psi_k(r) = e^{ik\cdot r}u_k(r), \tag{4.2}$$

where $u_k(r)$ has the periodicity of the lattice. An electron with $\psi_k(r)$ is called a *Bloch electron*. One can show that $u_k(r)$ satisfies

$$\left(\frac{1}{2m}(\mathbf{p} + \mathbf{k})^2 + V(\mathbf{r})\right)u_k(\mathbf{r}) = \varepsilon_k u_k(\mathbf{r}), \tag{4.3}$$

where $V(r)$ satisfies the periodic condition:

$$V(\mathbf{r} + \mathbf{a}) = V(\mathbf{r})$$

of the lattice.

The Bloch state given by Eq. (4.2) can be superposed to include many ions. If the ions are at $\mathbf{R}_s$ $(s = 1, 2, \ldots)$, the corresponding wavefunction is

$$\Psi_k(\mathbf{r}) = \sum_s \exp(i\mathbf{k}\cdot\mathbf{R}_s)u(\mathbf{r} - \mathbf{R}_s). \tag{4.4}$$

Note that $\psi_k(r)$ satisfies

$$\psi_k(\mathbf{r} + \mathbf{a}) = \exp(i\mathbf{k}\cdot\mathbf{a})\psi_k(r).$$

In a lattice with unit vectors $\mathbf{a}_1$, $\mathbf{a}_2$, and $\mathbf{a}_3$ in the x, y, and z directions, one can write

$$\mathbf{R}_s = s_1\mathbf{a}_1 + s_2\mathbf{a}_2 + s_3\mathbf{a}_3.$$

Also, $\mathbf{k}$ is such that

$$\mathbf{k} = \frac{n_1}{N_1}\mathbf{b}_1 + \frac{n_2}{N_2}\mathbf{b}_2 + \frac{n_3}{N_3}\mathbf{b}_3,$$

where n_i $(i = 1, 2, 3) = 0, 1, 2, \ldots, N_{i-1}$. The N's are the numbers of ions. Since the vector $\mathbf{k}$ is specified by n (n_1, n_2, n_3), it is appropriate to denote it as $\mathbf{k}_n$.

The function defined by

$$W(\mathbf{r} - \mathbf{R}_s) = \frac{1}{N^{1/2}}\sum_n \exp(-i\mathbf{k}_n\cdot\mathbf{R}_s)\psi_{k_n}(\mathbf{r}) \tag{4.5}$$

is called the *Wannier function*. The Wannier function $W(\mathbf{r} - \mathbf{R}_s)$ is almost the inverse transformation of Eq. (4.4), but not exactly equal to $u(\mathbf{r} - \mathbf{R}_s)$, because $\sum_n \exp[-i(\mathbf{k}_n - \mathbf{k})\cdot\mathbf{R}_s]$ is not exactly a delta function. As one can easily see, the Wannier function is large within the lattice cell at $\mathbf{R}_s$ and decreases away from the cell.

The properties of Bloch electrons are well known. In particular, for small magnitude of **k**, the energy can be expressed in a quadratic free-electron form with an effective mass. A wavenumber **k** can be associated with a Bloch state similarly to the case of a plane wave state. Nevertheless, the original Schrödinger problem must be solved to obtain explicit results. This is not necessarily a simple problem, and several approximate models have been developed. Of these, a *nearly-free-electron model* and a *tight-binding model* are two extremes. In the former model, a Bloch state is approximated by a small number of plane waves. The relative weights of these waves are determined so as to minimize the energy. This process is performed for the first Brillouin zone. For the second zone, a wave function that is orthogonal to the first is used. In the latter tight-binding model, atomic wavefunctions, instead of plane waves, are used to construct a Bloch state. If the atoms forming a lattice are not very close to each other, this model is well defined. The potential energy $V(x, y, z)$ is no longer a small perturbation. The Bloch function is expressed by Eq. (4.4). We can take a spherically symmetrical atomic function $u(r)$ and define the overlap integral between the nearest-neighboring sites:

$$t = \int u^*(\mathbf{r} - \mathbf{R}_n) H u(\mathbf{r} - \mathbf{R}_{n+1}) \, d\mathbf{r}. \tag{4.6}$$

The energy of the electron depends on a lattice structure. For the simple cubic lattice with unit vectors $(\pm a, 0, 0)$, $(0, \pm a, 0)$, and $(0, 0, \pm a)$, the energy is given by

$$\varepsilon = \varepsilon_0 - 2t(\cos k_x a + \cos k_y a + \cos k_z a), \tag{4.7}$$

where ε_0 is the energy in a single well. For small k, the cosine functions can be expanded to yield a quadratic energy with an effective mass.

The above two models have been well investigated as can be found in almost any solid-state text books. Much less known is the case when a Bloch electron is subject to a strong magnetic field. On the other hand, some very unusual phenomena have been discovered recently in 2D electron systems with a strong perpendicular magnetic field. Therefore, that case deserves our attention.

The problem of a Bloch electron in a magnetic field is peculiar because the periodicity of the lattice structure and the periodicity of the electron's cyclotron motion are primarily determined independently of each other. Hence, the problem of treating a Bloch electron in a magnetic field will depend on the commensurability of these two periods.

Since the motion of electrons is not affected by an external magnetic field in the parallel direction, let us study the case of a 2D square lattice with spacing a in the presence of a perpendicular magnetic field H. The problem depends only one one parameter α defined by

$$\alpha = \frac{eHa^2}{ch}. \tag{4.8}$$

This is the magnetic flux through a unit cell in units of the flux quantum defined by

$$\phi_0 = \frac{ch}{e}. \tag{4.9}$$

It is also the ratio of the two characteristic periods: one is the period given by a^2m/h of the motion of an electron in a state with crystal momentum h/a, and the other is the reciprocal of the cyclotron frequency eH/mc.

The energy is expected to be a function of the parameter α [3]. In order to find how the energy varies with α, let us start with a tight-binding form of the energy given by

$$W(k) = 2t(\cos k_x a + \cos k_y a). \tag{4.10}$$

For investigating the effect of a magnetic field, we construct an effective Hamiltonian by replacing $\hbar k$ in this expression by the operator $\mathbf{p} - (e/c)\mathbf{A}$. $\mathbf{A}$ is the vector potential, which is given in the Landau gauge by $H(0, x, 0)$. The effective Hamiltonian thus obtained includes a translational operator $\exp(ap_y/\hbar)$. The Schrödinger problem associated with this new effective Hamiltonian is

$$t[\psi(x + a, y) + \psi(x - a, y) + \exp(-ieHax/c\hbar)\psi(x, y + a)$$
$$+ \exp(ieHax/c\hbar)\psi(x, y - a)] = E\psi(x, y). \tag{4.11}$$

Note that the wavefunction $\psi(w, y)$ is linked with its four nearest neighbors in the lattice.

Let us introduce new parameters

$$x = ma, \qquad y = na, \qquad E/t = \varepsilon,$$

so that s and y are measured in units of a, and the energy in units of t. The coefficients in Eq. (4.11) involves only x. Therefore, it is reasonable to assume plane-wave behavior in the y direction and to write

$$\psi(ma, na) = e^{ivq}g(p). \tag{4.12}$$

The Schrödinger problem is then reduced to a one-dimensional difference equation:

$$g(m + 1) + g(m - 1) + 2\cos(2\pi m\alpha - v)g(m) = \varepsilon g(m) \tag{4.13}$$

This equation can be rewritten in a matrix form:

$$\begin{pmatrix} g(m+1) \\ g(m) \end{pmatrix} = \begin{pmatrix} \varepsilon - 2\cos(2\pi m\alpha - v) & -1 \\ 1 & 0 \end{pmatrix} \begin{pmatrix} g(m) \\ g(m-1) \end{pmatrix}$$

$$= A(m) \begin{pmatrix} g(m) \\ g(m-1) \end{pmatrix} \qquad (4.14)$$

In view of the cosine function, let us assume that the matrices $A(m)$ are periodic in m with period q. This requirement is reduced to the condition that there should be an integer p such that

$$\alpha = p/q. \qquad (4.15)$$

The problem is now dependent on the rationality of α. This is a somewhat bizarre result because the parameter α was originally a continuous function of H. However, the result is based on the assumption on $A(m)$.

An energy spectrum obtained from Eq. (4.14) for rational values of α is illustrated in Fig. 2.5. The ordinate represents ε and the abscissa is α. The center vertical line corresponds to $\alpha = 1/2$ and the horizontal line is $\varepsilon = 0$. The energy bands are plotted for all rational values of α with denominator less than 50. For $\alpha = p/q$, the Bloch band breaks up into q distinct energy bands. The graph shows a recursive structure in which energy bands cluster into groups, which themselves also cluster into larger groups. It is continuous as the magnetic field varies and shows certain symmetries. Irrational values of α appear in correspondence to an uncountable but measure-zero set of points. Small variations in the magnitude of α are expected to produce fluctuations in the value of the denominator q. These fluctuations would blur the graph, but the smeared graph appears to preserve a similar recursive character.

2.5. Hubbard and Anderson models

In a monovalent metal, in which each atom has one valence electron, the valence electrons are delocalized and form a conduction band to move about the lattice. Usually, such a metallic state can be described well in terms of one-electron band theory. However, this description is not always successful. Conceptually, there is a basic problem underlying one-electron theory, as we will discuss below. Also, there are metals that have a very narrow s-band. for which the one-electron approximation is not very good.

One-electron band theory is actually subject to the structure of the lattice. One can see this by artificially increasing the lattice constant, because then the band should be narrowed down. At sufficient separation the atoms will be completely isolated from each other. They become neutral with all the electrons localized, and the system is no longer metallic. Hence, the original

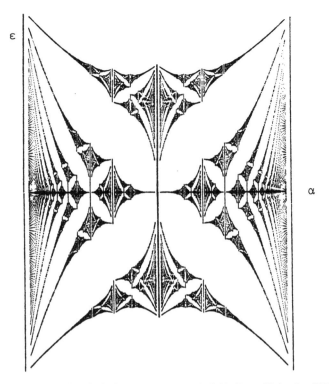

FIG. 2.5. Energy bands of a Bloch electron in a magnetic field. (From Hofstadter [33])

band model ceases in its applicability. The failure of the Bloch band theory is obvious in the case of a large number of transition-metal oxides; the band theory implies that these are metallic, whereas they are actually insulators.

Similarly, an electron-gas model breaks down if the density of electrons is reduced sufficiently at low temperatures. As the Fermi energy decreases, the electrons will form a crystal (*Wigner crystallization*) in order to minimize the electrostatic energy. In summary, both models become invalid as electron localization takes place. Indications are that electron–electron interaction should be taken into consideration in such cases.

Among models of electrons in a lattice that include electron–electron interaction, the *Hubbard model* [4] is often used. The corresponding Hamiltonian consists of

$$H = H_0 + H_1. \tag{5.1}$$

The first term allows the electrons to hop from one lattice site to another. If electrons hop from site j to site i, the kinetic energy can be expressed in terms of the creation and annihilation operators as follows:

$$H_0 = \sum_{(ij)} [w_{ij} a_{i\sigma}^\dagger a_{j\sigma} + \text{c.c.}], \tag{5.2}$$

where w_{ij} represents the strength of hopping, which is determined by the overlapping integral between the two sites. If this strength depends on the difference $|i - j|$, H_0 can be expressed in k-space as follows:

$$H_0 = \sum_{k\sigma} w(k) a_{k\sigma}^\dagger a_{k0}, \tag{5.3}$$

where

$$w(k) = \sum_j w_j e^{ikj}; \qquad a_{k\sigma} = \sum_j e^{-ikj} a_{j\sigma}. \tag{5.4}$$

Without spin indices, the interaction Hamiltonian between the electrons can be expressed as

$$H_1 = \tfrac{1}{2} \sum U_{ijkl} a_j^\dagger a_k^\dagger a_l a_i. \tag{5.5}$$

Here,

$$U_{ijkl} = \int dr_1 \, dr_2 \psi_j^*(r_1) \psi_i(r_1) \phi(|r_1 - r_2|) \psi_k^*(r_2) \psi_l(r_2), \tag{5.6}$$

with $\phi(r) = e^2/r$. If the site i is at R_i, the wavefunction is

$$\psi_i(r) = \psi(r - R_i).$$

In general, the parameter U_{ijkl} depends on overlapping of the wavefunctions at four sites. If such overlapping is small, U can be approximated by two-site Coulomb interaction in which $k = l$ and $i = j$. Even in this case, the strongest interaction is expected to take place among nearest-neighbour sites.

In the Hubbard model, only the Coulomb interaction on the same site is taken into consideration. The strength of this interaction is usually denoted by U. The corresponding Hamiltonian is of the form:

$$H = w \sum_i n_{i\sigma} + t \sum_{(ij)} [a_{i\sigma}^\dagger a_{j\sigma} + \text{c.c.}] + U \sum_j n_{j\uparrow} n_{j\downarrow} \tag{5.7}$$

with number density operators such as $n_{i\downarrow}$ or $n_{j\uparrow}$ and (ij) pairs that can be nearest neighbors. The parameters w, t, and U, assumed to be constant, determine the strength of the respective term.

In the limit $t \to 0$ the electrons do not hop but localize. In the limit, $U \to 0$, the model is reduced to the tight-binding model. In the opposite limit, $U \to \infty$, no two electrons can be at the same site unless their spins are antiparallel. If U is large but finite it is possible to have virtual processes in which electrons with opposite spins exchange their sites and then go back to their original sites. As a result the system will have effectively an antiferromagnetic interaction. If $t = 0$ each site is independent of other sites. Thus, the Hamiltonian given by Eq. (5.7) is expected to be applicable to a variety of cases.

Prior to Hubbard, Anderson[5] investigated localized magnetic states in metals based on a similar Hamiltonian. In dilute solutions of iron, cobalt, and to a lesser extent, nickel in nonmagnetic metals, localized magnetic moments are observed. Such localized moments cannot be satisfactorily described by one-electron theory. Moreover, in terms of a Hartree-Fock theory it is difficult to understand why the states of opposite spin on the magnetic ion can be empty while the parallel spin states are full. Anderson analyzed the conditions necessary for the occurence of localized magnetic moments based on the Hamiltonian:

$$H = H_0 + H_{0d} + H_c + H_{sd}, \tag{5.8}$$

where

$$H_0 = \sum_{\mathbf{k}\sigma} \varepsilon(k) n_{\mathbf{k}\sigma}; \qquad H_{0d} = \varepsilon \sum_{\sigma} n_{d\sigma}; \qquad H_c = U n_{d\uparrow} n_{d\downarrow};$$

$$H_{sd} = \sum_{\mathbf{k}\sigma} V_{dk}(a_{\mathbf{k}\sigma}^{\dagger} a_{d\sigma} + \text{c.c.}). \tag{5.9}$$

The Hamiltonians H_0 and H_{0d} are the unperturbed energies of the conduction (s) and localized (d) electrons respectively. H_c represents the Coulomb interaction between the localized electrons, for which a constant-strength U is assumed. H_{sd} is the interaction energy between the s and d electrons. It was found that the magnetic state depends not only on the s–d admixture matrix elements but also on the Coulomb correlation integral in the d shell.

Associated with such magnetic impurities is the *Kondo effect* [6], which is characterized by a rise in the electrical resistivity toward low temperatures. This effect is due to spin-flip scattering processes of conduction electrons at the magnetic impurities. Hence, its theoretical description is based on a Hamiltonian that includes an s–d interaction energy. This description can be made without the interaction energy H_c, and is essentially dependent on second-order terms linear in V_{dk} such as those proportional to $V_{dk}V_{dk'}$ when the terms linear in V_{dk} are eliminated from the Hamiltonian of Eq. (5.8) by a canonical transformation [7]. Further discussion of the Kondo effect will be given in Section 7.5.

REFERENCES

1. E. Batke, D. Heitmann and C. W. Tu, *Phys. Res. B* **34**, 6951 (1986).
2. A. Isihara, *Physica* **71**, 83 (1974). A. Isihara and C. A. Ten Seldam, *Physica* **76**, 153 (1974).

3. D. R. Hofstadter, *Phys. Rev. B* **14**, 2239 (1976). G. H. Wannier, *Phys. Stat. Sol.* (*b*) **88**, 757 (1978).
4. J. Hubbard, *Proc. Roy. Soc. London Ser. A* **276**, 238 (1963).
5. P. W. Anderson, *Phys. Rev.* **124**, 41 (1961).
6. J. Kondo, *Prog. Theor. Phys.* **32**, 37 (1964).
7. J. R. Schrieffer and P. A. Wolff, *Phys. Rev.* **149**, 491 (1966).

3

ELECTRON CORRELATIONS

Most properties of metals and semiconductors are due to free charge carriers. However, the carriers do have various interactions. Mutual interaction of electrons through the Coulomb potential is fundamentally important in understanding metallic binding. Electron–phonon interaction, indirect electron interaction via phonons, electron–hole interaction, interaction between conduction electrons and localized impurities, and so on, cause interesting phenomena. In this chapter we discuss only some aspects of electron correlations via the Coulomb potential based on the jellium and other models. Some other interaction effects will be addressed in later sections.

3.1. Correlation energy

Among many important roles played by electron correlations, perhaps that in metallic binding is most fundamental. The strong metallic cohesion depends not only on ions but also on electrons. The calculation of the so-called *cohesive energy* is not simple and requires several steps for evaluation. It is given in reference to the system of free atoms at infinite separation, that is, the ionization energy ε_I per atom.

For monovalent alkali metals one can start effectively with the *Wigner–Seitz method*. In this method a given metal is divided into polyhedral centered about each lattice point. These polyhedra are electrically neutral and are considered independent of others. Each polyhedron has one electron, which interacts with the ion at the center. The corresponding Schrödinger problem is solved under the boundary condition that the derivative of the wavefunction vanishes at the polyhedral surface. The energy eigenvalue ε_{WS} at vanishing wave number is then obtained.

The next step is to evaluate the electron's *ground-state energy* ε_g. Its evaluation is usually made based on the jellium model. The energy consists of the kinetic energy, exchange energy, and the so-called *correlation energy*. That is, it is expressed as

$$\varepsilon_g = \varepsilon_0 + \varepsilon_x + \varepsilon_c. \tag{1.1}$$

It is customary to express these energies in Rydbergs and per electron.

The kinetic energy is given by

$$\varepsilon_0 = \frac{2.21}{r_s^2}. \tag{1.2}$$

This energy is proportional to r_s^{-2} in Rydberg units. Indeed, the Hamiltonian of an electron system can be written in these units as

$$H = \frac{me^4}{\hbar^2} \frac{1}{r_s^2} \left[\sum_i \frac{p_i'^2}{2} + r_s \cdot \sum_{i<j} \frac{1}{r_{ij}'} \right], \tag{1.3}$$

where the first factor $me^4/\hbar^2$ is the Rydberg energy, and the quantities in the square brackets are dimensionless.

Because of the form of Eq. (1.3), the interaction energy can be evaluated for small r_s by perturbation. That is, one can take a high-density approach to evaluate the interaction energy.

The next term in Eq. (1.1) is the exchange energy given by

$$\varepsilon_x = -0.916 \frac{1}{r_s}. \tag{1.4}$$

Since the kinetic and exchange energies are obtained in the Hartree–Fock approximation, the correlation energy represents the difference between the true ground-state energy and that evaluated in the Hartree–Fock approximation.

The evaluation of ε_c can be based on the OCP model. In this case two corrections must be introduced into the result; one is the self-energy of the positive background and the other is the interaction energy of the electrons with the uniform positive background. The sum turns out to be $1.2/r_s$. When this is combined with the exchange energy, we obtain $0.284/r_s$. The cohesive energy is then obtained as the difference between the ionization theory and all of these energies.

There are several contributions to the correlation energy. For high densities these contributions can be classified in terms of the interaction parameter e^2. They start with terms of order e^4 since the kinetic and exchange energies are of order $(e^2)^0$ and e^2 respectively. The $(e^2)^2$ terms include the ring-diagram contribution and the second-order exchange contribution. Adding still higher-order contributions, which include the third-order exchange energy, we get [1]:

$$\varepsilon_c = -0.0937 + 0.062 \ln r_s + 0.086 r_s - 0.013 r_s \ln r_s - \frac{24}{\pi^3} r_s X_3. \tag{1.5}$$

The first constant term includes the positive contribution from the regular second-order exchange diagrams first obtained by Onsager et al. [2],

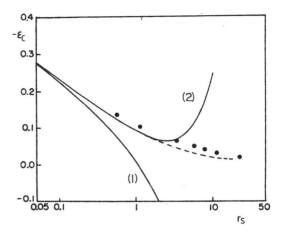

FIG. 3.1. The correlation energy of an electron gas as a function of r_s. Curves (1) and (2) are theoretically possible extremes. The dashed curve is an interpolated result. The solid circles represents numerical results obtained by Stevens and Pokrant [5] for low densities. (From Kojima and Isihara [1])

$$\varepsilon_{2x} = 0.04836,$$

and also the negative term from the ring diagrams. The contribution from the anomalous second-order exchange diagrams is canceled out by the corresponding ring diagram contribution [3]. The terms that are proportional to r_s and $\ln r_s$ are due to the ring diagrams. The last term, which is proportional to r_s, has an unknown function X_3 that includes the contributions from the third- and higher-order exchange diagrams.

For r_s of order 1 the correlation energy is not very large. Precisely for this reason, each term in the above series has been accurately evaluated. However, it should be pointed out that r_s of real metals is not small. In fact in alkali metals, r_s values of Li, Na, and K are respectively 3.22, 3.96, and 4.87.

It is important to note that the correlation energy is negative. Figure 3.1 illustrates the correlation energy as a function of r_s. The dashed line represents the results obtained by interpolating the high-density result with a Wigner lattice result based on a Padé approximation [4]. The dots are the numerical results of Steven and Pokrant [5] based on a variational method with a Jastrow-type trial function. Curves (1) and (2) correspond to the two extreme cases in which the ring-diagram contribution $0.086r_s$ in Eq. (1.5) is fully included and completely neglected. That is,

$$(1): \quad O(r_s) = 0.086r_s - 0.013r_s \ln r_s,$$
$$(2): \quad O(r_s) = -0.013r_s \ln r_s.$$

Since the third-order exchange diagram is expected to cancel at least a part of the term $0.086r_s$, the correct correlation energy must be between these two extreme cases.

3.2. Para- and diamagnetic susceptibilities

Metals show paramagnetism due to the spin and diamagnetism due to the orbital motion of the electrons. In an ideal electron gas, the paramagnetic susceptibility known as Pauli susceptibility is three times the absolute magnitude of Landau's diamagnetic susceptibility. In the presence of Coulomb interaction, both become functions of r_s. The case of paramagnetism has already been discussed in Section 2.1 based on a Fermi liquid approach. If we want to analytically evaluate the Coulomb effects on both susceptibilities, the energy must be evaluated in the presence of a magnetic field as a function of r_s. Its precise calculation is difficult because the correlation energy must be evaluated in the presence of a magnetic field.

Nevertheless, a systematic calculation based on an OCP model is possible because the eigenvalues and eigenfunctions of the relevant Hamiltonian,

$$H = \sum_I \left[\left(\mathbf{p}_i + \frac{e}{c}\mathbf{A} \right)^2 - \tfrac{1}{2}g\mu_B\boldsymbol{\sigma}_i\cdot\mathbf{H} \right] + \sum_{i<j} \frac{e^2}{r_{ij}}, \tag{2.1}$$

are known. Here, g is Landé's factor, $\boldsymbol{\sigma}$ corresponds to the spin, and μ_B is the Bohr magneton. We assume that the magnetic field is uniform and is in the z direction, and choose the vector potential $\mathbf{A}$ such that

$$\mathbf{A} = (-Hy, 0, 0). \tag{2.2}$$

The eigenfunction is given by

$$\psi_{n\sigma} = \frac{A_n}{2\pi} \exp\left[i(p_x x + p_z z) - \frac{a^2(y-y_0)^2}{2} \right] H_n[b(y-y_0)]|\sigma\rangle. \tag{2.3}$$

Here,

$$y_0 = -\frac{p_x}{a^2}; \qquad A_n^2 = \frac{a^2}{\pi^{1/2}2^n n!}; \qquad a^2 = \frac{eH}{2mc} = \frac{\omega_c}{2}. \tag{2.4}$$

Note that

$$a^2\hbar = \mu_B H,$$

and that ω_c is the cyclotron frequency. The eigenvalues are

$$\varepsilon_{n\sigma} = (n + \tfrac{1}{2})\hbar\omega_c + \frac{p_z^2}{2m} \pm \tfrac{1}{4}g\hbar\omega_c. \tag{2.5}$$

In order to obtain the magnetic susceptibilities at absolute zero, it is sufficient to evaluate the ground-state energy to order H^2. In grand-ensemble theory, the susceptibility is obtained from

$$\chi = \frac{kT}{HV}\left(\frac{\partial \ln \Xi}{\partial H}\right)_z,$$

where H is the magnetic field, Ξ is the grand partition function, and z is the absolute activity. This evaluation has been performed [6] by extending the calculation of the correlation energy in the absence of a magnetic field. The exchange and ring-diagram contributions to the grand partition function must be treated in combination in order to avoid a divergence. The combinations is given by a formula that modifies the first term of Eq. (A.19) of the Appendix. The eigenvalues λ_j in that formula can be obtained as functions of magnetic field by using Eqs. (2.3) and (2.5). After obtaining the grand partition function, the Fermi momentum must be renormalized as a function of magnetic field and r_s in accordance with the rule of grand ensemble theory. This approach is convenient because both para- and diamagnetic susceptibilities can be obtained simultaneously.

The Fermi momentum decreases almost linearly with $a^4\hbar^2/\varepsilon_F^2$, where ε_F is the Fermi energy in the absence of a magnetic field. However, the slope depends on r_s. An increase in r_s or in the magnetic field causes a decrease in the Fermi momentum as the ideal gas behavior is destroyed.

For small r_s, the ground-state energy can be expressed in an r_s series. Its correction due to a magnetic field can be expressed such that

$$\Delta\varepsilon = -\frac{5}{8}\left(\frac{\mu_B H}{\varepsilon_F}\right)^4 \frac{2.21}{r_s^2}\left[\frac{g^2}{2} - \frac{2}{3} + \frac{4}{\pi}\left(\frac{4}{9\pi}\right)^{1/3} r_s\left(-0.0697 + \frac{g^2}{8} - \frac{1}{36}\ln r_s\right) + \cdots\right].$$

(2.6)

The right-hand side series looks complicated, but note that the correction starts with a^4. For $r_s = 1$ and $g = 2$, we find that the ratio of $\delta\varepsilon$ to the total ground-state energy is almost equal to the ratio $(\mu_B H/\varepsilon_F)^4$:

$$\frac{\Delta\varepsilon}{\varepsilon} = -0.982\left(\frac{\mu_B H}{\varepsilon_F}\right)^4.$$

(2.7)

In general, the ratio $(\mu_B H/\varepsilon_F)$ is small. Therefore, $\Delta\varepsilon/\varepsilon$ is very small. Nevertheless, this is the source for the magnetic response. In the expression for $\Delta\varepsilon$, the terms with Landé's g-factor correspond to paramagnetism, those without to diamagnetism. The para- and diamagnetic susceptibilities χ_p and χ_d are given by [6]

$$10^6\chi_p = 2.589\,\frac{1}{r_s}\frac{m^*}{m} + 0.429 - 0.071r_s + 0.0356r_s\ln r_s + \cdots,$$

(2.8)

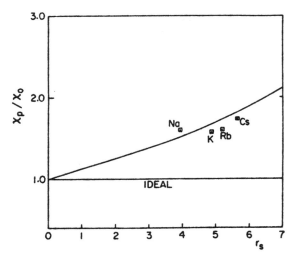

FIG. 3.2. Paramagnetic susceptibility of alkali metals relative to the Pauli susceptibility. The data points represent average values of several data. (From Kojima and Isihara [6])

$$10^6 \chi_d = -0.863 \frac{1}{r_s} \frac{m}{m^*} + 0.0598 + 0.02385 \ln r_s - 0.0246 r_s + 0.0127 r_s \ln r_s$$

$$+ \cdots, \quad (2.9)$$

where χ_p corresponds to the case $g = 2$ and an effective mass m^* has been introduced to the ideal gas term in order to include electron–phonon, lattice and other neglected effects.

The above results show that both susceptibilities are enhanced due to electron–electron interaction. Figure 3.2 shows the ratio χ_p/χ_0 as a function of r_s. The curve is based on Eq. (2.8) and the data represent averages. The ideal susceptibility is given by

$$\chi_0 = \frac{m}{\hbar^2} \frac{k_F}{\pi^2} \mu_B^2 = \frac{3}{2} \frac{n \mu_B^2}{\varepsilon_F}$$

$$= \frac{1}{4\pi^2 \alpha r_s} \left(\frac{1}{137.04}\right)^2, \quad (2.10)$$

where $\alpha = (4\pi/9)^{1/3} = 0.52106$.

Figure 3.3 illustrates the paramagnetic susceptibility χ_p given by Eq. (2.8) in comparison with the ideal gas curve. The theoretical curve includes effective mass corrections. Except for the case of lithium, the theoretical curve is reasonably close to the black dots representing average experimental points. The vertical bars are their ranges. Figure 3.4 shows the diamagnetic

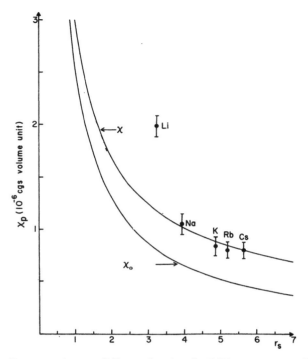

FIG. 3.3. Paramagnetic susceptibility as a function of r_s. Solid curve represents Eq. (2.8). (From Kojima and Isihara [6])

susceptibility without effective mass corrections. Note, in both cases, the differences from the ideal gas curve increase with r_s. Experimentally, the susceptibilities are determined by separating out the ion contribution from the measured total susceptibility and also the dia- and paramagnetic parts from each other. Therefore, the determination of each quantity is not simple. On the other hand, a perturbational approach becomes increasingly poor as r_s increases. Nevertheless, a reasonable agreement between theory and experiment has been achieved.

3.3. Specific heat of metals

According to Fermi liquid theory, the electronic specific heat of metals varies linearly with temperature as in the ideal case, although the proportionality constant is determined by an effective mass. This effective mass is expected to depend on electron–electron interaction or the density of electrons. For small r_s, this dependence can be evaluated in principle as a function of r_s by a perturbation method. The evaluation is difficult, but the temperature

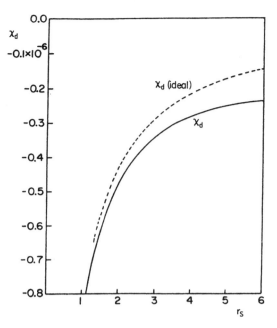

FIG. 3.4. Diamagnetic susceptibility given by Eq. (2.9) without effective mass corrections. (From Kojima and Isihara [6])

dependency has been at least partially determined with accuracy in the case of the correlation energy at absolute zero.

The result of such an evaluation depends on a dimensionless parameter given by

$$\eta_0 = \beta \varepsilon_F, \tag{3.1}$$

where $\beta = 1/kT$, and ε_F is the ideal Fermi energy. The calculation is based on the assumption that this parameter is large. We remark that the Fermi momentum p_F deviates from its ideal gas value p_0 at absolute zero in such a way that

$$\frac{p_F}{p_0} = A(r_s) - \frac{\pi^2}{24\eta_0^2} B(r_s, T). \tag{3.2}$$

The first term, representing the effect of Coulomb interaction at absolute zero, decreases almost linearly for small r_s as shown in Fig. 3.5. Here, Curves (1) and (3) are based on an iterative determination of p_F by keeping the terms to order e^6 and e^4 respectively in the grand partition function, except that in the former the third-order exchange contribution is not included. Curves (2) and (4) are obtained numerically without employing the iterative process

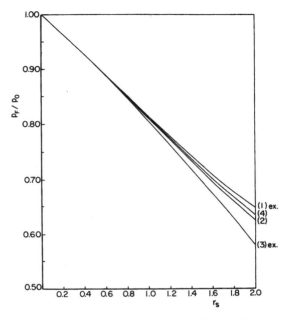

FIG. 3.5. The r_s dependence of the Fermi momentum p_F relative to its value p_0 for the ideal case. (From Isihara and Kojima [1])

and by keeping the terms to order e^6 and e^4 respectively in the grand partition function. Note that their differences start appearing at around $r_s = 1$. The coefficient $B(r_s, T)$ of the second term is shown in Fig. 3.6 as a function of T. For small r_s it is almost constant because the ideal gas term is proportional to T^2. The coefficient is negative and increases its magnitude and temperature variation with r_s.

The kinetic energy is given by

$$\varepsilon_0(T) = \frac{2.21}{r_s^2}\left(1 + \frac{5\pi^2}{12\eta_0^2} + \cdots\right). \tag{3.3}$$

The second term yields the well-known Sommerfeld specific heat. The first-order exchange contribution is expressed as

$$\varepsilon_{1x}(T) = -\frac{0.916}{r_s}\left(1 - \frac{\pi^2}{12\eta_0^2}(0.9503 - 2\ln\eta_0) + \cdots\right). \tag{3.4}$$

Note that a logarithmic term appears. The difference in the kinetic and exchange energies from their zero-temperature values depends on T and r_s. For $\alpha = (4/9\pi)^{1/3} = 0.521$, this dependence is given by

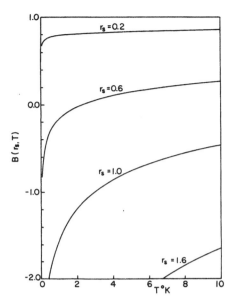

FIG. 3.6. The coefficient $B(r_s, T)$ for the temperature-dependent part of p_F. (From Isihara and Kojima [1])

$$\Delta\varepsilon_{0+x}(T)\left(\frac{e^4 m}{\pi\hbar^2\alpha k}\right)^2 = T^2 r_s^2\{1 + r_s[2.123 + 0.166\ln(Tr_s^2)]\}. \quad (3.5)$$

The correlation energy is given by

$$\varepsilon_c(T) = \varepsilon_c(0) - \frac{\pi^2}{12\eta_0^2}[1.427 - 0.152\ln r_s - 0.616\ln\eta_0 - 0.304(\ln\eta_0)^2]$$
$$+\cdots. \quad (3.6)$$

This expression can be rewritten in a more convenient and explicit form as follows:

$$\Delta\varepsilon_c(T)\left[\frac{e^4 m}{\pi\hbar^2\alpha k}\right]^2 = T^2 r_s^4\{5.458 - 0.786\ln(r_s^2 T)$$
$$+ 0.0755[\ln(r_s^2 T)]^2 + 0.0138\ln r_s + \cdots\}. \quad (3.7)$$

In Eqs. (3.5) and (3.7) the temperature T is expressed in kelvins and the energy is in Rydbergs. That is, their left-hand sides are in units of (Rydbergs $\cdot$ K^2).

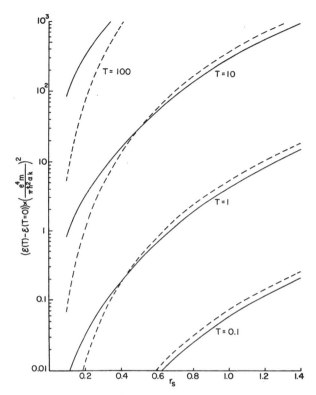

FIG. 3.7. The differences between the internal energies at T and at 0 K. (From Isihara and Kojima [1])

Figure 3.7 shows the increment of the internal energy (solid curve) per electron and that of the correlation energy (dashed curve) over their zero-temperature values as functions of r_s for a given temperature. The dashed curves cross their respective solid curves, and there is a trend that the crossing point r_s shifts towards higher values for higher temperatures. The difference between the solid and dashed curves is smaller on the right-hand side (larger r_s) of each crossing point, indicating increasing contributions from the correlation energy. On the left-hand side of the crossing point, the kinetic and first-order exchange energies contribute significantly.

The temperature variation of the correlation energy is illustrated in Fig. 3.8 for several values of r_s less than 1. There is similarity in these curves, but the difference in the correlation energy becomes larger and increases more rapidly with temperature as r_s increases.

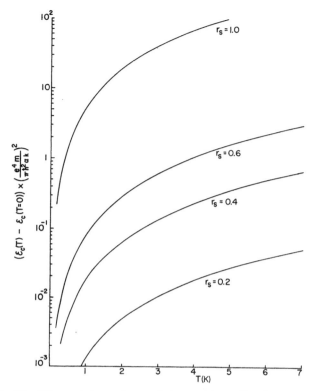

FIG. 3.8. The difference between the correlation energies at T and 0 K. (From Isihara and Kojima [1])

The specific heat c_v at constant volume is obtained from the internal energy, but we must use the energy per volume rather than the energy per electron. To order r_s^2, its ratio to the ideal gas specific heat is given by [7]

$$\frac{c_V}{c_v^0} = 1 + 0.162 - 0.166r_s \ln \eta_0 - 0.157r_s^2 \ln r_s$$
$$+ r_s^2[0.0282 \ln \eta_0 + 0.0275(\ln \eta_0)^2] + \cdots. \quad (3.8)$$

Here, the ideal gas term is

$$c_v^0 = \frac{k^3 T}{3}\left(\frac{2m}{\hbar^2}\right)k_F = \frac{\pi^2 k^2 T}{2\varepsilon_F}\, n. \quad (3.9)$$

Note that the last term with $(\ln \eta_0)^2$ is exactly the square of the first logarithmic term $-0.166r_s \ln \eta_0$. Therefore, they might be a part of a

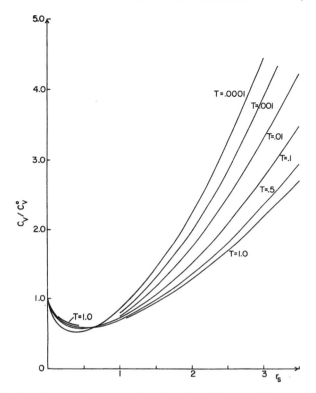

FIG. 3.9. Specific heat ratio as a function of r_s. (From Isihara and Kojima [7])

geometric series. The specific heat ratio is illustrated in Fig. 3.9 as a function of r_s for several temperatures. The specific heat c_v corresponds to the case in which the geometric series has been summed to infinite order. The ratio is small for small r_s, but exceeds 1 for r_s of order 1.3 or larger. Hence, the specific heat effective mass can be larger than the bare-electron mass depending on temperature. However, the approximate results obtained by perturbation become increasingly poor.

The electronic specific heat is often associated with the density of states $g(\varepsilon_F)$ at the Fermi energy by the formula:

$$c_v = \tfrac{1}{3}\pi^2 k^2 T g(\varepsilon_F). \tag{3.10}$$

This formula can be derived by expressing the average energy as

$$\langle \varepsilon \rangle = \int_0^\infty \varepsilon f(\varepsilon) g(\varepsilon)\, d\varepsilon. \tag{3.11}$$

Here, $f(\varepsilon)$ is the Fermi distribution function. When the average energy is expressed by such a one-electron distribution function, the density of states can be a complicated function of r_s and temperature. Neglecting this temperature dependence, and assuming that $f(\varepsilon)$ is a step function, one arrives at the above one-electron formula.

3.4. Electron–hole droplets

By laser light illumination valence electrons in germanium or silicon can be excited by the conduction band, with holes left behind. An excited electron and a hole attract each other through the Coulomb potential and can form a neutral particle called an *exciton*. This particle is similar to the hydrogen atom, or more closely to positronium because the mass difference between the positive and negative charges is small. However, unlike positronium, the lifetime of excitons is of order $1\,\mu s$, that is, 10^4 times longer. Therefore, in a sense they are the simplest atoms. The lifetime is determined by the recombination, which causes emission of luminescence light. This luminescence is observable. A schematic diagram showing such a recombination transition is illustrated in Fig. 3.10.

As in the case of any other atoms, excitons are in a gaseous state when the density is small. As the density increases the excitons can fall into a liquid state. Because the total number of excitons is not large, the latter occurs in small droplets. Moreover, at a certain density and relatively high temperature electrons and holes can be unpaired, and the system is metallic. That is, a phase transition from an insulating phase to a metallic phase takes place.

A phase diagram of electrons and holes in silicon, originally due to Shah, Combescot, and Dayem [8], is illustrated in Fig. 3.11. Above the coexistence curve are two gas phases; one is insulating and the other metallic. The dashed curve represents the Mott criterion for the metal–insulator transition. In the classical regime this criterion states:

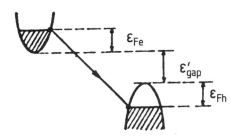

FIG. 3.10.   Schematic energy diagram for recombination

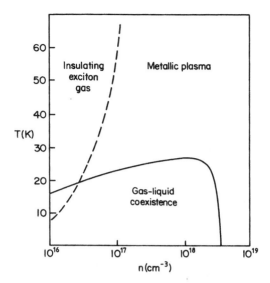

FIG. 3.11. Phase diagram of electron–hole liquid in silicon. (After Sah et al. [8])

$$n_M(T) = \frac{kT}{16\pi\varepsilon_{ex}a_0^{*3}}$$
$$= 1.5 \times 10^{15} T \text{ cm}^{-3}.$$

Here the excitons have the effective Bohr radius $a_0^* = 44$ Å, and the binding energy $\varepsilon_{ex} = 14.7$ meV. At $T = 0$ the criterion predicts

$$n_M = 1.3 \times 10^{14} \text{ cm}^{-3}.$$

The critical temperature is given by

$$T_c = 27 \pm 1 \text{ K},$$

and the critical density is around

$$n_c \sim 1.1 \times 10^{18} \text{ cm}^{-3}.$$

On the high-density side of the coexistence curve the liquid density varies as T^2. The limiting density at 0 K is given by

$$n_0 = 3.3 \times 10^{18} \text{ cm}^{-3}.$$

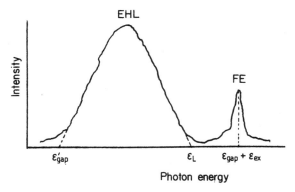

FIG. 3.12. Luminescence spectrum in the coexistence regime.

Concerning luminescence spectra, a sharp luminescent peak is observed at relatively high temperatures. This corresponds to free excitons. As the temperature is lowered the intensity of this line decreases, and at a certain low temperature and on a lower-frequency side a much broader liquid-line appears. A schematic graph of these two luminescence spectra is shown in Fig. 3.12.

In actuality, some of the excitons may form excitonic molecules as in the case of the hydrogen molecule. Figure 3.13 represents the luminescence data from ultrapure silicon under uniform stress obtained by Gourley and Wolfe [9]. The free exciton peak on the extreme right is very sensitive to the stress. The broad liquid-line disappears above around 29 kg/mm^2. In between, the luminescence line EM corresponding to excitonic molecules changes its strength with the stress. At higher stresses the EHL binding energy is reduced and the excitonic gas density increases.

A particularly interesting aspect of exciton gases is that a Bose condensation can be expected at a low temperature since they obey Bose statistics. In this respect, bound electron–hole pairs have several observational advantages. First, their mass is much smaller than that of an atom. Second, the exciton gas density can be controlled over a large range by optical excitation. Third, a recombination luminescence line reflects the kinetic energy distribution.

The study of Bose–Einstein condensation has been conducted in the semiconductor Cu_2O because the interaction between the excitons is repulsive at low densities and because the hole mass is only $0.7m_0$. In this semiconductor, electrons in the lowest conduction band and holes in the highest valence band can form excitons. Exchange interaction splits the $n = 1$ level of the so-called yellow series into a triply degenerate *orthoexciton* and a singly degenerate *paraexciton*. The paraexcitons have a rather long lifetime of around 10 μs at 15 K. Such a long lifetime is an additional advantage for

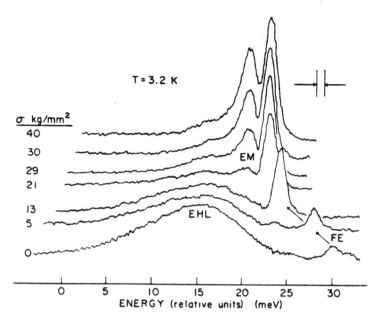

$T = 3.2$ K

σ kg/mm^2

40

30

29

21

13

5

0

EM

EHL

FE

0   5   10   15   20   25   30

ENERGY (relative units) (meV)

FIG. 3.13. Stress dependence of the luminescence spectrum of silicon. (From Gourley and Wolfe [9])

observing Bose–Einstein condensation. Time-resolved spectral and spatial measurements of photoluminescence in Cu_2O at low temperatures show that the gas density follows the expected $T^{3/2}$ variation of the ideal Bose gas. Moreover, the paraexcitons are found to undergo Bose–Einstein condensation at high densities [10].

Thus, the electron–hole systems show a variety of interesting phenomena. The interaction potential between the electrons and holes is Coulombic, but its magnitude is reduced considerably by the dielectric constant of the medium. For instance, the dielectric constant is 15.4 in Ge, 11.4 in Si, and 12.35 in GaAs. Although the effective masses of the electron and hole, m_e and m_h respectively, are small, the excitonic Bohr radius is large. It is defined by

$$a_0^* = \frac{\kappa \hbar^2}{\mu_0 e^2},$$ (4.1)

where μ_0 is the reduced mass given by

$$\frac{1}{\mu_0} = \frac{1}{m_e} + \frac{1}{m_h}.$$ (4.2)

The actual values of the effective Bohr radius are 177 Å in Ge, 49 Å in Si, and 140 Å in GaAs. Due to such large values of the effective Bohr radius, the binding energy of an exciton becomes very small. It is 0.01–0.001 of that of a hydrogen atom. Also, the dimensionless interaction parameter r_s becomes relatively small. This smallness is helpful in making a high-density theoretical approach to the exciton energy.

Let us now discuss the conditions under which electron–hole pairs are formed at absolute zero. The energy of N electron–hole pairs in a volume V can be expressed as

$$E = N\varepsilon(n) + N\varepsilon'_{\text{gap}}, \tag{4.3}$$

$\varepsilon'_{\text{gap}}$ being the effective energy gap between the valence and conduction bands, which includes the correction due to the correlations of the particles.

The width of the liquid-line depends on the momenta of electrons and holes, which range from zero to the maximum corresponding to their chemical potential or the actual Fermi energy. The upper edge, that is, the violet boundary, of the liquid-line is given by

$$\hbar\omega_v = \varepsilon'_{\text{gap}} + \mu_L, \tag{4.4}$$

where μ_L is the chemical potential of the liquid measured from the gap edge $\varepsilon'_{\text{gap}}$. It is the sum of the actual Fermi energies of the electrons and holes and is equal to [11]

$$\mu_L = (N + 1)\varepsilon\left(\frac{N + 1}{V}\right) - N\varepsilon\left(\frac{N}{V}\right).$$

$$= \varepsilon(n) + \frac{n \, \partial\varepsilon}{\partial n}. \tag{4.5}$$

At an equilibrium density n_0, the derivative vanishes:

$$\left(\frac{\partial\varepsilon}{\partial n}\right)_{n_0} = 0. \tag{4.6}$$

Hence, for equilibrium

$$\mu_L = \varepsilon(n_0). \tag{4.7}$$

The low-frequency edge, that is, the red boundary, of the liquid line is essentially determined by the band gap. However, this energy must be renormalized in consideration of the effects of collective interaction. This renormalized band gap $\varepsilon'_{\text{gap}}$ is given by

$$\hbar\omega_r = \varepsilon'_{\text{gap}} = \varepsilon_{\text{gap}} + \varepsilon(n_0) - (\varepsilon_{Fe} + \varepsilon_{Fh}). \tag{4.8}$$

The difference between Eqs. (4.4) and (4.8) is

$$\hbar(\omega_v - \omega_r) = \varepsilon_{Fe} + \varepsilon_{Fh}. \qquad (4.9)$$

The equilibrium density of droplets at absolute zero corresponds to the minimum of the function $\varepsilon(n)$. Condensation takes place when the minimum energy per particle, $\varepsilon(n_0)$, is lower than the excitonic binding energy ε_{exc}:

$$\varepsilon(n_0) < \varepsilon_{exc}. \qquad (4.10)$$

The cohesive energy of an electron–hole liquid can be evaluated similarly to the ground-state energy calculation of an OCP in consideration of the structure of the conduction and valence bonds. For instance, in the case of silicon, the 6-fold degenerate conduction band and the doubly degenerate valence band must be taken into consideration. Accordingly, several band parameters have to be introduced. Otherwise, a high-density approach can be used. The large effective Bohr radius or the small r_s is favorable to this approach, but an exact treatment of low densities has not been given as yet. Therefore, theoretical descriptions of the coexistence curve resort to certain approximations.

3.5. Heavy fermions

It was discussed in the previous chapter that the specific heat of actual metals can be expressed in the ideal gas form by using an effective mass. Concerning the specific heat constant γ that determines the effective mass, we remark that there are metals with γ values that are extraordinarily large; their γ is of order 500–1000 mJ/mole·K^2 in contrast to mJ/mole·K^2 of ordinary metals. For example, γ is 450 mJ/mole·K^2 in UPt$_3$ and 1100 mJ/mole·K^2 in UBe$_{13}$. Correspondingly, their paramagnetic susceptibility is also large. It is 7 in UPt$_3$ and 15 in UBe$_{13}$ in units of 10^{-3} emu/mole. Therefore, the electrons in these metals are classified as *heavy fermions*.

There are three known groups of heavy fermion systems. In Type 1, the ordinary linear specific heat is observed although the proportionality constant γ is very large. U$_2$Zn$_{17}$ and UCd$_{11}$ belong to this category. In Type 2 such as UPt$_3$, the specific heat c follows

$$\frac{c}{T} = \gamma + \beta T^2 + \delta T^2 \ln T, \qquad (5.1)$$

where γ, β, and δ are constants. The second term on the right represents the lattice contribution. The appearance of the third logarithmic term is characteristic of this group. A similar logarithmic term has been found in liquid ^{3}He; the term heavy fermion is not restricted to electrons. In Type 3, c/T varies as $T^{-1.6}$ or as $T^{-0.88}$–$T^{-1.8}$. CeCu$_2$ and UBe$_{13}$ show the former variation, and $CeCu_6$ and $CeAl_3$ the latter. However, UBe_{13} becomes superconductive, causing complications.

The above metals show a nearly constant spin susceptibility at low temperatures that is also strongly enhanced over the ordinary Pauli susceptibility as we mentioned earlier. The susceptibility follows the Curie–Weiss law at temperatures above a certain temperature T^* due to the localized f electrons (4f in cerium and 5f in uranium). It is considered that the conduction electrons interact strongly with these localized electrons. Below T^*, the f electrons are delocalized to form a Fermi liquid together with the conduction electrons. Therefore, the heavy fermion phenomenon has been investigated in connection with the Kondo effect. Their relationship can be seen from the resistivity of heavy fermion systems such as $CeCu_6$, which increases with decreasing temperature.

The Kondo effect is due to the interaction between the conduction electrons and the localized magnetic moments of magnetic impurities. These impurities are usually dilute and cause spin-flip scattering. The susceptibility follows the Curie–Weiss law at high temperatures due to the localized magnetic moments and becomes constant below a certain temperature characteristic of a metal when the magnetic moments are shielded by conduction electrons.

Two distinctive phenomena take place as the concentration of magnetic impurities increases. First, in the case of transition elements with a relatively wide d-orbital, the Kondo effect disappears as a result of the overlapping of these d-orbitals. Consequently, a magnetically ordered state or a spin-glass state occurs. Second, in the case of rare-earth atoms such as cerium or ytterbium with very localized f orbitals, the overlapping of the f-orbitals may not take place. Hence, the rare-earth impurities form a Kondo lattice. In such a lattice each impurity atom scatters conduction electrons as in the dilute case.

However, below a certain temperature it can be energetically advantageous if the f electrons admix with the conduction electrons to form a Fermi liquid. The susceptibility increases toward low temperatures following the Curie–Weiss law until such a point is reached. Therefore, the susceptibility can increase significantly if this point is very low. The characteristic susceptibility enhancement can then be expected. Although actual heavy-fermion systems show a variety of properties, the disappearance of the magnetic moment seems to give a clue to understanding the basic mechanism.

The temperature variations of the magnetic susceptibility and resistivity of cerium compounds are illustrated in Fig. 3.14 and 3.15 respectively [13]. At relatively high temperatures, the susceptibility varies as $1/T$ in $CeCu_6$, $CeAl_3$, and $CeCu_2Si_2$. The resistivity of these compounds increase logarithmically as the temperature is decreased, and therefore cerium atoms in these compounds are considered to form Kondo lattices. There are other cerium compounds such as CeG_6 or $CePb_3$ in which magnetic ordering destroys the Kondo effect. Note in general that the lower the Kondo temperature,

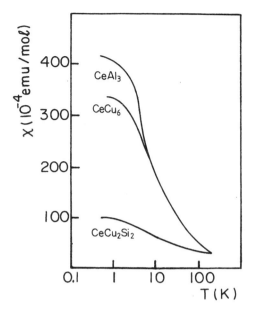

FIG. 3.14. Temperature variation of the magnetic susceptibility of cerium compounds that show the Kondo effect. (From Onuki and Komatsubara [13])

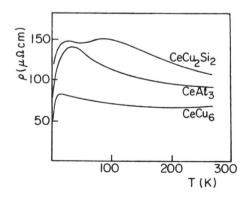

FIG. 3.15. The temperature variation of the resistivity ρ of cerum compounds.

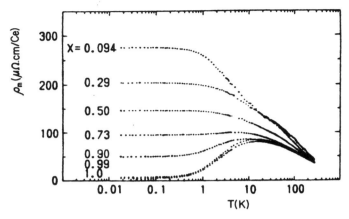

FIG. 3.16. The temperature variation of the relative resistivity per mole cerium of $Ce_xLa_{1-x}Cu_6$. (From Onuki and Komatsubara [13])

the higher the susceptibility. This provides one possible mechanism for the appearance of a large effective mass.

The relative resistivity of $Ce_xLa_{1-x}Cu_6$ is illustrated in Fig. 3.16 [13]. The ordinate represents

$$\rho_m = \frac{\rho_x - \rho_0}{x},$$

where the suffix x represents the concentration of cerium atoms. The current is along the b axis. Around 50 K, the resistivity increases toward low temperatures. However, when the concentration is above 80%, the resistivity drops after a maximum. The Kondo temperature at $x = 0.094$ is estimated to be 3.7 K. The residual resistivity per mole of cerium varies as $320x(1 - x)$ $\mu\Omega\cdot$cm. The spin–spin correlation between the cerium atoms is varied by lanthanum atoms.

A single ion with an f orbital can be treated in terms of the Anderson Hamiltonian given by Eq. (5.8) in Section 2.5. We use a suffix f for the f orbital so that ε in that equation is replaced by ε_f and separate out the energy H_0, which is concerned only with the conduction electrons. The coupling between the f and conduction electrons may be assumed to depend only on the magnitude of the k-vector and is characterized by a constant interaction amplitude V. The number operator of the f electrons in the mth state is denoted by $n_m^f = f_m^\dagger f_m$, where m varies from 1 to $v_f = 2J + 1$, which represents the degeneracy of the f orbital. As a consequence, we obtain a new effective Hamiltonian given by [14]

$$H = H_0 + \sum_{km} \varepsilon(k) c_{km}^{\dagger} c_{km} + \varepsilon_f \sum_n n_m^f + \frac{U}{2} \sum_{m \neq m'} n_m^f n_{m'}^f$$
$$+ V \sum_{km} (f_m^{\dagger} c_{km} + c_{km}^{\dagger} f_m). \tag{5.2}$$

We are concerned with two types of spin states, spin-zero singlet states and magnetic multiplet states. The energies associated with these two states are represented by ε_s and ε_m respectively, and their difference by ε:

$$\varepsilon_s = \varepsilon_m + \varepsilon. \tag{5.3}$$

The energy ε_m can be expressed as

$$\varepsilon_m = \varepsilon_0 + \varepsilon_f, \tag{5.4}$$

where ε_0 is the energy of the ground state $|\phi_0\rangle$, which corresponds to a completely filled Fermi sea.

A trial wavefunction for the singlet state is constructed by a linear combination of the ground state and the states $|f_m^{\dagger} c_{km}|\phi_0\rangle$ in which a single conduction electron is transferred to the localized state m. This linear combination is expressed as

$$|\psi\rangle = a \left[1 + \sum_k b(k) f_m^{\dagger} c_{km} \right] |\phi_0\rangle. \tag{5.5}$$

The energy of this singlet state is given by

$$\varepsilon_s = \frac{\langle \psi | H | \psi \rangle}{\langle \psi | \psi \rangle}. \tag{5.6}$$

This energy is required to be stationary with respect to variations in a and the hybridization parameter $b(k)$. Eliminating the hybridization parameter from two variational equations thus obtained, we arrive at a single equation that determines ε:

$$\varepsilon = -\varepsilon_f + v_f V^2 \sum_k \frac{1}{\varepsilon + \varepsilon(k)}. \tag{5.7}$$

The solutions of this equation can be obtained by plotting both sides against ε. We find then for small V that there is a solution ε that is negative. That is, the system can gain energy by entering the singlet state, which is nonmagnetic. The normalization constant a of the singlet state can be expressed in terms of the f-level occupancy n_f such that

$$|a|^2 = 1 - n_f. \tag{5.8}$$

When a magnetic field is applied, the energy ε_f becomes $\varepsilon_f - g_J\mu_B mH$, with m varying between $-J$ and J. The k-sum in the above equation will be changed in such a way that

$$v_f \sum_k \rightarrow \sum_{km}.$$

Otherwise, the corresponding energy is given in a similar form. The magnetic susceptibility can then be obtained by the second derivative of the energy in the following form:

$$\chi = (g_J\mu_B)^2 \frac{J(J+1)}{3} \frac{1}{v_f\Gamma} \frac{n_f^2}{1-n_f}, \tag{5.9}$$

where $\Gamma = \pi g(\varepsilon_F)V^2$ with $g(\varepsilon_F)$ is the density per spin of the conduction electrons on the Fermi surface. Corresponding to Eq. (5.9), the specific heat constant γ is given by

$$\gamma = \tfrac{1}{3}\pi^2 k^2 \frac{1}{v_f\Gamma} \frac{n_f^2}{1-n_f}. \tag{5.10}$$

These results show that both χ and γ increase as n_f approaches 1.

Some heavy fermion systems become superconductors in spite of their large effective masses. The critical temperature T_c is 0.7 K in $CeCu_2Si_2$, 0.9 K in UBe_{13}, 0.5 K in UPt_3, and 1.5 K in URu_2Si_2. The temperature variation of the specific heat is anomalous at the superconductive transition point, and the ratio c/T reaches a peak. In the case of UPt_3, T_c appears in the Fermi liquid range, but this is not the case in Be_{13} and $CeCu_2Si_2$. According to the BCS theory, $\Delta c/\gamma T_c = 1.43$, where Δc is the specific heat jump. In $CeCu_2Si_2$, this ratio is 1.2, whereas UBe_{13} has 2.5, and UPt_3 about 0.6. Below T_c, the specific heat of heavy fermion systems does not decrease exponentially following $\exp(-\Delta/kT)$. In fact, in $CeCu_2Si_2$, it varies as $T^{2.4}$, in UBe_{13} as $T^{2.9}$, and in UPt_3 as T^2. Note that the specific heat in liquid ${}^3He$ is proportional either to T^3 or to T^2 depending on the superfluid states.

The ultrasonic attenuation coefficient α in the BCS-type superconductors decreases sharply below T_c and their NMR relaxation time T_1 increases toward a maximum just below T_c and then decreases to zero. On the other hand, the attenuation coefficient in UPt_3 and UBe_{13} decreases slowly in proportion to T^2. Also, T_1 of both $CeCu_2Si_2$ and UBe_{13} does not increase below T_c but decreases in proportion to T^3. Therefore, these heavy fermions do not seem to become S-wave BCS superconductors. Instead, they can be

P-wave superconductors. The case of $CeCu_2Si_2$ appears to be in a spin singlet state with an anisotropic energy gap that may vanish in a certain direction. When impurities are added to expand the lattice, this material favors antiferromagnetic ordering [12]. A similar antiferromagnetic ordering occurs in $U_{0.95}Th_{0.05}Pt_3$.

Let us try to understand the relation between such an antiferromagnetic spin configuration and superconductivity based on the strong coupling limit of the Hubbard Hamiltonian [15]. Omitting the first term in Eq. (5.7) in Chapter 2, we express the Hamiltonian as

$$H = t \sum_{(ij)} (a_{i\sigma}^\dagger a_{j\sigma} + \text{c.c.}) + U \sum_i n_{i\uparrow} n_{i\downarrow}, \quad (5.11)$$

where the first sum is over (ij) pairs and spin σ. When U is very large, this Hamiltonian reduces to an antiferromagnetic Heisenberg model in a band that is half-filled. Each site in this case has one electron. From this limit, the hopping term is considered to cause perturbation. To first order in t the kinetic energy may be expected to have the form

$$H_1 = t \sum_{(ij)} (c_{i\sigma}^\dagger c_{j\sigma} + \text{c.c.}). \quad (5.12)$$

Here the new operators $c_{i\sigma}$ hop single electrons from site to site. In contrast to $a_{i\sigma}$ they are not true fermion operators since they do not allow double occupancy. This means

$$c_{i\uparrow} c_{i\downarrow}^\dagger = 0. \quad (5.13)$$

These operators describe spinless fermions in one dimension. Their role is unknown in higher dimensions. We assume that the system is still describable by spinless fermions.

Perturbation to first order in t/U allows virtual transitions into states with doubly occupied sites. The effective Hamiltonian that acts only on the states with singly occupied sites can be expected to be characterized by energy that is proportional to $t(t/U)$:

$$V_{ij} = 2 \frac{t^2}{U} [(\sigma_i \cdot \sigma_j) - \tfrac{1}{4}]. \quad (5.14)$$

This interaction energy corresponds to superexchange between nearest-neighbor spins. It is antiferromagnetic in that antiparallel spin pairs have lower energies by $2t^2/U$. If two electrons with opposite spins are favored in nearest-neighbor sites, they will have an effective attraction, which can be a cause of superconductivity. A similar attraction results from the *Anderson lattice model* with d and f electron states:

$$H = t \sum_{\substack{(ij) \\ \sigma}} (d_{i\sigma}^\dagger d_{j\sigma} + \text{c.c.}) + V \sum_{i\sigma} (d_{i\sigma}^\dagger f_{i\sigma} + \text{c.c.})$$

$$+ \varepsilon_d \sum_{i\sigma} n_{di\sigma} + \varepsilon_f \sum_{i\sigma} n_{fi\sigma} + U \sum_i n_{fi\uparrow} n_{ni\downarrow}. \tag{5.15}$$

Perturbation to lowest order in $V(\varepsilon_d - \varepsilon_f)$ reveals that the energy of the singlet state for $U \to \infty$ is lowered from the case $U = 0$ by the Kondo coupling characterized by

$$\Delta E = \frac{2V^2}{\varepsilon_d - \varepsilon_f}. \tag{5.16}$$

Thus, the mechanism for lowering energy is somewhat different from the earlier case.

REFERENCES

1. A. Isihara and D. Y. Kojima, *Z. Phys.* B **21**, 33 (1975). D. Y. Kojima and A. Isihara, *Z. Phys.* B **25**, 167 (1976).
2. L. Onsager, L. Mittag, and M. J. Stephen, *Ann. Phys.* **18**, 71 (1966).
3. M. Gell-Mann and K. A. Brueckner, *Phys. Rev.* **106**, 364 (1957). P. Nozières and D. Pines, *Phys. Rev.* **111**, 442 (1958).
4. A. Isihara and E. W. Montroll, *Proc. Nat. Acad. Sci.* **68**, 3111 (1971).
5. F. A. Stevens, Jr. and M. A. Pokrant, *Phys. Rev.* A **8**, 990 (1973).
6. A. Isihara and D. Y. Kojima, *Phys. Rev.* B **11**, 710 (1975). D. Y. Kojima and A. Isihara, *Phys. Rev.* B **20**, 489 (1979).
7. A. Isihara and D. Y. Kojima, *Physica* **77**, 469 (1974).
8. J. Shah, M. Combescot, and A. H. Dayem, *Phys. Rev. Lett.* **38**, 1497 (1977). C. D. Jeffries and L. V. Keldysh (eds.), *Electron–Hole Droplets in Semiconductors* (North-Holland, Amsterdam, 1983).
9. P. L. Gourley and J. P. Wolfe, *Phys. Rev.* B **10**, 6338 (1982).
10. D. W. Snoke, J. P. Wolfe, and A. Mysyrowicz, *Phys. Rev. Lett.* **64**, 2543 (1990).
11. M. Combescot and P. Nozières, *J. Phys. C* **5**, 2369 (1972).
12. G. R. Stewart, *Rev. Mod. Phys.* **56**, 755 (1984). F. Steglich, in *Theory of Heavy Fermions and Valence Fluctuations*, ed. by T. Kasuya and T. Saso (Springer-Verlag, New York, 1985), p. 23. A. I. Goldman, G. Shirane, G. Aeppli, B. Batlogg, and E. Bucher, *Phys. Rev.* B **34**, 6564 (1986).
13. Y. Onuki and T. Komatsubara, *J. Mag. & Mag. Mater.* **63**, 281 (1987); *Butsuri* **42**, 732 (1987).
14. P. Flude, J. Keller, and G. Zwicknagl, in *Solid State Physics*, ed. by H. Ehrenreich and D. Turnbull (Academic Press, New York, 1988), **41**, p. 1. C. M. Varma and Y. Yafet, *Phys. Rev.* B **13**, 2950 (1976).
15. J. E. Hirsch, *Phys. Rev. Lett.* **54**, 1317 (1985).

4

TWO-DIMENSIONAL ELECTRON SYSTEMS

Traditionally, condensed matter physics has dealt with three-dimensional systems. However, several two-dimensional electron systems have been realized since about twenty years ago. They have attracted considerable attention in recent years because unusual and important phenomena have been discovered. In what follows, we address some of these properties.

4.1. Realization of two-dimensional electron systems

Although thin metallic foils can be two-dimensional, other clean and important 2D electron systems have been fabricated. In what follows, we introduce three typical such systems.

In the first place, electrons can be trapped two-dimensionally on the surface of liquid ^{4}He. The reason for this is that liquid helium is a very good insulator, with its surface providing an energy barrier of more than 1 eV. On the other hand, electrons near the surface are attracted by their own image charges. As a consequence, these electrons are trapped near the surface. Figure 4.1 shows schematically the barrier and image potentials. The electron wavefunction is also shown schematically.

Electron motion perpendicular to the surface is now restricted, causing the energy to split, while electron motion parallel to the surface is free. Thus, this electron system is two-dimensional if no excitation to higher energy levels takes place.

The density of the electrons on the surface of liquid ^{4}He is typically 10^9 cm^{-2}. Hence, they are classical at around a few kelvin. An important aspect of these electron systems is that they are generally free from impurities.

In the second place, electrons are trapped two-dimensionally in the so-called *inversion layer or accumulation layer* in a silicon MOSFET (metal–oxide–silicon field-effect transistor). A typical silicon MOSFET is a parallel plate condenser consisting of a metallic plate (*gate*), an SiO$_2$ dielectric layer, and another plate made of silicon. As a semiconductor, silicon can be doped into either a p-type or an n-type. If p-type silicon is used as a condenser plate, as in Fig. 4.2, its charge carriers are positively charged. These charges will be pushed into the interior of the silicon plate by the charges near the bottom of the SiO$_2$ layer when a positive voltage is applied to the gate. This results in space charges confined to a narrow inversion

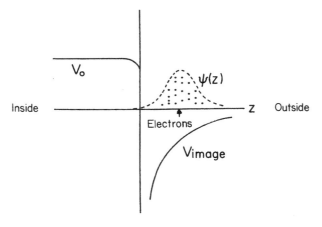

FIG. 4.1. Barrier and image potentials of electrons on the surface of liquid ^{4}He.

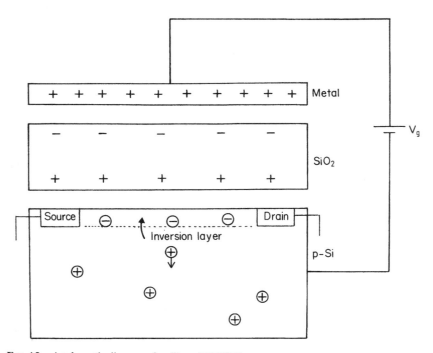

FIG. 4.2. A schematic diagram of a silicon MOSFET.

layer, so-called because the charge carriers are negatively charged. Although not shown in Fig. 4.2, the electron source and drain are electrically connected.

If n-type silicon is chosen instead of p-type, an accumulation layer appears in which the sign of the induced charge carriers is the same as that of the semiconductor. Typically, the thickness of inversion or accumulation layers is of order 100 Å or less. The charges in these layers can move freely in the plane parallel to the condenser plates. However, their perpendicular motion is restricted, causing the energy to split. If excitations to high energy levels do not take place, they are again two-dimensional. Typically, their density is of order 10^{12} cm^{-2} so that at low temperatures they are quantum-mechanically degenerate. The induced charges can easily be driven by an applied field from one condenser to another if they are electrically connected. In the [100] direction of silicon, the effective mass of inversion layer electrons is $0.19m_0$, m_0 being the bare mass.

In the third place, electrons are induced in the so-called *superlattice* or *heterostructure* that is formed by molecular-beam epitaxy such that thin layers of GaAs and GaAlAs alternate. The aluminum atoms do not alter the lattice structure of the GaAs lattice but alter the band structure as shown in Fig. 4.3. As a result, a one-dimensional Kronig–Penney-type periodic alternation of the conduction and valence bands can be produced. Upon doping the GaAlAs layers with donor impurities, electrons can be introduced.

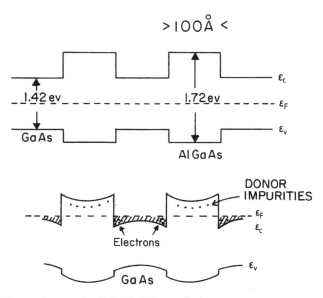

FIG. 4.3. Energy diagram of a GaAs/GaAlAs superlattice.

These electrons fall into a wedgelike narrow potential well at the interface in the GaAs layers. Again, these electrons are two-dimensional for low energies in that their motion in parallel with the interface is not restricted while in the perpendicular superlattice direction it is restricted. Their density is typically 10^{12} cm^{-2}. The electrons in the interface are highly mobile, as reflected in their very small effective mass of $0.068m_0$.

These superlattices are classified as Type I in contrast to some other types. In the Type-II superlattice consisting of InAs–GaSb, for example, the conduction band of InAs is close to the valence band of GaSb, resulting in strong band interaction. InGaAs/GaSbAs superlattices also belong to this category.

In polytype superlattices, a third component is added. For instance, AlSb can be added as a third component to Type II. In the so-called NIPI superlattices, n(Si)- and p(Be)-doped GaAs layers are separated from each other by a thin intrinsic (i) layer of the same semiconductor material. Such a structure provides tunability of conductivity, absorption coefficient, and luminescence. In yet another type, spin-superlattices, the electron g-factor is modulated spatially. For example, $Hg_{0.99}Mn_{0.01}Se/Hg_{0.98}Cd_{0.02}Se$ belongs to this category. In the absence of a magnetic field, the two semiconductors have matching bands. In the presence of a magnetic field, the spins in manganese will cause spin splitting, resulting in an energy variation.

In two dimensions, the Coulomb potential is logarithmic. However, in the above systems the ordinary 3D Coulomb potential can be used because the 2D electron systems have a small but finite thickness. Only their motion at low energies is two-dimensional. Thus, the Fourier transform $u(q)$ of their Coulomb potential in two dimensions is

$$u(q) = \int \frac{e^2}{r} d^{i\mathbf{q}\cdot\mathbf{r}} d\mathbf{r}$$

$$= \frac{2\pi e^2}{q}. \tag{1.1}$$

Note that in three dimensions,

$$u(q) = \frac{4\pi e^2}{q^2}$$

so that it decreases faster with q.

Near absolute zero, the Fermi wavevector k_F can be introduced. It is given by

$$k_F^2 = 2\pi n, \tag{1.2}$$

where n is the 2D density.

Another important quantity is the dimensionless density parameter r_s defined by

$$r_s = \frac{1}{a_0^*} \frac{1}{(\pi n)^{1/2}}, \tag{1.3}$$

where a_0^* is an effective Bohr radius. In silicon inversion layers, the dielectric constant $\kappa = 11.5$, and in GaAs $\kappa = 12$ can be used for a_0^*.

The dimensionless r_s parameter is used as in the 3D case, and the Hamiltonian of 2D electrons can be expressed as in Eq. (1.3) of Chapter 3. However, since r_s for two dimensions is inversely proportional to the square root of number density instead of the cube root, relatively more significant Coulomb effects can be expected for two than for three dimensions at low densities, where in general the interaction term is dominant in the Hamiltonian. In fact, we shall see in the next section that the kinetic energy is much less, while the exchange and correlation energies are much larger than the 3D case for the same r_s. This fact and the wide available density range make the above 2D systems fundamentally important.

The 2D character of the inversion layer and superlattice electrons has been confirmed in several ways. As discussed in Section 2.2, the plasmon dispersion curve approaches zero in two dimensions and a constant in three dimensions in the long-wavelength limit. This feature has been observed clearly in silicon inversion layers and in GaAs/GaAlAs superlattices. Also, a tilted magnetic field has been effectively utilized to identify the two-dimensionality. In a magnetic field that is tilted from the vertical direction, electrons' orbital motion depends only on the perpendicular component. Experiments on the Shubnikov–de Haas oscillation and on the effective g-factor of these systems have shown that this is indeed the case.

4.2. Phase transitions in two-dimensional electron systems

We discuss below two interesting phase transitions due to interaction.

4.2.1. *Valley occupancy phase transition in silicon inversion layers*

The conduction band of silicon is characterized by the ellipsoidal equal-energy surfaces in the six crystal directions. The projection of these surfaces on to the [100] plane produces two equivalent circles and four ellipses that correspond respectively to the ground and higher-energy states. Figure 4.4 illustrates this projection. In the [110] and [111] directions, the state is 4-fold and 6-fold degenerate respectively. These states in which electrons nest are called *valleys*. Thus, the ground state in the [100] direction shown by the black circle in Fig. 4.4 is doubly degenerate. These two states are expected to be equally populated by electrons. However, as in the case of an Ising lattice, this symmetry may be broken by interaction. The basic mechanism

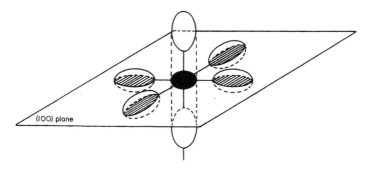

Fig. 4.4.   Projection of the equal energy surfaces of silicon onto the [100] plane.

for this symmetry breaking can be understood by comparing the energy difference between one- and two-valley occupancies [1, 2].

   For this purpose let us first comment on the case of a single valley. The ground-state energy can be evaluated as in the 3D case starting from the high-density limit. Per electron, it can be expressed as

$$\varepsilon_g = \frac{1}{r_s^2} - \frac{1.2004}{r_s} + 0.2287 + \varepsilon_r, \tag{2.1}$$

where the first and second terms correspond respectively to the kinetic and exchange energies. The constant term is the second-order exchange energy, and the last term is the ring-diagram contribution. This contribution can be evaluated by using Eq. (A.19) of the Appendix. The eigenvalues λ_j in this formula can be derived by introducing the 2D version of Eq. (A.13) into Eq. (A.15). Fortunately, by making coordinate transformations, the eigenvalue expression becomes very simple. As a result, the ring energy is given by a single angle integral in a finite domain. From this integral, the following result is obtained [3]:

$$\varepsilon_r = \begin{cases} -0.6137 - 0.1726 r_s \ln r_s + 0.8653 r_s + O(r_s^2 \ln r_s) & (r_s \leqq \sqrt{2}); \\[2mm] -\dfrac{1.2935}{r_s^{2/3}} + \dfrac{1.2004}{r_s} - \dfrac{0.14018}{r_s^{4/3}} + O\left(\left(\dfrac{\ln r_s}{r_s}\right)^2\right) & (r_s \geqq \sqrt{2}). \end{cases} \tag{2.2}$$

Here the first equation is subject to the condition that r_s is less than $\sqrt{2}$, while the second equation is valid for $r_s \geqq \sqrt{2}$. The latter is convergent for $r_s \to \infty$. Note in the latter high-density series that the second term cancels exactly the first-order exchange contribution in Eq. (2.1). Hence, in principle the Hartree–Fock approximation is not valid around r_s of order 1.5.

   In silicon inversion layers, the parameter r_s must be expressed in terms of the effective Bohr radius a_0^*, which in turn depends on the dielectric constant. If the average dielectric constant $\bar{\kappa}$ is used it is given by

$$\pi r_s^2 \left(\frac{m_0 a_0 \bar{\kappa}}{m} \right)^2 = \frac{1}{n}, \tag{2.3}$$

where the effective mass $m = 0.19 m_0$ in the [100] silicon inversion layers. The average dielectric constant can be 7.8. The Fermi wavenumber is given by

$$k_F = \left(\frac{2\pi n}{g_v} \right)^{1/2}, \tag{2.4}$$

where g_v is the valley degeneracy. In the [100] direction, $g_v = 2$.

For a given n, one can then compare the energy difference $\Delta\varepsilon$ between the one- and two-valley states. The kinetic energy

$$\varepsilon_0 = \frac{1}{2r_s^2} \tag{2.5}$$

of the two-valley equally occupied state can be used as units of the energy. The energy difference for low densities is given by [2]

$$\frac{\Delta\varepsilon}{\varepsilon_0} = 1 - 0.14 r_s^{2/3} - \cdots. \tag{2.6}$$

This shows that for small r_s the difference is positive, but it decreases as r_s increases.

A more precise calculation shows that $\Delta\varepsilon = 0$ at $r_s = 8.011$ as shown in Fig. 4.5. At this point what is called *valley occupancy phase transition* takes place. For an average dielectric constant of 7.8, this translates into 1.2×10^{11} cm^{-2}. This is the right order of magnitude observed in experiments. Below this density the electron–electron correlation causes the one-valley state to be more favorable energetically even though the kinetic energy increases. Further reduction of the density would favor a spin-polarized state, but the system is expected to crystallize beforehand.

The above calculation is for absolute zero. The temperature dependency of the kinetic and exchange energies as well as the ring energy at high densities are known. Although logarithmic terms appear, the energy varies almost quadratically with temperature. In particular, the exchange specific heat has been obtained exactly as follows [4]:

$$\frac{c_v}{c_v^0} = 1 + r_s(0.1997 - 0.2251 \ln \eta_0), \tag{2.7}$$

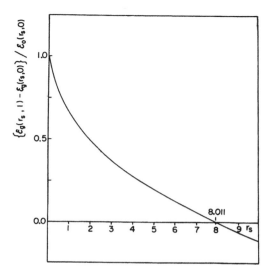

FIG. 4.5. The energy difference between the one- and two-valley states in an [100] silicon inversion layer. (From Isihara and Ioriatti [2])

where $\eta_0 = \varepsilon_F/kT$ and

$$c_v^0 = \frac{\pi^2 k^2 T}{3\varepsilon_F} \tag{2.8}$$

is the ideal specific heat.

From the temperature-dependent energy, the specific heat effective mass m^* has been derived in the following form [5]:

$$\frac{m^*}{m} = 1 + 0.168 n^{-1/2} \times 10^6. \tag{2.9}$$

Thus, the effective mass increases as the density is reduced. According to Eq. (1.3), the corresponding r_s will increase. Hence, the valley occupancy transition might be expected to take place at a larger r_s or a lower density. However, a finite thickness of actual inversion layers is expected to shift the critical r_s to a lower value. To some extent, therefore, these two effects cancel each other.

4.2.2. Wigner crystallization

In 1934 Wigner predicted that an electron gas would make a phase transition into a crystal at a sufficiently low density near absolute zero. At high densities, the Fermi energy is large and the electrons are in a gaseous state. As the density is reduced, their Coulomb interaction becomes more and more important. Finally, they form a crystal in order to reduce the electrostatic energy due to their Coulomb repulsion. However, despite considerable efforts over many years no Wigner crystal was found. The reason was simply that in traditional electron systems such as metals it was difficult to vary the electron density widely enough.

The 2D electron systems such as introduced in Section 4.1 are advantageous in this respect because their electron density can be varied widely. This basic aspect caused renewed interest in Wigner crystallization, and finally Grimes and Adams [6] succeeded in 1979 in observing crystallization of electrons on the surface of liquid helium. These electron systems are very pure and are suitable to such observations. In contrast, it is difficult to obtain low density and high purity in semiconductors.

The experimental observation of Grimes and Adams was based on the coupling of the surface waves of liquid helium and the collective oscillations in the 2D electron system above the surface, that is, ripplon–plasmon coupling. The electron motion in this coupling is not vertical but rather horizontal, because the former costs more energy. At low temperatures, the electron system is a crystal and this coupling was observed. It disappeared as the temperature was increased from around 0.42 K to 0.46 K at density 4.4×10^8 cm^{-2} as shown in Fig. 4.6. More precisely, melting took place at 0.475 K at a density 4.4×10^8 cm^{-2}.

The data points obtained by Grimes and Adams at various values of electron density n are given in Fig. 4.7. This graph shows that the measured melting points are proportional to the square root of electron density n. The linearity of the data points suggests that the dimensionless parameter

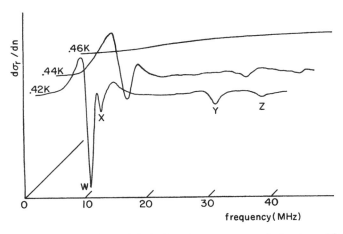

FIG. 4.6. Disappearance of coupled ripplon–plasmon resonances due to melting. (From Grimes and Adams [6])

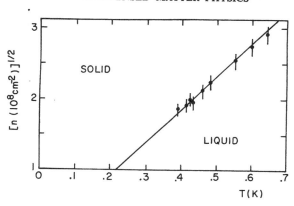

FIG. 4.7. Melting curve of a Wigner crystal. (From Grimes and Adams [6])

$$\Gamma = \frac{\pi^{1/2}e^2 n^{1/2}}{kT} \qquad (2.10)$$

can be used to determine the phase boundary. This is given by the constancy of Γ:

$$\Gamma_m = 131 \pm 7. \qquad (2.11)$$

For a classical OCP, critical Γ values of similar magnitudes have been predicted based on Monte Carlo and molecular dynamics.

Melting of a 2D electron system has also been treated based on the Kosterlitz and Thouless theory. As in Eq. (5.8) of Chapter 1, the melting point is determined by the parameter J, which includes the Lamé's constants λ and μ, and absolute magnitude of the Burgers' vector $\mathbf{b}$. However, the bare value of λ is infinite for a Wigner lattice. Hence, if we use μ instead, the melting point is given by

$$T_m = \frac{\rho b^2 c^2}{4\pi k}, \qquad (2.12)$$

where c is the phonon velocity, and ρ is the mass density. This T_m is found to yield $\Gamma_m = 78.7$. Considering the temperature dependence of the electric constant, Morf [7] then obtained an improved value $\Gamma_m = 128.2$, which is very close to the experimental value of 131.

Two-dimensional phase transitions differ from those in three dimensions since a true long-range order does not exist. If the melting corresponds to a first-order transition, the associated entropy change is estimated to be of order of $0.3k$ per electron. However, Glattli et al. [8] obtained an entropy upper limit of $0.02k$. Hence, it appears that the transition is a continuous type.

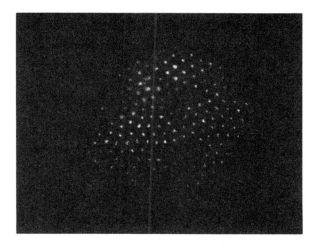

FIG. 4.8. Dimple electron crystal. Bright spots correspond to the center of dimples. (Courtesy of P. Leiderer. See [9])

The surface of liquid helium is not absolutely flat. Electrons are easily trapped into shallow depressions on the surface. These depressions have been found to form a hexagonal lattice called *dimple crystal*. By using total reflection of light from the dimples, such a lattice has been photographed [9]. Figure 4.8 is one example at 3.5 K. (See the frontispiece also.) Each bright spot in this picture corresponds to the center of a dimple. A theoretical electron density, and the surface profile $z(r)$ of a dimple are illustrated in Fig. 4.9. Each dimple contains 5×10^6 electrons in an electric field of 3400 V/cm at a helium temperature of 2.5 K.

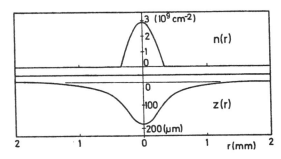

FIG. 4.9. Electron density $n(r)$ and surface profile $z(r)$ of a dimple. (Courtesy of P. Leiderer. See [9]).

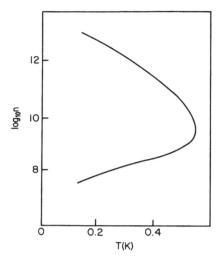

FIG. 4.10. Schematic melting curve of a Wigner lattice.

In the above, we have used the classical parameter Γ. Near absolute zero, the appropriate dimensionless parameter is r_s. A schematic phase diagram is illustrated in Fig. 4.10. Note that for very small r_s and low temperatures, reentrance into the liquid state takes place. According to molecular dynamics calculations [10], reentrance appears to take place above r_s of around order 34.

Evidence for the Wigner crystal in the extreme quantum limit is mounting on the basis of observations on the 2D electron system in GaAs/GaAlAs heterostructures in magnetic fields, and a phase diagram has been obtained with the Landau-level filling factor ν as a relevant variable. For instance, Andrei et al. [6] observed strong resonance absorption at certain frequencies that they associated with those of the normal modes of the electron solid. Jiang et al. [6] found a drastic change in the temperature dependences of the resistance when the filling factor was slightly increased; the dependences corresponded to an incompressible quantum fluid at $\nu = 1/5$, but to an insulating phase such as a pinned Wigner crystal at $\nu = 0.21$. Goldman et al. [6] reported observations of an electric-field threshold conduction and of related ac voltage generation that can be interpreted as evidence for formation of a pinned Wigner crystal.

4.3. Effective g-factor

The g-factor of electrons in bulk silicon is 2. In silicon inversion layers the g-factor is strongly enhanced toward low densities, as was first observed by Fang and Stiles [11]. Such a density dependence can be attributed to

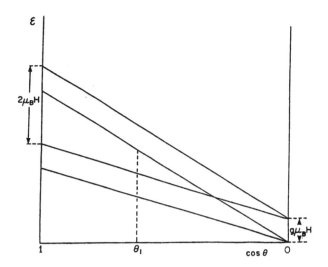

FIG. 4.11.   Landau levels in a tilted magnetic field.

Coulomb interaction that becomes more effective at low densities. Their experiment was conducted in a magnetic field that was tilted by angle θ from the direction perpendicular to an n-type silicon surface. The temperature range was 1.3 to 4.2 K. It is important to recognize that this tilted magnetic field experiment not only determined the effective g-factor but also demonstrated the two-dimensionality of the electron system.

The Landau splitting of a 2D electron depends only on the perpendicular component of the magnetic field, while the spin splitting is determined by the total field. The Landau levels are given by

$$\varepsilon_n = (n + \tfrac{1}{2})2\mu_B H \cos\theta \pm \frac{g}{2}\mu_B H, \qquad (3.1)$$

where μ_B is the Bohr magneton. When plotted against $\cos\theta$, the levels for a given n are represented by a pair of parallel lines, as shown in Fig. 4.11. At a particular angle θ_1, the energy difference between the $n = 1$ and 0 levels is equal to that between the two spin states of the ground level. Thus, at this point

$$\hbar\omega_c \cos\theta_1 = 2g\mu_B H, \qquad (3.2)$$

where on the left-hand side we have used the cyclotron frequency because the orbital motion is characterized by an effective mass m. In contrast, the right-hand side, which is due to spin, is given by the Bohr magneton. Hence, the effective g-factor is given by

$$g = (m_0/m) \cos \theta_1. \tag{3.3}$$

At the angle θ_1 the phase of the Shubnikov–de Haas oscillation is reversed. This enables an accurate determination of θ_1.

The effective g-factor can be derived from the spin-dependent part of the ground-state energy [12]. If p_F is the Fermi momentum in the unpolarized state, the Fermi momenta of the spin up and down states are given by

$$p_F^\uparrow = (1 + P)^{1/2} p_F; \qquad p_F^\downarrow = (1 - P)^{1/2} p_F, \tag{3.4}$$

where P is a polarization parameter determining the ratio of the two components. Since the experimentally determined effective g-factor does not show any oscillation, one can make a small-P approach. To order P^2, which is the relevant order, the difference in the ground-state energies of the polarized and unpolarized states is given per electron by

$$\varepsilon_g(r_s, P) - \varepsilon_g(r_s, 0) = \frac{e^2 m}{2\kappa^2 a_0 m_0} A(r_s) P^2 - \mu_B H P, \tag{3.5}$$

where r_s is determined by

$$\pi r_s^2 \left(\frac{m_0 a_0 \kappa}{m} \right)^2 = \frac{1}{n}. \tag{3.6}$$

κ is the average dielectric constant and a_0 is the Bohr radius. The factor $A(r_s)$ can be given by the sum of the contributions from the kinetic, exchange, and correlation energies. For instance, the kinetic and exchange contributions are given by

$$A^0(r_s) = \frac{1}{g_v r_s^2}, \tag{3.7}$$

$$A^x(r_s) = -\frac{2^{1/2}}{\pi g_v^{1/2} r_s}, \tag{3.8}$$

where g_v is the valley degeneracy. The correlation contribution is more complicated, but since the second-order exchange energy is independent of P it is given by the contribution from the ring diagrams. It turns out that the exchange contribution given by Eq. (3.8) is exactly canceled by a term from the ring contribution. The ring-diagram contribution $A^c(r_s)$ can be obtained as a function of P since it is a function of the Fermi momentum. After all these contributions are obtained, the polarization parameter is

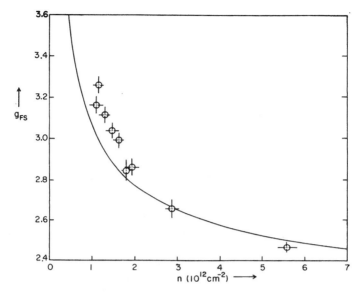

FIG. 4.12. Effective g-factor of a 2D electron system. Data, Fang and Stiles [11]; solid curve, Isihara and Ioriatti [12].

determined so as to minimize the energy. We then obtain

$$P = \mu_B H a_0 \frac{m_0}{m} \frac{\kappa^2}{e^2 A(r_s)}. \tag{3.9}$$

Note that P is inversely proportional to $A(r_s)$.

The magnetic moment is given by

$$M = n\mu_B P. \tag{3.10}$$

Hence, the paramagnetic susceptibility is given by

$$\frac{\chi}{\chi_0} = \left(1 - \frac{(2g_v)^{1/2}}{\pi} + g_v r_s^2 A^c(r_s)\right)^{-1}, \tag{3.11}$$

$$\chi_0 = \frac{g_v m}{\pi \hbar^2} \mu_B^2. \tag{3.12}$$

The effective g-factor is obtained simply by forcing χ to take the form of Eq. (3.12).

The theoretical result [12] thus obtained with the above effective mass correction is illustrated in Fig. 4.12. In the [100] direction, $m = 0.19m_0$, but for a wide density range its density dependence due to electron correlation must be taken into consideration. An experimental relation [13]

$$\frac{m}{m^*} = 1 - 0.071 r_s^*$$ (3.13)

may be adopted for this purpose. In this relation,

$$r_s^* = me^2 \hbar^{-2} \kappa_{sc}^{-1} (\pi n)^{-1/2}.$$ (3.14)

For the present case, $\kappa_{sc} = 12$ and $m^* = 0.195 m_0$. The solid curve in Fig. 4.8 includes the above effective mass correction. Even though there is room for theoretical improvements, the overall agreement without any artificial theoretical adjustment is remarkable. The increase of the g-factor is a correlation effect since the first-order exchange contribution is canceled out and the second-order exchange energy does not contribute.

When the field strength is increased, the effective g-factor is expected to show oscillations corresponding to the de Haas–van Alphen effect. For high densities, such an oscillating effective g-factor is given by [14]

$$g^* = 2 + \frac{\pi}{\alpha \beta n} \sum_s (-)^s \frac{\cos(gs/2)}{\sinh(\pi^2 s/\alpha)} \coth(\pi^2 s/\alpha) \cos(\pi s/\gamma_0),$$ (3.15)

where

$$\alpha = \frac{\mu_B H}{kT}, \qquad \beta = \frac{1}{kT}, \qquad \gamma_0 = \frac{\mu_B H}{\varepsilon_F}.$$ (3.16)

The above result has been obtained in consideration of the contributions from the first-order exchange and ring diagrams. The period of oscillation depends on the ideal Fermi energy ε_F, which in turn is determined by the density of electrons. For high densities where such oscillations are strong, the effective g-factor does not have any other r_s dependence. The reason is simply that g^* given above was obtained only from the oscillating part of the magnetic response, which is actually related to electrons' orbital motion rather than spin.

4.4. Integral quantized Hall effect

When a magnetic field is applied vertically to the direction of a current in a slab, a *Hall voltage* appears perpendicularly to both. The motion of an electron in the slab may be described in a single relaxation time approximation by

$$m \frac{d\mathbf{v}}{dt} + \frac{m\mathbf{v}}{\tau} = -e \left[\mathbf{E} + \frac{1}{c} \mathbf{v} \times \mathbf{H} \right].$$ (4.1)

In a steady state $dv/dt = 0$. The current vector is given by

$$\mathbf{j} = -ne\mathbf{v}$$
$$= \sigma \cdot \mathbf{E}, \tag{4.2}$$

where σ is the conductivity tensor. Its inverse is the resistivity tensor ρ:

$$\mathbf{E} = \rho \cdot \mathbf{J} \tag{4.3}$$

These two tensors are related with each other as follows:

$$\sigma_{xy} = -\frac{\rho_{xy}}{\rho_{xx}^2 + \rho_{xy}^2}; \qquad \sigma_{xx} = \frac{\rho_{xx}}{\rho_{xx}^2 + \rho_{xy}^2}. \tag{4.4}$$

In a steady state, $dv/dt = 0$, and we find

$$\sigma_{xx} = \sigma_{yy} = \frac{\sigma_0}{1 + (\omega_c \tau)^2}; \qquad \sigma_{xy} = -\sigma_{yx} = -\frac{\sigma_0 \omega_c \tau}{1 + (\omega_c \tau)^2}. \tag{4.5}$$

Here,

$$\sigma_0 = \frac{ne^2 \tau}{m}, \tag{4.6}$$

and ω_c is the cyclotron frequency:

$$\omega_c = \frac{eH}{mc}. \tag{4.7}$$

The electric current may be allowed to flow only in the x direction. Under the condition

$$j_y = 0, \tag{4.8}$$

the *Hall conductivity* is given by

$$\sigma_H = \frac{j_x}{E_y} = -\frac{nec}{H}$$
$$= \frac{1}{\rho_{yx}}. \tag{4.9}$$

Note that in contrast to σ_0 this conductivity is dependent on the sign of the charge. This helps to identify the type of charge carriers.

Instead of the condition given by Eq. (4.8), if the condition

$$E_y = 0 \tag{4.8'}$$

is imposed, the Hall conductivity is given by

$$\sigma_{yx} = \frac{j_y}{E_x} = -\sigma_H. \tag{4.10}$$

Equation (4.4) shows that in 2D if $\sigma_{xx} = 0$ then $\rho_{xx} = 0$:

$$\sigma_{xx} = 0 \rightarrow \rho_{xx} = 0.$$

According to Eq. (4.5), σ_{xx} does not vanish for a finite density. However, if

$$\sigma_{xx} = 0, \tag{4.11}$$

then σ_{xy} is given by

$$\sigma_{xy} = \sigma_H = \frac{j_x}{E_y} \tag{4.12}$$

$$= -\frac{nec}{H}. \tag{4.13}$$

The last equality is based on the classical equation of motion. It shows that the absolute magnitude of σ_{xy} is proportional to electron density n.

The Hall coefficients of bulk metals, especially alkali metals, agree generally with the theoretical expectation. Measurements of the magneto-conductivity of surface electrons started around 1966 and those of the Hall conductivity around 1975. These measurements supplied important information such as the dimensionality and Coulomb effects. While interest in 2D electron systems was rapidly growing, a very unusual phenomenon of the electrons in silicon MOSFETs was discovered in 1980 by von Klitzing, Dorda, and Pepper [15] during the course of studying electron localization: the Hall resistance took on plateau values for finite intervals of gate voltage where the magnetoresistance dropped to zero. Note that electron density is changed by gate voltage in silicon MOSFETs. The plateau values correspond to the quantization of the resisticity component ρ_{xy} in units of h/e^2 such that

$$\rho_{xy} = \frac{h}{e^2 i}, \tag{4.14}$$

where i is an integer. The magnetoconductivity σ_{xx} in the plateau region stays nearly zero: it can be 10^{-7} times smaller than its maximum value. The phenomenon is now called the *integral quantized Hall effect*.

The integral quantized Hall effect (IQHE) is illustrated schematically in Fig. 4.13. The upper graph illustrates the Hall conductivity σ_{xy} as a

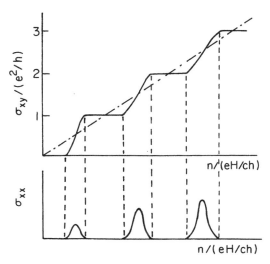

FIG. 4.13. Schematic representation of the IQHE.

function of electron density n. The lower curve illustrates the magneto-conductivity. This graph may be considered to represent the density of states plotted against the Fermi energy. The dashed line represents the classical expectation.

The IQHE is very important particularly because the Hall plateaus enable very accurate determination of the fine-structure constant:

$$\alpha = \frac{e^2}{h} \frac{\mu_0 c}{2}, \tag{4.15}$$

which is a fundamental constant. Here, $\mu_0 = 4\pi \times 10^{-7}$ H/m is the vacuum permeability and c is the velocity of light. Note that the velocity of light is known with great accuracy. At the same time, the IQHE provides a new resistance standard because the Hall resistance can be determined with accuracy of order 0.03 parts per million.

The resistivity ρ_{xx} shows an activated behavior at low but finite temperatures, indicating the existence of an energy gap. As the temperature is increased, the width of Hall plateaus decreases. In silicon MOSFETs the width is determined by the density of electrons, or the gate voltage. In GaAs/GaAlAs heterostructures, a magnetic field is used to change the electron filling factor of Landau levels. In an InGaAs/InP heterostructure, the magnetoresistivity is found to show an interesting temperature variation [16]; if ρ_{xx} vanishes in an interval of magnetic field with a half-width Δ, a power law holds:

$$\Delta \sim T^{\kappa}, \tag{4.16}$$

where

$$\kappa = 0.42 \pm 0.04. \tag{4.17}$$

At the same time, the maximum slope of ρ_{xy} with respect to the magnetic field varies as

$$\left(\frac{\partial \rho_{xy}}{\partial H}\right)_{\text{max}} \sim T^{-\kappa}. \tag{4.18}$$

The IQHE has been interpreted in several ways. It is important to recognize that σ_{xx} vanishes in the plateau region. Although the integral plateau values coincide with the classical values at the center of a broadened Landau level, the coincidence occurs only at such a particular point. That is, the crux of the IQHE is in the appearance of "plateaus" rather than "points" where $\sigma_H/(e^2/h)$ is an integer. These facts suggest that electron states are extended only near the center of each broadened Landau level and are localized at the outskirts.

The classical relation between σ_{xx} and σ_{xy} in Eq. (4.5) can be expressed as

$$\sigma_{xy} = -\frac{nec}{H} + \frac{\sigma_{xx}}{\omega_c \tau}. \tag{4.19}$$

Hence, if $\sigma_{xx} = 0$ for an interval of density between a pair of adjacent Landau levels with index i and $i + 1$,

$$\sigma_{xy} = -\frac{nec}{H}. \tag{4.20}$$

When all i Landau levels are filled by electrons, no current flows and

$$n = i\left(\frac{eH}{ch}\right). \tag{4.21}$$

Hence,

$$\sigma_{xy} = -i\frac{e^2}{h}.$$

This result does not explain the IQHE but indicates that the existence of a localized region is essential to understand the phenomenon.

Let us now try to interpret the IQHE [17]. Since this effect is characterized by the fundamental constant α, the effect of impurities may be considered only to broaden the Landau levels. Let us assume that each broadened Landau level has an extended region near its center and localized

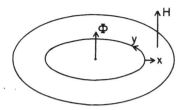

FIG. 4.14. A Corbino disk geometry.

regions at its edges on both sides. If the Fermi energy lies in the ith localized region below which all the levels are filled, no current can flow.

Let us use a Corbino geometry as shown in Fig. 4.14. An external electric field is in the x direction with component $E_x = E$. In the perpendicular y direction, we use the condition $E_y = 0$. A magnetic field is applied perpendicularly to the disk. The size of the disk is assumed to be so large that x and y constitute a rectangular coordinate system. If an electron in an extended state circulates the disk once at $y = R$, the magnetic flux $\pi R H$ should be an integral multiple of the flux quantum defined by

$$\phi_0 = \frac{ch}{e}. \tag{4.22}$$

Let us now imagine sending in a magnetic flux Φ through a very thin solenoid at the center of the disk. The flux is confined in the solenoid and does not produce a magnetic field to affect the motion of the electron. We can require rot $\mathbf{A} = 0$, where $\mathbf{A}$ is the vector potential of the flux. Hence, the vector potential is given by a gradient of a single-valued function χ:

$$\mathbf{A} = \text{grad } \chi(r). \tag{4.23}$$

The vector potential in the Schrödinger equation can be eliminated by a gauge transformation:

$$\psi \to \psi \exp\left[i \frac{e}{ch} \chi(r) \right]. \tag{4.24}$$

When the electron circulates the disk once with all other electrons fixed in space, ψ will change such that

$$\psi \to \psi \exp\left[i \frac{e\Phi}{ch} \right]. \tag{4.25}$$

Hence,

$$\Phi = \text{integer} \times \phi_0, \tag{4.26}$$

and the total flux given by the sum $\pi R^2 H + \Phi$ must also be an integral multiple of ϕ_0:

$$\pi R^2 H + \Phi = j\phi_0. \tag{4.27}$$

For a given integral value of J, increasing Φ causes decreasing R. If Φ is increased by ϕ_0, R will be reduced such that it corresponds to $(j-1)$ in the above equation. Other than this change, no effect of the change of Φ is expected. However, in the presence of an electric field E_x, the energy changes

$$\Delta \varepsilon = e E_x L$$

should take place. Here, L is the distance between the outer and inner edges of the disk. The current density j_y due to the electron is

$$j_y = -\frac{c}{L} \frac{\Delta \varepsilon}{\Delta t} \frac{\Delta t}{\Delta \Phi}$$

$$= -\frac{c}{L} \frac{\Delta \varepsilon}{\phi_0}. \tag{4.28}$$

That is,

$$j_y = -\frac{e^2}{h} E_x. \tag{4.29}$$

The same process can be repeated in all i Landau levels. The total current will become i times that given by Eq. (4.26), so that

$$\sigma_{yx} = -\sigma_H = \frac{e^2}{h} i. \tag{4.30}$$

This demonstrates the IQHE.

The IQHE has been interpreted in other ways based in general on the existence of extended and localized states in each Landau level. The former states are located about the center of each impurity-broadened Landau level and correspond to those electron orbits that cover the entire system. The latter states are located at the outskirts of the Landau levels. In real space they may be expected around impurities.

4.5. Fractional quantized Hall effect

The *fractional quantized Hall effect* (FQHE) was discovered by Tsui, Störmer, and Gossard [18] two years after the discovery of the IQHE. They observed that the Hall resistivity of high-mobility GaAs/GaAlAs heterostructures takes on a plateau value as a function of H:

$$\rho_{xy} = 3 \frac{h}{e^2} \qquad (v = 1/3), \tag{5.1}$$

when the filling factor of the lowest Landau level is 1/3. If the density of electrons is n, the filling factor is defined by

$$v = \frac{n}{(eH/ch)}. \tag{5.2}$$

In the plateau interval, the resistivity ρ_{xx} drops to zero. Therefore, the IQHE and the FQHE are similar, except that the FQHE occurs basically when the lowest Landau level is only fractionally occupied.

Since the discovery, the FQHE has been observed at $v = 1/3, 2/3, 1/5, 2/5, 3/5, \ldots$. It occurs primarily when

$$v = \frac{p}{q}, \tag{5.3}$$

where q is an odd integer. However, the effect is actually not restricted to the lowest Landau level. In fact, the cases $v = 4/3, 7/3, 7/5, 8/5$, and so on, have been observed. Moreover, the case of an even denominator at $v = 5/2$ has also been reported [19]. However, this is exceptional, and is attributable to a spin-unpolarized state in distinction to the ordinary spin-polarized fractionally quantized states.

At finite temperatures, the magnetoresistivity shows an activated behavior, as in the case of the IQHE, except for very low temperatures where a variable-range hopping behavior is observed. The activation energy is of order $0.02(e^2/\kappa l)$, where κ is the dielectric constant and l is the magnetic length. This fact, and low electron densities, suggest that the FQHE is due to electron–electron interactions in a magnetic field. Also at finite temperatures, the slope of $d\rho_{xy}/dH$ and the width ΔH of ρ_{xx} depend on temperature. It has been found [20] that

$$\left(\frac{d\rho_{xy}}{dH}\right)_{max} \sim T^{-\kappa},$$

$$\frac{1}{\Delta H} \sim T^{-\kappa}. \tag{5.4}$$

Interestingly, the exponent κ is the same as in the case of the IQHE:

$$\kappa = 0.43 \pm 0.02. \tag{5.5}$$

The width ΔH corresponds to the field between the two extrema of $d\rho_{xx}/dH$ observed at $v = 2/5$ and $1/3$ for $22\,\text{mK} < T < 1.2\,\text{K}$ in GaAs/GaAlAs heterostructures.

Since the FQHE is observed primarily in the lowest Landau level, the electron state may be constructed by a linear combination of the ground Landau states:

$$\phi_l(z) = (z^{l+1}\pi l!)^{-1/2} z^l \exp(-|z|^2/4), \tag{5.6}$$

where, in units of the magnetic length, the coordinate of an electron in two dimensions is expressed as

$$z = x + iy,$$

where l is an angular momentum quantum number that specifies the states in the ground level. Since only the ground Landau level is involved, the momentum dependency of the wavefunction and the kinetic energy can be ignored for the interpretation of the FQHE.

If a total of N electrons are in the ground Landau level, they must be distributed among the angular momentum states such that the Pauli principle is satisfied. Laughlin [21] proposed the following wavefunction for the $v = 1/m$ state, where m is an odd integer:

$$\psi_m = \prod_{i<j}^{N} (z_i - z_j)^m \exp\left(-\sum_{i=1}^{N} |z_i|^2\right). \tag{5.7}$$

The total angular momentum determined by the power of z's in the polynomial part is given by

$$L = mN(N-1)/2. \tag{5.8}$$

Note that this polynomial part can be expressed by the van der Mond determinant:

$$\prod_{i<j}^{N} (z_i - z_j) = \begin{vmatrix} z_1^0 & z_1^1 & z_1^2 & \cdots & z_1^{N-1} \\ z_2^0 & z_2^1 & z_2^2 & \cdots & z_2^{N-1} \\ \vdots & & & & \vdots \\ z_N^0 & z_N^1 & z_N^2 & \cdots & z_N^{N-1} \end{vmatrix} \tag{5.9}$$

This determinant shows clearly how the electrons are distributed in conformity with the Pauli principle.

Moreover, ψ_m has an interesting electrostatic analog: the square of the absolute magnitude of the wavefunction can be expressed in the form:

$$|\psi_m|^2 = \exp(-\beta\Phi), \tag{5.10}$$

where

$$\beta = 1/m, \tag{5.11}$$

$$\Phi = -\sum_{i<j} 2m^2 \ln|z_i - z_j| + \frac{m}{2} \sum_{i=1}^{N} |z_i|^2. \qquad (5.12)$$

When β is considered to be $1/kT$, Eq. (5.10) represents the Boltzmann factor of an OCP of quasiparticles with charge m because the corresponding electrostatic potential is given by Eq. (5.12). The first term in Eq. (5.12) is the mutual repulsion and the second term is the attraction due to a cloud of neutralizing charges that are distributed uniformly in the system with a charge density $1/(2\pi)^{1/2}$.

According to Monte Carlo studies, such an OCP is a fluid if

$$\Gamma = (\pi n)^{1/2} e^2 \beta \qquad (5.13)$$

is less than the critical value 128 and is a crystal above this value. In the present case,

$$\Gamma = (2m)^{1/2}. \qquad (5.14)$$

For the observed FQHE, the Γ values are much less than the critical value. Therefore, the OCP is a liquid.

Moreover, this liquid state has a lower ground-state energy than the charge-density wave (CDW) state. The energy for $v = 1/3$ in the limit $N \to \infty$ has been found numerically to be [22]

$$\varepsilon_g = -0.4100 \qquad (5.15)$$

in units of $e^2/\kappa l$. This is lower than the corresponding CDW state, which has energy -0.389. Furthermore, it has been found numerically that the ground-state energy as a function of m reaches cusp-type minima at odd integers. Thus, one can conclude that the electrons fall into liquidlike fractional states at odd-integral $m = 1/v$.

How can numerous other fractional states be interpreted? We will discuss shortly the fact that the state $v = 1 - 1/q$ can be considered to be the electron–hole conjugate of the state $1/q$. For instance, the 2/3 state is conjugate to the 1/3 state. Note that not only electrons but also holes can fall into liquid states. We will also show that the state ψ_m for $m = 1/v$ can be considered as the parent state of the fractional states at $v = p/q$.

In order to see a hierarchy of the fractional states, let us consider perturbing the $1/m$ state by expanding the system very slightly at a fixed magnetic field. Since this state is stable, the electrons prefer to remain in the same state. The expansion will cause a charge deficiency that would cost energy. In order to create a quasihole at a certain position z_0 we pierce the system at z_0 with an infinitesimally thin solenoid and send in a flux quantum

ϕ_0 adiabatically. The state with one quasihole at z_0 can be represented by

$$\psi_m^+ = \prod_{i=1}^{N} (z - z_0)\psi(z_1, z_2, \ldots, z_N). \tag{5.16}$$

In fact, if we use the same electrostatic analog as expressed by Eq. (5.10), we have

$$|\psi_m^+|^2 = \exp(-\Phi^+/m). \tag{5.17}$$

$$\Phi^{+1} = -2m^2 \sum_{i<j} \ln|z_i - z_j| + \frac{m}{2}\sum_i |z_i|^2 - 2m \sum_i \ln|z_i - z_0|. \tag{5.18}$$

In comparison with Eq. (5.12), this electrostatic potential has an extra term which is due to an additional phantom charge 1 at z_0. In order to screen this charge, the plasma will accumulate charge -1, which is equivalent to a real charge $-1/m$.

If, on the other hand, a quasiparticle is added to the system at position z_0, the wavefunction is given by

$$\psi_m^{-1} = \prod_{i=1}^{N} \left(2\frac{\partial}{\partial z_i} - z_0^*\right)\psi_m(z_1, z_2, \ldots, z_N), \tag{5.19}$$

where the prefactor operates only on the polynomial part of ψ_m. That ψ_m^{-1} is the correct wavefunction can be seen on the basis of the same electrostatic analog as before.

If there are many quasiholes or quasiparticles in the state ψ_m, they can form their own liquid states. If M quasiholes are created, the wavefunction is given by

$$\psi_m^{+M} = \prod_{i<k}^{M} (Z_i - Z_k)^\theta \exp\left(-\sum_k |Z_k|^2/4l_h^2\right) \prod_j^{N} (z_j - Z_k)\psi_m(z_1, z_2, \ldots, z_N), \tag{5.20}$$

where Z_k are the coordinates of the quasiholes, while z_j are those of the electrons. The quasiholes have magnetic length l_h. θ can be 0 (Bose statistics) or 1 (Fermi statistics) so that the quasiholes obey *anyon statistics*.

The zeros of the wavefunction can be associated with particles. The original wavefunction ψ_m is characterized by the m-fold zeros. This represents the strongest approach to zero for a given $\nu = 1/m$. In a magnetic field corresponding to the magnetic length l there are $1/(2\pi l^2)$ zeros per unit area. We call these zeros *vortices*. Since we have kept the magnetic field constant in creating quasiholes, the number of electron vortices is given by

$$mn + n_h = \frac{eH}{ch}$$

$$= \frac{1}{2\pi l^2}$$

$$= g_e, \tag{5.21}$$

where the density of quasiholes is denoted by n_h. For given H the electron density is reduced from g_e/m to $(g_e - n_h)/m$.

The number of the zeros associated with the quasiholes is given by

$$n + \theta n_h = \frac{1}{2\pi l_h^2}$$

$$= g_h. \tag{5.22}$$

At a certain density these quasiholes form a new liquid state. In analogy to the original wavefunction ψ_m, the corresponding wavefunction is given by

$$\psi_{p_1+\theta} = \prod_{i<k} (Z_i - Z_k)^{p_1+\theta} \exp\left(-\sum_k \frac{|Z_k|^2}{4l_h^2}\right).$$

Since the quasihole statistics are determined by θ, p_1 must be an even number. The filling factor of the quasiholes is

$$\nu_h = \frac{n_h}{g_h}$$

$$= \frac{1}{p_1 + \theta}. \tag{5.23}$$

On the other hand, Eqs. (5.21) and (5.22) yield

$$n_h = g_e - mn, \qquad g_h = n + \theta(g_e - mn).$$

Using these in Eq. (5.23) we find that the corresponding filling factor of electrons is given by

$$\nu = \frac{n}{g_e}$$

$$= \frac{1}{m + 1/p_1}. \tag{5.24}$$

This result is independent of θ. If $m = 1$ and $p_1 = 2$, $\nu = 2/3$. That is, the fractional state $2/3$ is the electron–hole conjugate state of $\nu = 1/3$.

One can work on the case of quasiparticles in a similar manner. We arrive at

$$\nu = \frac{1}{m - 1/p_1}. \tag{5.25}$$

Repeating the same process for either the quasihole liquid or quasiparticle liquid, we can generate a hierarchy of the fractional states.

It is important to recognize that according to the electrostatic analog of Eq. (5.10), ψ_m is associated with particles with fractional charge $m = 1/\nu$. In general, a 2D conductivity σ_{xx} is proportional to e^2/h so that the conductivity at the $\nu = p/q$ fractional state is expected to scale with $(1/q^2)$. From its activated form

$$\rho_{xx}(T) = \rho^c_{xx} \exp(-\Delta/2kT) \tag{5.26}$$

the classical limit ρ^c_{xx} can be determined. This in turn yields a temperature-independent limiting magnetoconductivity σ^c_{xx}. Experimental evidence for such fractional charges has been obtained recently [23]. Figure 4.15 illustrates σ^c_{xx} as a function of $1/q$ obtained by Clark et al. [23]. Fits (a) and (b) correspond respectively to $c = 0.91$ and 1.0 in

$$\sigma^c_{xx} = \frac{c(e/q)^2}{h}, \tag{5.27}$$

while (c) represents $\sigma^c_{xx} = (1.07/q^{2.1})e^2/h$. The graph shows that the quasiparticles in the $1/q$ FQHE for $\nu = p/q$ have charge e^* given by

$$e^* = \pm\frac{e}{q}. \tag{5.28}$$

Even more convincing evidence for such a fractional charge is provided by the plot in Fig. 4.16 for the case $q = 3$.

Laughlin's wavefunction ψ_m given by Eq. (5.7) describes the FQHE well but is not a true ground state. In an effort to approach the true ground state, Takano and Isihara [24] proposed a new variational wavefunction for $m = 1/\nu = 2p + 1$ as follows:

$$\psi = \prod_{i,j} (z_i - z_j) f_{ij} f_{ji} \exp\left[-\sum_i |z_i^2|/4\right], \tag{5.29}$$

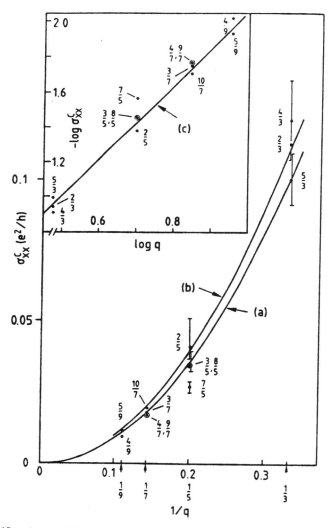

FIG. 4.15. $\sigma_{xx}^c = \sigma_{xx}(1/T = 0)$ indicating fractional charges e^* in the FQHE. Curves a and b correspond respectively to $c = 0.91$ and 1 in Eq. (5.27), and curve c to $(1.07/q^{2.1})e^2/h$. (From Clark et al. [23]).

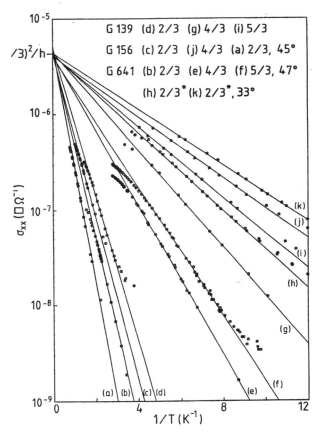

FIG. 4.16. Activation data for $q = 3$ FQHE in GaAs/GaAlAs heterostructures, showing the quasiparticle charge $e^* = \pm e/3$. (From Branch et al. [23])

where

$$f_{ij} = z_i - z_j + c \sum_k{}' z_k. \qquad (5.30)$$

In contrast to ψ_m, the wave-function ψ includes a variational parameter c. If $c = 0$, ψ is reduced to ψ_m. The c term in Eq. (5.3) takes account of the presence of the other electrons than the chosen pair ij. This variational wavefunction is found to produce ground-state energies lower than Laughlin's for small but finite c values. Moreover, their numerical results show that the ground state of the $1/3$ fractional state is triply degenerate. Since only one-third of angular momentum states are occupied in this case, the true wavefunction is expected to have 3-fold degeneracy. Therefore, ψ is an improved wavefunction. Note that ψ is constructed within the lowest Landau state but f_{ij} defined by Eq. (5.30) is not translationally invariant.

It has been shown that Laughlin's ground state is associated with a peculiar off-diagonal long-range order, although the one-body density matrix evaluated directly in term of Eq. (5.7) is short-ranged [25]; a gauge transformation that in effect replaces the polynomial part $(z_i - z_j)^m$ of ψ_m by $|z_i - z_j|^m$ yields a new density matrix that decays algebraically [25]:

$$\rho(z, z') \sim |z - z'|^{-m/2}. \tag{5.31}$$

We say in such a case that there exists *off-diagonal long-range order*. The transformation is associated with the singular gauge field used in the study of anyons [26]. The FQHE can be described as Bose condensation based on anyon statistics [27]. Note that the above replacement of the polynomial part removes the requirement of Fermi statistics.

REFERENCES

1. W. L. Bloss, L. J. Sham, and B. Vinter, *Phys. Rev. Lett.* **43**, 1529 (1979).
2. A. Isihara and L. C. Ioriatti, Jr., *Phys. Rev. B* **25**, 5534 (1982).
3. A. Isihara and L. C. Ioriatti, Jr., *Physica A* **103**, 621 (1980). L. C. Ioriatti, Jr. and A. Isihara, *Z. Phys. B* **44**, 1 (1981).
4. L. Ioriatti, Jr. and A. Isihara, *Phys. Stat. Sol.* (B) **97**, k65 (1980).
5. Isihara and T. Toyoda, *Phys. Rev. B* **21**, 3358 (1980).
6. C. C. Grimes and G. Adams, *Phys. Rev. Lett.* **42**, 795 (1979). E. Y. Andrei et al., *Phys. Rev. Lett.* **60**, 2765 (1988). H. W. Jiang et al., *Phys. Rev. Lett.* **65**, 633 (1990). V. H. Goldman et al., *Phys. Rev. Lett.* **65**, 2189 (1990).
7. R. Morf, *Phys. Rev. Lett.* **43**, 931 (1979).
8. D. C. Glattli, E. Y. Andrei, and F. I. B. Williams, *Phys. Rev. Lett.* **53**, 588 (1984).
9. W. Ebner and P. Leiderer, *Phys. Lett. A* **80**, 277 (1980). P. Leiderer, W. Ebner, and V. B. Shikin, *Surf. Sci.* **113**, 405 (1982).
10. M. Imada and M. Takahashi, *J. Phys. Soc. Jpn* **53**, 3770 (1984).
11. F. F. Fang and P. J. Stiles, *Phys. Rev.* **174**, 823 (1968).
12. A. Isihara and L. Ioriatti, Jr., *Physica B* **113**, 42 (1982).
13. J. L. Smith and P. J. Stiles, *Phys. Rev. Lett.* **29**, 102 (1972).
14. A. Isihara and D. Y. Kojima, *Phys. Rev. B* **19**, 846 (1979).
15. K. von Klitzing, G. Dorda, and M. Pepper, *Phys. Rev. Lett.* **45**, 494 (1980). R. Prange and S. M. Girvin (eds.), *The Quantum Hall Effect* (Springer-Verlag, New York, 1987).
16. H. P. Wei, D. C. Tsui, M. A. Paalanen, and A. M. Pruisken, *Phys. Rev. Lett.* **61**, 1294 (1988).
17. R. B. Laughlin, *Phys. Rev. B* **23**, 5632 (1981). B. I. Halperin, *Phys. Rev. B* **25**, 2185 (1982).
18. D. C. Tsui, H. L. Störmer, and A. C. Gossard, *Phys. Rev. Lett.* **48**, 1559 (1982).
19. R. Willett, J. P. Einsenstein, H. L. Störmer, D. C. Tsui, A. C. Gossard, and J. H. English, *Phys. Rev. Lett.* **59**, 1776 (1987). R. C. Clark, R. J. Nicholas, A. Usher, C. T. Foxon, and J. J. Harris, *Surf. Sci.* **170**, 141 (1986).

20. L. Engel, H. P. Wei, D. C. Tsui, and M. Shayegan, in *Electrical Properties in Two-Dimensional Systems*, Grenoble, 1989, to be published.

21. R. Laughlin, *Phys. Rev. Lett.* **50**, 1395 (1983); *Surf. Sci.* **142**, 163 (1984).

22. J. M. Caillol, D. Levesque, J. J. Weiss, and J. P. Hansen, *J. Stat. Phys.* **28**, 325 (1982). R. Morf, *Phys. Rev. Lett.* **43**, 931 (1979).

23. R. G. Clark, J. R. Mallett, S. R. Haynes, J. J. Harris, and C. T. Foxon, *Phys. Rev. Lett.* **60**, 1747 (1988). J. V. Branch, R. G. Clark, C. Andrikidis, S. J. Collocott, G. J. Griffiths, J. J. Harris, C. T. Foxon, and J. T. Chalker, *Phys. Rev. Lett.*, to be published. J. A. Simmons, H. P. Wei, L. W. Engel, D. C. Tsui, and M. Shayegan, *Phys. Rev. Lett.* **63**, 1731 (1989). A. M. Chang and J. E. Cunningham, *Solid State Commun.* **72**, 651 (1989).

24. K. Takano and A. Isihara, in *Anderson Localization*, ed. by T. Ando and H. Fukuyama (Springer-Verlag, New York, 1988), p. 268. A. Isihara, in *Solid State Physics*, ed. by H. Ehrenreich and D. Turnbull (Academic Press, New York, 1989), **42**, p. 271.

25. S. M. Girvin and A. H. MacDonald, *Phys. Rev. Lett.* **58**, 1252 (1987). N. Read, *Phys. Rev. Lett.* **62**, 86 (1988).

26. F. Wilczek, *Phys. Rev. Lett.* **49**, 957 (1982).

27. D. H. Lee and M. O. A. Fisher, *Phys. Rev. Lett.* **63**, 903 (1989). D. J. Lee and C. L. Kane, *Phys. Rev. Lett.* **64**, 1313 (1990). S. C. Zhang, T. H. Hansson, and S. Kivelson, *Phys. Rev. Lett.* **62**, 82 (1989).

5

QUASI ONE-DIMENSIONAL SYSTEMS

A one-dimensional electron system was considered for a number of years a highly theoretical object, a mathematically convenient model but rather remote from reality. However, since the early 1960s many quasi-one-dimensional systems have been realized. These include charge-transfer salts such as TTF-TCNQ, conducting polymers such as polyacetylene, platinum salts called KCP, and trichalcogenides such as $NbSe_3$. These systems are one-dimensional in the sense that electrical conduction is high in one direction and is very low in all other directions. Moreover, two-dimensional systems such as discussed in the previous chapter have been squeezed into one dimension. Experiments on these systems have not only confirmed basic theoretical expectations but also unveiled unexpected properties such as superconductivity. The physics of one-dimensional condensed matter requires a book of its own for detailed discussions. Therefore, in what follows we shall address only a few characteristic properties.

5.1. Squeezing two-dimensional systems

In a silicon MOSFET 2D electrons are induced under its metal gate. Hence, by reducing the width of the gate a quasi-1D system can be realized. Similarly, by depositing a split gate on top of the GaAlAs layer of a GaAs/GaAlAs heterostructure, one can realize a 1D channel. Figure 5.1 illustrates a split gate device. The gap between the split gates is about 1 μm. Applying negative voltage to the gates removes electrons from beneath the metal except for the gap region. These electrons can be forced to pass through the narrow channel by a potential difference. The result is a ballistic 1D electron transport to be discussed below.

The drift velocity of an electron due to the potential difference V is given for low temperatures by eV/mv_F, where v_F is the Fermi velocity. The current passing through the channel of width w is given by

$$I = ne\left(\frac{eV}{mv_F}\right)w.$$

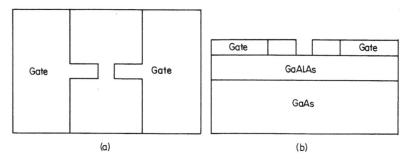

FIG. 5.1. Point contacts geometry of a heterostructure: (a) top view, (b) side view.

Here, n is the density of electrons that participate in the current. It is given by

$$2\frac{k_F\langle \cos\theta\rangle(2k_F)}{(2\pi)^2} = \frac{k_F^2\langle \cos\theta\rangle}{\pi^2},$$

where the average $\langle \cdots \rangle$ must be taken for positive values of $\cos\theta$ representing the direction of flow. Thus, the conductance is given by

$$G = \frac{e^2\langle k_x\rangle w}{\pi^2\hbar} \tag{1.1}$$

where the average $\langle k_x\rangle$ is taken in the domain of positive k_x. It can be evaluated from

$$\langle k_x\rangle = \frac{1}{2\pi k_F}\int d\mathbf{k}|k_x|\,\delta(k-k_F)\frac{2\pi}{w}\sum_{n=1}^{n_{\max}}\delta\left(k_F - \frac{n\pi}{w}\right). \tag{1.2}$$

We obtain

$$G = \sum_{1}^{n_{\max}}\frac{e^2}{\pi\hbar}, \tag{1.3}$$

where the summation is up to the largest integer that is smaller than

$$n_{\max} = k_f w/\pi.$$

The quantized conductance of ballistic transport in a GaAs/GaAlAs heterostructure has been measured by van Wees et al. [1] by a point contact method as illustrated in Fig. 5.2. The conductance clearly shows steps at multiples of $e^2/\pi\hbar$. They observed up to sixteen steps by changing the gate from 0 to 360 nm.

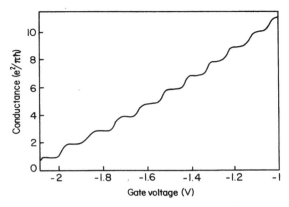

FIG. 5.2. Quantized conductance of point contacts in a 2D electron gas. (From van Wees et al. [1])

When a 2D electron system is confined in a system of a finite width w and is subject to a perpendicular magnetic field, its energy dispersion generally has three distinctive regions corresponding to the edge, Landau, and 1D discrete states. One example is shown in Fig. 5.3 [2]. Here, the ordinate is the reduced electron energy $\varepsilon/\hbar\omega_c$. The width is given by

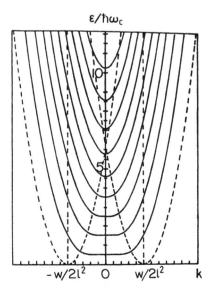

FIG. 5.3. Energy dispersion of a confined 2D electron in a magnetic field. The width is $w = 5l$. (From Isihara and Ebina [2])

$w = \sqrt{5\pi}l$, l being the magnetic length. The dashed parabolas represent

$$\varepsilon = \frac{\hbar^2}{2m}\left|k \pm \frac{w}{2l}\right|^2. \tag{1.4}$$

The dashed lines illustrate

$$k = \frac{\pm w}{2l}.$$

For small k and energies below

$$\varepsilon_{cr} = \frac{\hbar^2}{2m}\left(\frac{w}{2l^2}\right)^2 \tag{1.5}$$

there is a triangular region formed by the two dashed parabolas. In this region the energy curves are flat, representing magnetic quantization. The number of Landau states in this triangular region is given by the integer that is smaller than and closest to $[(w/2l)^2 + 1]/2$. In the present case it is 6.

On top of this region the energy varies parabolically due to spatial quantization. The energy levels are given by

$$\varepsilon_n = \varepsilon_g^*[(n + 1)^2 + k^2], \tag{1.6}$$

where ε_g^* is the reduced ground-state energy defined by

$$\varepsilon_g^* = \frac{(\pi\hbar/w)^2/2m}{\hbar\omega_c/2}. \tag{1.7}$$

In the present case $\varepsilon_g^* = 0.2$.

The density of states corresponding to the above case is illustrated in Fig. 5.4. The ordinate is $g(\varepsilon)/(eH/ch)$. The density of states is saw-toothlike with its minima and mutual distances gradually increasing towards larger values of the abscissa. In the present case the peaks are due mainly to magnetic quantization. The number of peaks for a given interval of ε decreases when the width is decreased. The curve that connects the minima is parabolic. The area under this parabolic curve decreases slowly when the width is increased.

The magnetoconductivity of narrow 2D systems depends on the width. When the width is finite but not very narrow, the system is quasi-two-dimensional. The experimental magnetoconductivity corresponding to such a case has been given by Skocpol et al. [3] as illustrated in Fig. 5.5. The conductivity increases with the gate voltage in a similar way. The regular Shubnikov–de Haas oscillation visible in the top curve decreases as the field strength is reduced. However, the bottom curve for zero magnetic field still

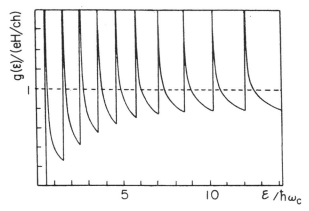

FIG. 5.4. Density of states of a narrow 2D electron system. The width is $w = \sqrt{5\pi}l$. (From Isihara and Ebina [2])

shows oscillations. These oscillations may be due to electron hopping or to width fluctuations. We shall discuss shortly the sharp irregular conductance fluctuations that take place in quasi-one-dimensional systems. The oscillations in the bottom curve can be precursors to such fluctuations even though they are rather wide.

In order to evaluate the magnetoconductivity, let us adopt a soft-wall model in which the electron is confined in a parabolic potential well given

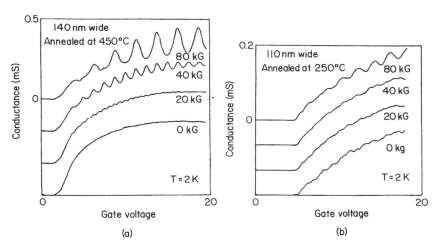

FIG. 5.5. Magnetoconductance of a narrow silicon inversion layer: (a) width 140 nm, (b) width 110 nm. (From Skocpol et al. [3]).

by [4]

$$v(y) = \frac{m\Omega^2}{2} y^2. \tag{1.8}$$

Here, Ω is the angular frequency representing the curvature of the parabolic potential well, and y is the lateral direction in which the width is narrowed down. If the energy of an electron is ε, the "width" of the system is energy-dependent and can be defined by

$$w(\varepsilon) = \frac{2}{\Omega} \left(\frac{2\varepsilon}{m} \right)^{1/2}. \tag{1.9}$$

If x represents the extended direction, the wavefunction of the electron can be expressed as

$$\psi(x, y) = e^{ikx} u(y), \tag{1.10}$$

where $u(y)$ satisfies

$$\left[\frac{p_y^2}{2m} + \frac{m\tilde{\omega}^2}{2} (y - \alpha \tilde{l}^2 k)^2 + \frac{\hbar^2 k^2}{2m} \right] u(y) = \varepsilon u(y). \tag{1.11}$$

Here, the quantities with tildes are effective quantities defined by

$$\tilde{\omega}^2 = \omega_c^2 + \Omega^2; \qquad \tilde{m} = m/(1 - \alpha^2);$$

$$\tilde{l} = (\hbar/m\tilde{\omega})^{1/2} = \frac{l}{\alpha^{1/2}}.$$

The parameter α represents the ratio $\omega_c/\tilde{\omega}$ in which ω_c is the cyclotron frequency. It varies between 0 and 1 in accordance with the strength of the spatial confinement. Note that the above effective quantities enables us to use the conductivity formulas for wide systems.

According to linear response theory, the conductivity components of 2D systems are given by the velocity–velocity correlation functions as follows:

$$\sigma_{xx} = \frac{i\hbar e^2}{A} \langle \text{Tr}[v_x \, \delta(\varepsilon - H) v_x \, \delta(\varepsilon - H)] \rangle; \tag{1.12}$$

$$\sigma_{xy} = \frac{ec}{A} \frac{\partial N(\varepsilon, 0)}{\partial H} + \frac{i\hbar e^2}{2A} \langle \text{Tr}[v_x G^+(\varepsilon) v_y \, \delta(\varepsilon - H) - v_x \, \delta(\varepsilon - H) v_y G^-(\varepsilon)] \rangle. \tag{1.13}$$

Here, H is the Hamiltonian of the system and

$$N(\varepsilon, 0) = \int_{-\infty}^{\varepsilon} \delta(\varepsilon' - H)\, d\varepsilon', \tag{1.14}$$

and $G^{\pm}$ is the advance and retarded Green's functions given by

$$G^{\pm} = (\varepsilon - H \pm i0)^{-1}. \tag{1.15}$$

When averaged over impurity configurations, the Green's function G is expressed in terms of the Hamiltonian H_0 in the absence of impurities such that

$$\langle G(z) \rangle = (z - H_0 - \Sigma)^{-1}, \tag{1.16}$$

where

$$\Sigma = \Delta - i\Gamma \tag{1.17}$$

is the self-energy. The self-energy can be determined in a self-consistent way in terms of the impurity potential and the Green's function:

$$G_0 = (z - H_0)^{-1}. \tag{1.18}$$

One convenient formula for a delta function-type impurity potential is

$$\Sigma = cv + \frac{c(1 - c)v^2 K(z - \Sigma)}{1 + [\Sigma + (c - 1)v]K(z - \Sigma)}, \tag{1.19}$$

where v is the strength of the impurity potential and

$$K(z) = \frac{2\pi\hbar^2}{m\varepsilon_M A}\, \mathrm{Tr}[G_0(z)]. \tag{1.20}$$

ε_M is the cut-off energy ε_M. Note that the density of states is given by

$$g(\varepsilon) = -\frac{1}{\pi}\, \mathrm{Im}\, K(\varepsilon - \Sigma). \tag{1.21}$$

Finally, the conductivity at finite temperatures is given by

$$\sigma_{ij}(\varepsilon_F, T) = -\int_{-\infty}^{\infty} \frac{\partial f}{\partial \varepsilon}\, \sigma_{ij}(\varepsilon, 0)\, d\varepsilon, \tag{1.22}$$

where $f(\varepsilon)$ is the Fermi distribution function.

The above conductivity formulas can be applied to the present case of a finite width. The function K defined by Eq. (1.19) is

$$K(z - \Sigma) = \frac{\tilde{m}\tilde{\omega}\Omega}{h} \sum_{0}^{n_M} \frac{1}{z - \hbar\omega(n + \frac{1}{2}) - \Sigma}. \tag{1.23}$$

The conductivity is oscillatory. If the oscillation is small, it is given by

$$\sigma_{xx} = (1 - \alpha^2)\sigma_0\left(1 - \frac{\Delta g}{g_0}\right) + \frac{\alpha^2\sigma_0}{1 + \tilde{\omega}^2\tau_0^2}\left(1 + \frac{1 + 5\tilde{\omega}^2\tau_0^2}{2(1 + \tilde{\omega}^2\tau_0^2)}\frac{\Delta g}{g_0}\right); \tag{1.24}$$

$$\sigma_{xy} = -\alpha\sigma_0\frac{\omega\tau_0}{1 + \tilde{\omega}\tau_0^2}\left(1 - \frac{1}{2}\frac{7\tilde{\omega}^2\tau_0^2 + 3}{\tilde{\omega}^2\tau_0^2(1 + \tilde{\omega}^2\tau_0^2)}\frac{\Delta g}{g_0}\right). \tag{1.25}$$

Here

$$\sigma_0 = \frac{N_0 e^2 \tau_0}{m}.$$

$$\frac{\Delta g}{g_0} = \frac{1}{(2\varepsilon)^{1/2}} \sum_{1}^{\infty} (-)^s \left(\frac{\hbar\tilde{\omega}}{s}\right)^{1/2} \exp\left(\frac{-\pi s}{\tau_0\tilde{\omega}}\right) \frac{X_s}{\sinh X_s} \cos\left[\frac{2\pi s}{\hbar\tilde{\omega}}(\varepsilon - \Delta) - \frac{\pi}{4}\right]; \tag{1.26}$$

$$X_s = 2\pi^2 \frac{skT}{\hbar\tilde{\omega}}$$

σ_0 is the zero-field conductivity with an effective number N_0 of electrons and the relaxation time τ_0. $\Delta g/g_0$ is the ratio of the oscillatory and constant parts of the density of states. The relaxation time is determined by the imaginary part of the self-energy.

In reduced dimensions, localization at impurities is effective, and the energy dependence of the relaxation time becomes important. Under the assumption that a localized state lies just below the conduction band, and based on a localized state lies just below the conduction band, and based on a pseudopotential, its expression has been given as follows:

$$\tau_0 = a\left(\frac{\varepsilon_0\varepsilon}{\varepsilon_\sigma\varepsilon^2 + (\varepsilon_0^2 - 3\varepsilon_\tau^2)\varepsilon + \varepsilon_\sigma\varepsilon_\tau^2}\right), \tag{1.27}$$

where ε_σ (or ε_τ) represents the energy at which $\sigma(\varepsilon)$ (or $\tau_0(\varepsilon)$) reaches a maximum. Another parameter a is inversely proportional to the impurity concentration. These three constants originate from the form of the pseudo-potential representing the impurity effect. The parameters have been modified in the above form to facilitate their actual determination.

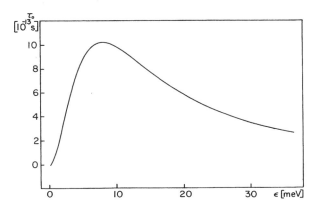

FIG. 5.6.   Energy dependence of the relaxation time. (From Smrčka et al. [4])

Figure 5.6 shows a theoretical relaxation time as a function of electron energy, which is the Fermi energy. The relaxation time increases first as the electron moves more freely and localization effects decrease. However, at high energies, electrons can get closer to scattering centres so that the scattering rate increases, causing the decrease in the relaxation time. For this graph the following values of the parameters in Eq. (1.26) are used:

$$a = 5.2 \times 10^{-10} \text{ meV}^2\text{s}; \qquad \varepsilon_\sigma = 17.5 \text{ MeV}; \qquad \varepsilon_\tau = 8 \text{ meV}$$

These values have been chosen in consideration of experimental data on silicon inversion layers including those in Fig. 5.5

The corresponding magnetoconductivity in arbitrary units is illustrated in Fig. 5.7. The four curves in each set are to be compared with those in Fig. 5.5. Graph (a) represents the case $\Omega = 1.2 \times 10^{12} \text{ s}^{-1}$ and (b) the case of $\Omega = 1.8 \times 10^{12} \text{ s}^{-1}$, which is more confined. Note that the zero-field case shows oscillations due to spatial quantization. These oscillations become irregular if the width fluctuates.

Because of Eq. (1.26), the amplitude of Shubnikov–de Haas (SdH) oscillations can be characterized by the exponential factor

$$\exp\left(\frac{-(\pi/\tau_0 + 2\pi^2 kT)}{\hbar\tilde{\omega}}\right).$$

The first term is temperature-independent while the second is temperature-dependent. When the logarithm of the oscillating part of σ_{xx} is plotted against $1/H$, it can be a straight line when the magnetic field is sufficiently strong. This has been confirmed by experiments on GaAs/GaAlAs wires with 0.4 μm width. However, the infinite-field limit was found not to vanish but rather is a constant that depends on width [5].

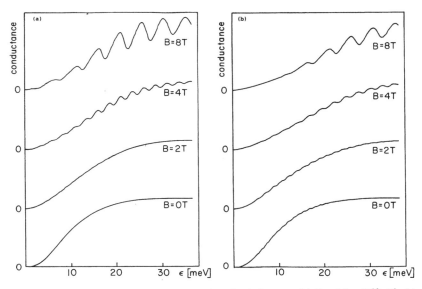

FIG. 5.7. Theoretical magnetoconductance of confined electrons: (a) $\Omega = 1.2 \times 10^{12}$ s^{-1}, (b) $\Omega = 1.8 \times 10^{12}$ s^{-1}. (From Smrčka et al. [4])

5.2. Aharonov–Bohm effect in quasi one-dimensional systems

The *Aharonov–Bohm effect* is manifested when an electron beam is split into two waves that propagate in opposite directions to encircle a cylinder and reunite on the opposite side; it represents quantum interference due to the wave character of electrons. In the presence of a magnetic field as in Fig. 5.8(a), the resultant electronic current measured at a point P can be related to the difference $\Delta\phi$ of the phases of the electron waves. If the vector potential is **A**, this difference is given by

$$\Delta\phi = \frac{e}{h} \int \mathbf{A} \cdot d\mathbf{s}. \qquad (2.1)$$

Here, s is the path encircling the magnetic flux. Due to quantum interference the magnetoresistance is expected to oscillate with a period h/e.

This interference takes place in vacuo. However, even in a disordered metallic ring there can be similar quantum interference if only elastic scattering is involved and phase coherence is preserved. Figure 5.8(b) illustrates quantum interference in a disordered metallic loop [6]. Here, two electron trajectories are shown. The leads on both sides are widened for clarity. Note that in addition to the major Aharonov–Bohm loop, there can

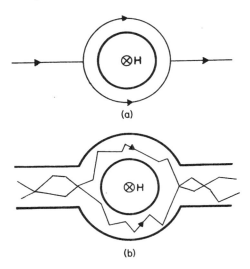

(a)

(b)

FIG. 5.8. Interference in an Aharonov–Bohm loop (a) in vacuo and (b) in a disordered metallic loop (b). In (b), two trajectories have closed loops of different sizes in the lead as indicated. (From Kaplan and Hartstein [6])

be many other smaller loops, as shown by two representative loops in the leads. Such small loops are expected to cause aperiodic structure.

The quantum interference in metallic loops was observed by Webb et al. in 1985 [7]. They used submicron-diameter gold rings and observed magnetoresistance oscillations. They observed periodic oscillations superimposed on a more slowly varying background. Figure 5.9(a) illustrates their resistance data for three temperatures, and (b) represents the corresponding Fourier spectra. The fundamental period h/e is clearly visible at $1/\Delta H = 131\ \mathrm{T}^{-1}$. The period of oscillation is h/e, in contrast to those previously known oscillations with period $h/2e$ due to flux quantization. This flux quantization is a one-electron phenomenon, while quantum interference involves many electrons. Near $1/\Delta H = 260\ \mathrm{T}^{-1}$, there is a much weaker one corresponding to $h/2e$. This may be due to self-interference or a harmonic content of the h/e oscillation.

As the temperature increases, all the magnetoresistance fluctuations shrink. At finite temperatures all of the states within kT of the Fermi energy contribute independently to the aperiodic fluctuations, and the amplitude decreases as $T^{-1/2}$. The h/e peak amplitude appears to have the same temperature variation. The high-frequency oscillation has a period $\Delta H = 0.007\,59\,T$. This period corresponds to the addition of a flux quantum to the area of the hole.

Fluctuations in conductance G at magnetic field H and Fermi energy

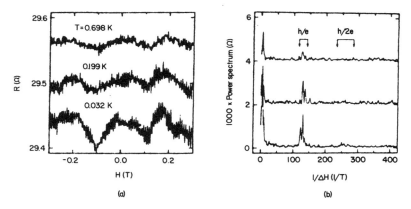

FIG. 5.9. Aharonov–Bohm effect in a gold loop. (From Webb et al. [7]).

ε_F may be expressed by the conductance autocorrelation function defined by

$$F(\Delta H, \Delta\varepsilon) = \langle G(H + \Delta H, \varepsilon_F + \Delta\varepsilon)G(H, \varepsilon_F)\rangle - \langle G(H, \varepsilon_F)\rangle^2. \qquad (2.2)$$

The root-mean-square fluctuation amplitude at $T = 0$ is given by

$$\Delta G = [F(0, 0)]^{1/2}. \qquad (2.3)$$

One very important feature of ΔG is that regardless whether a metal or an inversion or accumulation layer, its magnitude is universal:

$$\Delta G \sim \frac{e^2}{h}. \qquad (2.4)$$

The magnitude is essentially the same whether the electron energy or the applied magnetic field is varied. However, the phase coherence length L_ϕ must be of the order of the sample length L, which is much larger than the elastic mean free path.

The temperature scale over which the fluctuations are observed depends on electronic states which are correlated within bands of width $\varepsilon_\phi = hD/L_\phi^2$, where D is the diffusion constant. The band width depends on the time of traverse which is given by D/L_ϕ^2. If the thermal energy is much smaller than ε_ϕ, only a single band will contribute to the conductance and ΔG will be independent of temperature.

On the other hand, if the thermal energy is much larger than ε_ϕ fluctuations of uncorrelated energy bands will contribute, and the observed

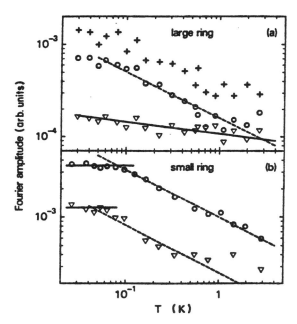

FIG. 5.10. Temperature dependence of conductance fluctuations in a gold loop. (From Washburn et al. [8])

fluctuations are the average over the uncorrelated patterns. Since the number of fluctuating patterns is given by $N = kT/\varepsilon_\phi$, the amplitude of the total conductance fluctuation which is proportional to $1/N^{1/2}$ varies as $1/T^{1/2}$.

Figure 5.10 depicts the data of Washburn et al. [8], showing the square root temperature dependency of ΔG in a gold loop with a diameter of 830 nm. Graph (a) corresponds to a larger ring. The crosses, triangles, and circles represent respectively the aperiodic fluctuations, $h/2e$ oscillations, and h/e oscillations. The $h/2e$ oscillation follows a weaker power than the h/e oscillations and crosses the $T^{-1/2}$ line near 1 K. For a smaller ring, both the h/e and $h/2e$ oscillations follow the $T^{-1/2}$ law above $\varepsilon_\phi = 0.03$ K. For temperatures less than 0.03 K, they are independent of temperature. The dependence of quantum fluctuations on size can be discussed in a similar way. When the sample length $L > L_\phi$, ΔG is the average of the contributions from the L/L_ϕ phase-coherent segments of the sample. Measurements on a string of silver loops separated by a distance of order L_ϕ such as shown in Fig. 5.11(a) [9] reveal that voltage fluctuations is proportional to

$$\Delta V \sim N^{1/2} \sim \left(\frac{L}{L_\phi}\right)^{1/2}. \tag{2.5}$$

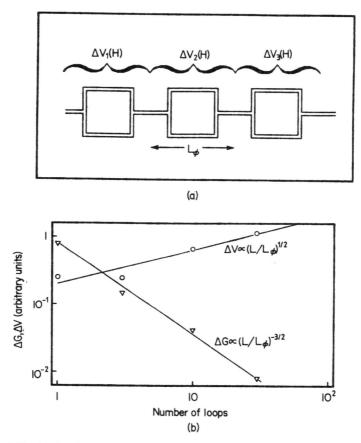

FIG. 5.11. (a) A string of silicon loops. (b) Size dependences of voltage and conductance fluctuations. (From Umbach et al. [9])

This is clearly a classical average of uncorrelated segments of the sample. The conductance fluctuations ΔG varies as

$$\Delta G \sim \frac{\Delta V}{I\langle R\rangle^2}$$

$$\sim \left(\frac{L}{L_\phi}\right)^{-3/2}. \qquad (2.6)$$

These variations are shown in Fig. 5.11(b). The temperature is 0.3 K. The circles are bare voltage fluctuations and the triangles are conductance fluctuations.

5.3. One-dimensional conductors

In 1955 Peierls suggested that a 1D metal is unstable abainst distortion of the lattice, so that it could actually never be metallic. However, he cautioned that his argument based on the adiabatic approximation may not necessarily be valid. This caution was appropriate, particularly in view of the 1954 work of Fröhlich concerning the possibility of superconductivity in one dimension. Fröhlich's work was based on a jellium model in which the positive ions forming an elastic continuum distort periodically with wavevector $2k_F$ while the electrons have a gap at the Fermi energy. However, since there is no preferred spatial location, the phase of the distortion is free to move. The electrons and the lattice distortion are coupled to from a sliding *charge-density wave* (CDW) carrying a current.

If a 1D system is made up of atoms separated by a lattice constant a, each giving one conduction electron, the electronic wavefunction that satisfies the periodic condition

$$\psi(x + a) = \psi(x)$$

is of the form

$$\psi(x) = u_k(x) \exp(ikx),$$

where $u_k(x)$ has the periodicity. In the tight-binding approximation and when measured from the Fermi energy, the electron with wavevector k has energy given by

$$\varepsilon_k = -2t \cos(ka). \tag{3.1}$$

where t is the overlapping integral between the nearest-neighboring sites. Figure 5.12(a) schematically illustrates the energy as a function of k. The wavevector k can range between $-\pi/a$ and π/a. The Fermi wavevector k_F corresponds to $\pi/2a$. The Fermi energy is at the middle, dividing the empty and filled levels. That is, the band is exactly half-filled at absolute zero.

However, Peierls showed that the lattice becomes more stable if it distorts itself as in the lower graph (b). This graph shows that a "dimerization" has taken place so that the lattice constant is now $2a$ instead of a. This costs energy, but the system gains energy because the electronic energy is lowered even more and a gap 2Δ opens up at the Fermi energy as shown.

At low temperatures, the electrons can be at either k_F or $-k_F$ so that they couple strongly with phonons with wavevector $2k_F$. This coupling depends on polarization function defined by

$$\chi(q, \omega) = \frac{2}{2\pi} \int \frac{f(\varepsilon_{k+q}) - f(\varepsilon_k)}{\varepsilon_{k+q} - \varepsilon_k + \omega + i0} \, dk, \tag{3.2}$$

where $f(\varepsilon)$ is the Fermi distribution function. In terms of this function, the RPA (random phase approximation) dielectric function is given by

(a)

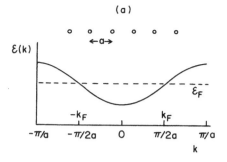

(b)

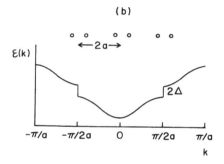

FIG. 5.12.   Enerby band and Peierls' distortion of a 1D system.

$$\varepsilon^{-1}(q, \omega) = 1 + u(q)\chi(q, \omega). \tag{3.3}$$

Here, $u(q)$ is the Fourier transform of the Coulomb potential.

For one dimension this Fourier transform is not well defined, but since in actual 1D systems electrons in different molecular stacks interact, the 3D form

$$u(q) = \frac{4\pi e^2}{q^2}, \tag{3.4}$$

must be modified in conformity with the anisotropy of the system. For low temperatures where many interesting phenomena are observed, q may be considered to be constant, with different values in different crystal directions. In any case, one can show that $\chi(q, \omega)$ diverges logarithmically at $q = 2k_F$ if the Fermi surface is simply given by $\pm k_F$. In contrast, the corresponding functions stay finite for two and three dimensions.

Figure 5.13 illustrates the static polarization function for one, two, and three dimensions. The 2D curve is continuous but the derivative is singular at $2k_F$. The 3D curve has an inflection point at $2k_F$. That is, the divergence

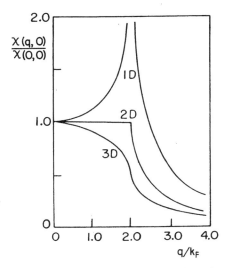

FIG. 5.13. Polarization function for one, two, and three dimensions.

is characteristic of one dimension. The Peierls instability is related to this divergence. Also related is the *Kohn anomaly*, which appears as a dip in the phonon mode at $2k_F$ because phonons are coupled with the electrons. This dip and the Peierls distortion also are of course dependent on temperature. They are strong near absolute zero, and as the temperature increases softening takes place, as schematically illustrated in Fig. 5.14 for the case of the Kohn anomaly.

The Peierls distortion and the resulting CDW occur in the chain direction. Experimental evidence has been provided by strong infrared absorption with polarized light in the chain direction.

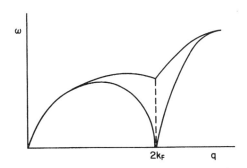

FIG. 5.14. Schematic Kohn anomaly at 0 K (lower curve) and at a finite temperature (upper curve).

The ideal 1D Fermi surface at $q = \pm k_F$ must be modified in actual quasi-1D organic conductors due to interchain coupling. Let us assume that molecular chains are stacked in the x direction, and choose the y and z axes perpendicularly. The system is highly anisotropic but is three-dimensional. For weak coupling between the chains, the tight-binding approximation may be adopted to express the electron energy in the form

$$\varepsilon(k) = \varepsilon_x(k) + \varepsilon_v(k), \tag{3.5a}$$

$$\sim v_F(|k_x| - k_F) + C(|k_x| - k_F)^2 + \varepsilon_v(k), \tag{3.5b}$$

where v stands for "vertical" and

$$\varepsilon_x(k) = -2t_a \cos k_x a,$$

$$\varepsilon_v(k) = -2t_v(\cos k_y b + \cos k_z c), \tag{3.6}$$

$$C = 2t_v a^2 \cos \frac{\pi n}{2}, \qquad v_F = 2t_v a \sin \frac{\pi n}{2}.$$

Here, the same transfer integral t_v is assumed in the vertical direction with lattice constants b and c, a is the longitudinal lattice constant, t_a is the corresponding transfer integral, and n is the number of electrons per unit cell. Accordingly, the Fermi surface deviates from $\pm k_F$ by

$$\Delta k_F = k_{Fx} \pm k_F$$

$$= \pm \frac{2\varepsilon_v(k_F)}{v_F}\left[1 + \left(1 - \frac{4C\varepsilon_v(k_F)}{v_F^2}\right)^{1/2}\right]^{-1}. \tag{3.7}$$

The solid curves in Fig. 5.15 illustrate the new Fermi surface at $k_z = 0$. The Fermi surface is open and periodic in the k_y direction, while it is bounded in the k_x direction. Both sides of the Fermi surface in the k_x–k_y plane are warped by an amount proportional to t_v. The shaded area is occupied by the electrons. The vector $\mathbf{Q}$ defined by

$$\mathbf{Q} = (2k_F, \pi/b, \pi/c) \tag{3.8}$$

transfers a point on the left Fermi surface to the right. It is determined by

$$-\varepsilon_-(k) = \varepsilon_+(\mathbf{k} + \mathbf{Q}). \tag{3.9}$$

This represents the *nesting condition* under which the denominator of the polarization function vanishes. Since the momentum integration is now three-dimensional, the polarization function may not diverge, but a sufficient disturbance may be expected still at this point. For $C \neq 0$, the above exact

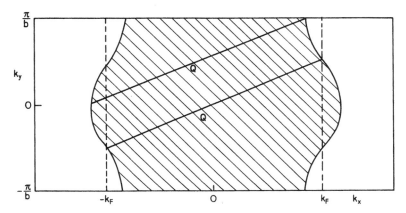

FIG. 5.15. Fermi surface of a weakly coupled chain system.

nesting condition is modified, but the modification is of order $(t_v/t_a)^2$, which is small. Hence, at a low temperature instability can be expected; the system may be in CDW or SDW states, depending on the form of the electron–electron interaction. While the form of the interaction Hamiltonian must be specified to find the criteria for the appearance of various states, backward scattering with momentum changes of order $2k_F$ (or forward scattering with very small momentum changes) is a dominant mechanism that is favourable to CDW (or SDW) states.

Let us now switch our attention to transport properties. By doping, the conductivity of 1D systems can be significantly increased. For instance, polyacetylene, which has the simple chemical structure $(CH)_n$, can be doped by sodium or potassium as donors and by AsF_6 as acceptors. Moreover, polyacetylene may have impurities as in any other semiconductors or metals, and also thermally created "topological defects" [10]. Figure 5.16 illustrates a mismatch of two degenerate configurations A and B. These two configurations have different sequences of double and single bonds but clearly correspond to the same energy. In order to satisfy the periodicity of the lattice, one mismatch, called a *soliton*, must be compensated by an antisoliton.

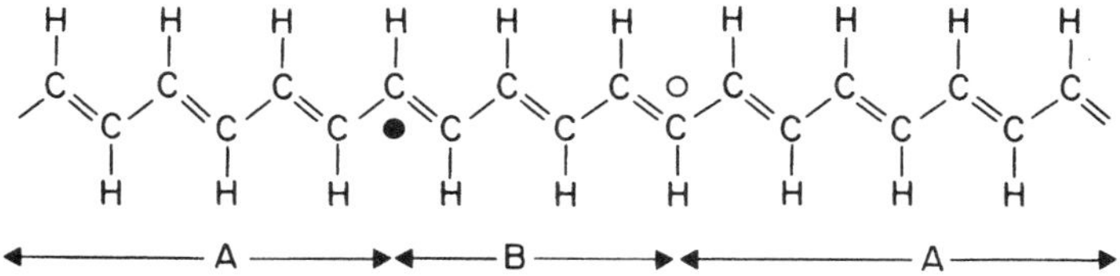

FIG. 5.16. Soliton (●) and antisoliton (○) in polyacetylene.

That is, a pair of solitons appears in the sequence *ABA*, as in the graph, or in the sequence *BAB*. Behaving like the case of lattice vibrations, these solitons can move in either left or right directions. In actual 1D conductors, the transition from one configuration to another may take place over several lattice distances.

A soliton in polyacetylene can accommodate up to two electrons, one spin-up and the other spin-down. With one electron at the site a soliton can be neutral and has spin 1/2. Hence, it is paramagnetic. If this electron is removed by acceptor doping, the soliton will be positively charged and nonmagnetic without spin. Donor doping may produce double occupancy, which corresponds to a negative soliton without spin. For dilute doping, the Curie-law contribution is found experimentally to decrease, indicating that the charge carriers are spinless solitons. The conductivity has also been interpreted in terms of the soliton concept. A soliton (black circle in Fig. 5.16) and antisoliton (white circle) distortion would induce localized mid-gap states that are screened by a charge deficit of $-e/2$ in the local valence-band charge density. If the single spinless Fermion removed from the latter is placed in the localized state at the black circle, leaving the state at the white circle empty, the net charge on the soliton is $e/2$ and that at the antisoliton is $-e/2$ [11].

5.4. Bechgaard salts

Among many 1D organic metals a relatively new family called Bechgaard salts has unusually interesting properties. The structure is given by the chemical formula $(TMTSF)_2X$, where X are anions such as PF_6, AsF_6, TaF_6, SbF_6 with hexahedral symmetry, or ClO_4, ReO_4, FSO_3, and so on, with tetrahedral symmetry. These salts show a variety of ground states that depend on the anions and pressure with their charge-transfer character and crystal structure preserved. They exhibit interplay between superconducting and insulating ground states.

The 2:1 TMTSF salts have the structure such as shown in Fig. 5.17. The planar cations of TMTSF (short for the long and difficult name of tetramethyltetraselenafulvalene) are stacked in a zig-zag way perpendicularly to the plane of selenium atoms shown by black circles. The stacking distance is approximately 3.6 Å in the direction of *a*. In the unit cell, there are four molecules in the *b* direction. With this structure the intermolecular Se–Se distance, which is of order 3.8 Å, is shorter than the intrastack separation. The conductivity is highly anisotropic; it is more than a factor of 20 larger in the *a* direction than in the *b* direction due to the large overlap integral in that direction. Figure 5.18 illustrates the phase diagram that represents a combination of several data [12] on slowly cooled $(TMTSF)_2ClO_4$ when a magnetic field is applied in the direction of the c^* axis along which stacks of TMTSF molecules are separated by anions X. This is the direction in which interchain coupling is weak. The phase diagram shows reentrance into

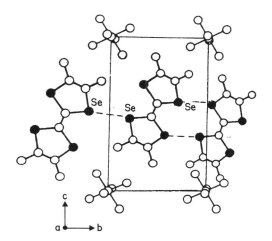

FIG. 5.17. The structure of $(TMTSF)_2X$.

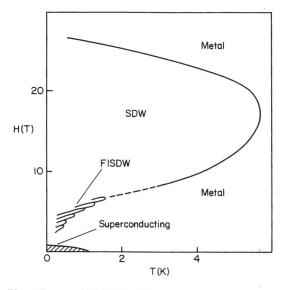

FIG. 5.18. Phase diagram of $(TMTSF)_2ClO_4$.

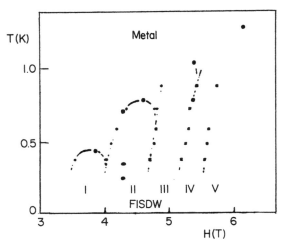

FIG. 5.19. Susceptibility of $(TMTSF)_2ClO_4$. (From Miljak et al. [13])

the metallic phase when the field is increased below around 5.6 K. The critical temperature is around 5.6 K at 18 T. When the field is increased at around 0.5 K the following changes take place. First, the superconducting phase appears for fields up to 0.03 T below around 1 K even without applying pressure. Then, a normal metallic state occurs up to around 3.5 T. Between 3.5 and 7.5 T, several different SDW semimetal phases are induced by the magnetic field as the Fermi level jumps between Landau level gaps. The appearance of these phases is a fascinating and unique feature of the phase diagram. The transitions between these *field-induced SDW states* (FISDW) are observed in several ways. For instance, the specific heat shows a series of jumps and peaks, indicating first- and second-order transitions.

Figure 5.19 provides more details in the same FISDW region than the phase diagram obtained by Pesty et al. [12] by specific heat measurements. The specific heat jumps at 3.5, 4.1, 4.3, and 4.9 T at 0.38, 0.49, 0.73, and 0.89 K respectively. The jumps separate the low-field normal metal from the high-field phases. The transitions are very abrupt; a 60% increase in the electronic specific heat is observed for a mere 3% field variation. The threshold field of the first specific heat jump exhibits a peculiar temperature variation. A weak increase is observed as long as the first anomaly is present. As the temperature increases, the first anomaly disappears and the threshold field is abruptly shifted to the next one. A plot of the electronic specific heat divided by T versus magnetic field shows a 20% decrease at 0.38 K and 0.73 K for a 2% field variation. This peculiar variation seems to originate from the finite width of the transitions. A series of specific-heat peaks is observed at high field, indicating phase transitions between different semimetallic states. They appear in the field region where the first jumps are observed. At low temperatures the size of the peaks and the electronic contribution decrease as the field increases. This evolution at constant temperature is probably caused by a corresponding increase of the gap.

Between 7.5 and 26 T of the phase diagram in Fig. 5.18 is a new state with the Fermi energy changing to keep a constant Landau-level filling. Finally, above 26 T, reentrance takes place into the normal metal nonmagnetic state.

The susceptibility of $(TMTSF)_2X$ for $X = ClO_4$ and PF_6 shows the Curie law at low temperatures in a field of 0.94 T [13]. The absolute magnitude of the spin susceptibility of these salts is around 1.3×10^{-4} emu/mole, although it depends on samples, crystal directions, and heat treatments. Despite differences in their ground states, the ClO_4 and PF_6 salts have similar magnetic susceptibilities above 20 K. From the susceptibility the density of states is found to be 1.02 states/eV/spin/TMTSF molecule. This value is close to the one determined from the specific heat, and it appears that superconducting fluctuations do not persist to high temperatures. The Fermi surface of the ClO_4 salts is open, since below 1 T there is no Landau diamagnetism characteristic of closed orbits. The magnetic susceptibility shows weak field dependence below 5 kOe due probably to magnetic ordering. The magnetic phase is antiferromagnetic.

If the on-site interaction U is strongly repulsive in the 1D Hubbard model, each site in the chain will be occupied by one electron. Either of two neighboring electrons may take the virtual process of hopping to the other's site and back if their spins are antiparallel, thus lowering the energy. If the spins are parallel, no such virtual hopping will take place. That is, antiparallel spin configurations are energetically favored.

In a tight-binding approximation the band energy for anisotropic crystals measured from the Fermi energy can be expressed as

$$\varepsilon(k) = -2[t_a \cos(ak_a) + t_b \cos(bt_b) + t_c \cos(ct_c)], \tag{4.1}$$

where the t's are transfer integrals, and $a = 3.6$ Å, $b = 7.7$ Å, and $c = 13.5$ Å, a being in the direction of the molecular stacks. The transfer integral t_a is the largest, and t_c is the smallest. Their orders of magnitude are

$$t_a = 0.35 \text{ eV}; \qquad t_b = 30 \text{ MeV}; \qquad t_c = 1 \text{ meV}.$$

The conductivity anisotropy in the a and b directions is given by [14]

$$\frac{\sigma_a}{\sigma_b} = \left(\frac{at_a}{bt_b}\right)^2 \tag{4.2}$$

$$\sim 30. \tag{4.3}$$

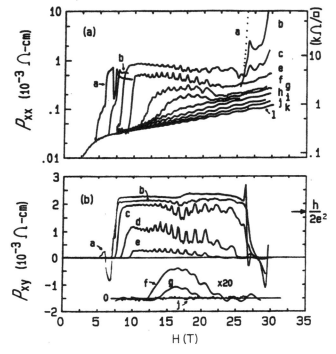

FIG. 5.20. ρ_{xx} and ρ_{xy} of (TMTSF)$_2$ClO$_4$. (From Naughton et al. [12]).

The magnetoresistivity components ρ_{xx} and ρ_{xy} of (TMTSF)$_2$ClO$_4$ measured by Naughton et al. [12] for fields up to 30 T are illustrated in Fig. 5.20. Above 6 K (curves i, j, k and l), ρ_{xx} increases smoothly with field, while ρ_{xy} stays small. The SdH oscillations are observable, indicating coherent transport in at least two directions to complete close orbits. The small Hall resistance is indicative of metallic behavior. At such high temperatures there is no sign of phase transitions. From 2.6 to 5.4 K (curves c–h), there is a single large-scale bump that develops in both ρ_{xx} and ρ_{xy}. This bump corresponds to a region of increased resistance and reduced density of carriers. It can be associated with the field-induced SDW semimetal phase. Below 2 K, ρ_{xx} and ρ_{xy} develop several steps characteristic of the sequence of FISDW transitions from 3 to 8 T, and the remarkably stable data from 8 to 26 T. This semimetallic regime develops systematically out of the metallic "background". Although not very clear in the graph, the variations of ρ_{xx} and ρ_{xy} are correlated; as ρ_{xy} increases from near zero to saturation, ρ_{xx} drops. Although these changes are not very sharp, they remind us of the quantum Hall effect. ρ_{xy} stays nearly constant at low temperatures between 7.5 and 26 T. This constancy corresponds to the FQHE at $\nu = 1/3$. The

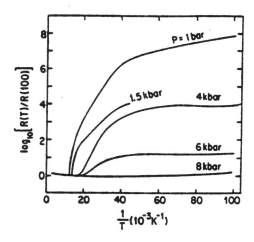

FIG. 5.21. Relative resistance of $(TMTSF)_2FSO_3$ at several values of pressure. (From Lacoe et al. [12])

Fermi level is apparently pinned at this filling factor for such a wide interval of magnetic field.

At low temperatures the Bechgaard salts become superconductors under pressure. The transition is rather sharp as observed in a sharp drop of the resistance. Figure 5.21 show the logarithm of the relative resistance of $(TMTSF)_2FSO_3$ as a function of reciprocal temperature for several values of applied pressure [15]. As the pressure is increased to about 6 kbar, the metal insulator transition point and the sharpness of the transition are reduced. The latter may be due to inhomogeneity in the pressure. The phase diagram of $(TMTSF)_2FSO_3$ is given in Fig. 5.22 [15]. The metal–insulator transition is suppressed almost linearly with pressure up to around 6 kbar and 40 K. At higher pressures this linearity no longer holds. Before reaching the superconducting phase there is a region where the metal–insulator transition is smeared out. This area is a glass phase, as indicated. A similar phase diagram has been obtained in yet another organic metal with the structure $(BEDT-TTF)_4(ReO_4)_2$.

5.5. FISDW and FQHE in Bechgaard salts

As shown in Figs. 5.18 and 5.19, a series of FISDW states appears in $(TMTSF)_2X$ salts when a magnetic field in the direction of c^* is increased. The appearance of such SDW states suggests that the metallic phase becomes unstable for several values of magnetic field. The instability of a metallic phase can be discussed based on *Stoner's criterion*. This criterion uses the susceptibility in the ladder approximation:

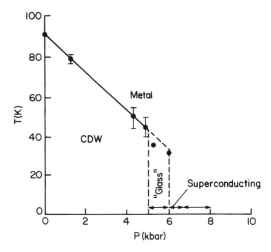

FIG. 5.22.   Phase diagram of $(TMTSF)_2FSO_3$. (From Lacoe et al. [12])

$$\chi_p = \frac{\chi(Q)}{1 - \lambda\chi(Q)}, \qquad (5.1)$$

where $\chi(Q)$ is the ideal polarization function. It states that a metallic phase is unstable when the denominator of the susceptibility becomes negative

$$1 - \lambda\chi(Q) \leqq 0, \qquad (5.2)$$

where λ is an approximate interaction parameter and $\chi(Q)$ is given by Eq. (3.2). Thus, the criterion depends on the behavior of the ideal polarization function $\chi(Q)$.

We note also that the nesting condition given by Eq. (3.9) is applicable to both CDW and SDW states, but in the presence of a magnetic field the two states are not equivalent to each other. While the SDW antiferromagnetic coupling is not affected by the field, the CDW pairing for parallel spins occurs under a new nesting condition:

$$\varepsilon(\mathbf{k} + \mathbf{Q}) + \varepsilon(\mathbf{k}) = 2\mu_B H. \qquad (5.3)$$

Hence, in an applied magnetic field CDW states are less likely to occur.

In order to find how a magnetic field induces a series of SDW states, let us introduce the Green's function defined by [16]

$$G_+(i\omega_n, \mathbf{k}; x, x') = \exp[ik_F(x - x')]g_+(i\omega_n, \mathbf{k}; x, x'), \qquad (5.4)$$

where g_+ is a slowly varying part of the Green's function near the right side of the Fermi surface and k is a 2D wavevector. Corresponding to the left side, a similar function g_- is defined. In terms of these Green's functions the susceptibility is given by

$$\chi(Q) = kT \sum_{\omega_n} \frac{1}{(2\pi)^2} \int d\mathbf{k} \int dx' g_+(i\omega_n, \mathbf{k}; x, x') g_-(i\omega_n, \mathbf{k} - \mathbf{Q}; x', x). \quad (5.5)$$

Here $\mathbf{Q}$ is the nesting vector, which depends on the form of one-electron energy. Therefore, let us use a simple energy expression similar to Eq. (3.5) based on the tight-binding approximation:

$$\varepsilon_\pm(\mathbf{k}) = \pm v_F(p_x \mp p_F) - \varepsilon_v^0(k_v); \quad (5.6)$$
$$\varepsilon_v^0(k_v) = 2t_b \cos(k_y b) + 2t_c \cos(k_z c^*).$$

A perfect nesting is expected in this case for $\mathbf{Q}$ of Eq. (3.8):

$$\mathbf{Q} = (2k_F, \pi/b, \pi/c^*)$$

so that the ground state is never metallic. Thus, a certain modification of the energy expression is necessary. One possible modification is to add the second harmonics such that

$$\varepsilon(\kappa_v) = \varepsilon_v^0(k_v) + 2t_b' \cos(2k_y b) + 2t_c' \cos(2k_z c^*). \quad (5.7)$$

In the presence of a magnetic field in the direction of the c^* axis, $\mathbf{k}$ is replaced by $\mathbf{k} - e\mathbf{A}/c$, where $\mathbf{A} = (0, Hx, 0)$, and $\varepsilon_v = \varepsilon_v(k_y - eHx/c, k_z)$. The Green's function equation for g_+ becomes

$$\left\{ i\hbar\omega_n + i\hbar v_F \frac{d}{dx} - \varepsilon_v \right\} g_+(i\omega_n, \mathbf{k}_v; x, x') = \delta(x - x'). \quad (5.8)$$

One can express g_+ in an integrated form in the interval from x' to x. Together with a similar expression for g_-, the resulting integral can be used in Eq. (5.5). The coupling parameter λ in Stoner's criterion can be converted to a temperature T_0 that corresponds to the case in which $t_b' = t_c' = 0$:

$$\frac{\lambda\hbar}{2\pi v_F bc} = \ln\left(\frac{v_F \hbar}{\pi k T_0 d}\right) = \zeta, \quad (5.9)$$

where d is a cut-off parameter that will appear below. When $t_c' \ll t_b'$, t_c' can be neglected. Since ε_v^0 is irrelevant, the Stoner criterion becomes

$$\frac{1}{\zeta} \geqq \int_d^\infty dx \, J_0\left[\frac{4l^2ct_b'}{\hbar v_F b}\sin\left(\frac{bx}{l^2}\right)\right]\frac{2\pi kT}{\hbar v_F \sinh(2\pi kTx/\hbar v_F)}, \tag{5.10}$$

where $l^2 = c\hbar/eH$.

The above integral is logarithmically divergent for large x. Using a formula

$$\frac{1}{\pi}\int_0^\pi J_0(z\sin\phi)\,d\phi = J_0^2(z/2),$$

and averaging over x, we find that the metallic phase is stable only for discrete values of the magnetic field corresponding to

$$J_0\left(\frac{2t_b'l_b^2}{\hbar v_F b}\right) = 0. \tag{5.11}$$

For large arguments, $J_0(x)$ oscillates:

$$J_0(x) \sim \left(\frac{2}{\pi x}\right)^{1/2}\cos\left(\frac{x-\pi}{4}\right).$$

Hence, a magnetic field is expected to induce a series of SDW states.

The energy expression plays an important role in the above consideration. It has been shown that reentrance to the metallic phase can be explained by introducing an anion gap [17]. Such a gap causes the electron states to split and become anisotropic in such a way that the states at even and odd chains have opposite energy corrections. As a result, a nesting becomes more difficult particularly when the interaction is local. Since the larger the magnetic field the smaller the electron orbit in the b direction, localization of electron orbits in the two inequivalent chains will be enhanced. Thus, at a sufficiently large magnetic field, reentrance to the metallic phase may take place.

Another mechanism that destroys a perfect nesting is given by a slight misalignment of the magnetic field from the c^* direction [16]. Let us take a coordinate system such that x and z are parallel to a and c^* axes respectively and y is perpendicular to the (a, c^*) plane. The Fermi surface is given by two slightly corrugated planes. The angle between the a and b axes is $\gamma = 71°$, $b = 7.7$ Å, and $c^* = 13.3$ Å.

Near the Fermi surface, we use Eq. (5.6) for the energy of electrons in these compounds. The magnetic field is considered to have a y-component in addition to the z-component. The classical equation of motion is

$$\hbar\frac{d\mathbf{k}}{dt} = -\frac{e}{c}[\mathbf{v}, \mathbf{H}]. \tag{5.12}$$

The electron is considered on the Fermi surface on which $\mathbf{v} = (v_F, 0, 0)$ is in the direction of the axis a. Accordingly, Eq. (5.12) is satisfied by

$$\hbar k_y = \frac{e}{c} v_F H_z t; \qquad \hbar k_z = -\frac{e}{c} v_F H_y t. \tag{5.13}$$

Since

$$\mathbf{v} = \frac{\partial \varepsilon}{\hbar \, \partial \mathbf{k}} \tag{5.14}$$

the electron motion in real space is described by

$$
\begin{aligned}
y &= \frac{2t_b c}{e v_F H_z} \cos\left(\frac{e v_F H_z b t \sin \gamma}{c}\right), \\
z &= \frac{2t_c c}{e v_F H_y} \cos\left(\frac{e v_F H_y t c^*}{c}\right).
\end{aligned}
\tag{5.15}
$$

Hence, the orbit depends on the field components. If the magnetic field satisfies the condition such that

$$\frac{H_y}{H_z} = \tan \phi = \frac{b \sin \gamma}{c^*} \frac{m}{n} \tag{5.16}$$

for integers m and n, the motion in the plane perpendicular to the chains is periodic. This will cause instability of the metallic phase and promote SDW pairing. It has been shown that the susceptibility exhibits a logarithmic divergence.

As a result of the periodicity due to magnetic field, new gaps may be expected. In order to find these gaps, we first note that the field components H_y and H_z specify magnetic lengths

$$l_y = \left(\frac{c\hbar}{eH_y}\right)^{1/2}; \qquad l_z = \left(\frac{c\hbar}{eH_z}\right)^{1/2}. \tag{5.17}$$

When coupled with the lattice constants b and c^*, these magnetic lengths introduce new characteristic wavevectors given by

$$q_y = \frac{b}{l_z^2}; \qquad q_z = \frac{c^*}{l_y^2}. \tag{5.18}$$

If their ratio is rational,

$$\frac{q_z}{q_y} = \frac{c^* H_y}{b H_z} = \frac{r}{q}, \tag{5.19}$$

where r and q are integers, the periods concerning the wavevectors are commensurate. If m, n, and p are integers one expects new gaps at

$$Q_x = 2k_F + \frac{mb}{l_z^2} + \frac{nc^*}{l_y^2}$$

$$= 2k_F + \frac{peH_z b}{(qch)}. \tag{5.20}$$

Each Landau band is split effectively into q subbands due to H_y. Between the Fermi level and the SDW gap at Q_x, there are p subbands. For a fixed value of q, a series of SDW states appear for different values of p. Each subband can accommodate eH_z/qch carriers per unit volume. The corresponding Hall conductivity for p subbands is given by [18]

$$\sigma_{xy} = -\frac{e^2}{h}\frac{p}{q}. \tag{5.21}$$

Note that q depends on the orientation of the field and p depends on its amplitude. Equation (5.21) represents the FQHE that is due to the fractional quantization of the density of carriers.

REFERENCES

1. B. J. van Wees, H. van Houten, C. W. J. Beenakker, J. G. Williamson, L. P. Kouwenhoven, D. van der Marel, and C. T. Foxon, *Phys. Rev. Lett.* **60**, 848 (1988). D. A. Wharam, T. J. Thornton, R. Newbury, M. Pepper, H. Ahmed, J. E. F. Frost, D. G. Hasko, D. C. Peacock, D. A. Ritshie, and G. A. C. Jones, *J. Phys. C* **21**, L209 (1988).
2. A. Isihara and K. Ebina, *J. Phys. C* **21**, L1079 (1988).
3. W. J. Skocpol, L. D. Jackel, R. E. Howard, H. G. Craighead, L. A. Fetter, P. M. Mankiewich, P. Grabbe, and D. M. Tennant, *Surf. Sci.* **142**, 14 (1984).
4. L. Smrčka, H. Havlova, and A. Isihara, *J. Phys. C* **19**, L457 (1986). A. Isihara, H. Havlova, and L. Smrčka, *J. Phys. C* **21**, 645 (1988).
5. A. D. C. Grassie, K. M. Hutchings, M. Lakrimi, C. T. Foxon, and J. J. Harris, *Phys. Rev. B* **36**, 4551 (1987-I).
6. S. B. Washburn, *IBM J. Res. Develop.* **32**, 335 (1988). S. B. Kaplan and A. Hartstein, *IBM J. Res. Develop.* **32**, 347 (1988)
7. R. A. Webb, S. Washburn, C. P. Umbach, and R. B. Laibowitz, *Phys. Rev. Lett.* **54**, 2696 (1985).
8. S. Washburn, C. P. Umbach, R. B. Laibowitz, and R. A. Webb, *Phys. Rev. B* **32**, 4789 (1985).
9. C. P. Umbach, C. van Haesendonck, R. B. Laibowitz, S. Washburn, and R. A. Webb, *Phys. Rev. Lett.* **56**, 386 (1986). S. Washburn, *IBM J. Res. Develop.* **32**, 335 (1988).
10. W. P. Su, J. R. Schrieffer, and A. J. Heeger, *Phys. Rev. Lett.* **42**, 1698 (1979).

11. M. J. Rice and E. J. Mele, *Phys. Rev. B* **25**, 1339 (1982).
12. T. Osada, N. Miura, and G. Saito, *Solid State Commun.* **60**, 441 (1986). J. P. Ulmet, A. Khmou, P. Auban, and L. Bachere, *Solid State Commun.* **58**, 753 (1986). P. M. Chaikin et al., *Physica* **143B**, 383 (1986). R. V. Chamberlin et al., *Jpn J. Appl. Phys.* **26-3**, 575 (1987). M. J. Naughton et al., *Phys. Rev. Lett.* **61**, 621 (1988). F. Pesty, P. Garoche, and K. Bechgaard, *Phys. Rev. Lett.* **55**, 2495 (1985). R. C. Lacoe, P. M. Chaikin, F. Wudl, and E. Aharon-Shalom, *J. de Phys. Colloq.* C3, Suppl. 6, 767 (1983).
13. M. Miljak, J. R. Cooper, and K. Bechgaard, *J. de Phys.* *C3*, **44**, 893 (1983).
14. G. Soda, D. Jérome, M. Weger, J. Alizon, J. Gallice, H. Robert, and J. M. Fabre, *Mol. Crys. & Liq. Crys.* **79**, 183 (1982).
15. R. C. Lacoe, P. M. Chaikin, F. Wudl, and E. Aharon-Shalom, *J. de Phys.* *C3*, **44**, 767 (1983).
16. L. P. Gorkov and A. G. Lebed, *J. de Phys. Lett.* **45**, 1433 (1984). A. G. Lebed, *J. de Phys. Lett.* **43**, 174 (1986). M. Héritier, G. Montambaux and P. Lederer, *J. de Phys. Lett.* **45**, 943 (1984).
17. A. G. Lebed and Per Bak, *Phys. Rev. Lett.* **63**, 1315 (1989).
18. G. Montambaux and P. B. Littlewood, *Phys. Rev. Lett.* **62**, 953 (1989).

6

HOPPING AND LOCALIZATION

The conductances of metals or semiconductors depend not only on impurities but also on dimensions. Even though these dependences were known for many years, their studies for high impurity concentrations and finite systems began to develop only relatively recently, since around 1979. This chapter deals with a new area concerning electron transport in disordered low-dimensional systems.

6.1. Hopping conduction

When an electron is placed in a periodic potential with wells so far apart that the overlap between the atomic wavefunctions on adjacent wells is small, the tight-binding approximation is good, and electron energy $\varepsilon(k)$ oscillates sinusoidally. The amplitude of this oscillation, that is, the width of electron's energy band, is determined by the overlap integral between the nearest neighbors. If the electron is subject to a potential that varies randomly in space or in well-depth, its wavefunction will be modulated randomly. However, the wavefunction will have large amplitudes at low-energy areas. Unless its energy is higher than the potential fluctuations, the electron will prefer to stay in low potential wells because the overlap integrals between such wells can be small. That is, the electron tends to localize. Figure 6.1 illustrates a random potential and two localized sites a and b that are farther apart from the original lattice constant. However, at finite temperatures the electron can be activated and may hop from one localized position to another, causing *hopping conduction*.

At a localized site the wavefunction of the electron can be considered to decrease exponentially as $\exp(-\alpha r)$, where α is a constant and r is the distance from the localized center. Hence, its probability to overlap with another localized wavefunction at a distance r is proportional to $\exp(-2\alpha r)$. At low but finite temperatures, the electron can hop to the next localized state depending on this probability and also the probability associated with the energy difference between the two sites. The conductivity due to hopping is expected to be proportional to the product of these two probabilities [1]:

$$\sigma \sim \exp\left(-2\alpha r - \frac{\Delta\varepsilon}{kT}\right). \tag{1.1}$$

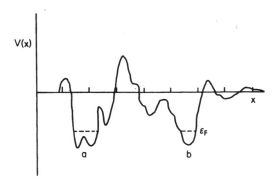

FIG. 6.1. Random potential $V(x)$.

In actuality, both r and $\Delta\varepsilon$ fluctuate, but for conduction the electron must move from one end to the other of the system. We replace both quantities by their averages. At low temperatures the average $\Delta\varepsilon$ may be estimated from the density of states $g(\varepsilon)$ per unit volume. If $g(\varepsilon)$ is given, the total density of states contained in the volume in a d-dimensional space with a characteristic length r is $g(\varepsilon)r^d$. An average energy difference $\Delta\varepsilon$ near the Fermi energy is then given by

$$\Delta\varepsilon = \frac{a}{g(\varepsilon)r^d}, \tag{1.2}$$

where a is an appropriate constant.

Since $\Delta\varepsilon$ is now expressed as a function of r, one can find a particular value of r that corresponds to a maximum of σ. The optimal condition is

$$\frac{d}{dr}\left[2\alpha r + \frac{a}{g(\varepsilon)r^d kT}\right] = 0.$$

Hence, the conductivity at low temperatures will vary with temperature as

$$\sigma \sim \exp[-(T_0/T)^{1/(d+1)}], \tag{1.3}$$

where T_0 is a constant. This formula, due to Mott [1], represents variable-range hopping conduction. Note that the temperature dependence varies with dimension. In particular, for three dimensions,

$$\sigma \sim \exp[-(T_0/T)^{1/4}]. \tag{1.4}$$

The temperature exponent is $1/2$ for one dimension and $1/3$ for two dimensions.

Such a variation has been observed experimentally. For instance, between 0.2 K and 3 K in silicon accumulation layers Fowler et al. [2] observed variable-range hopping. The temperature exponent was 1/2 or 1/3 depending on whether the gate voltage was below or above 6 V. This change indicates a transition in hopping from one-dimensional to two-dimensional.

What happens to the conduction if the disorder of a given system is significantly increased, for instance by introducing more impurities? Does the system continuously decrease its conductivity? In the classic article entitled "Absence of diffusion in certain random lattices" published in 1958 [3] Anderson pointed out that the system will become an insulator at a certain point. That is, there is a critical value of disorder at which the system changes its conductive behavior. This critical value is determined by the width of the potential fluctuation measured by the original overlap integral. It depends also on the lattice coordination number.

The existence of such a critical value of fluctuation is well demonstrated by a percolating threshold. For low magnitudes of disorder, electrons can percolate through a disordered lattice, but when randomness reaches a certain value, they cannot go through the entire system. Macroscopically, the system may be pictures as consisting of conductive and nonconductive domains, and its randomness may be represented by the probability p of finding the conductive domains. Experiment and computer simulation show that the conductivity varies as

$$\sigma(p) = \begin{cases} \sigma_0(p - p_c)^\beta & (p > p_c), \\ \sigma_0(\sigma_1/\sigma_0)^\gamma & (p = p_c), \\ \sigma_1(p_c - p)^{-\alpha} & (p < p_c), \end{cases} \tag{1.5}$$

where the threshold p_c, constants σ_0 and σ_1, and exponents α and β depend on the system and

$$\gamma = \frac{\alpha}{\alpha + \beta}. \tag{1.6}$$

Equation (1.5) means that $\sigma(p)$ near the threshold possesses a scaling behavior. The conductivity depends on parameters:

$$\varepsilon = p - p_c, \qquad x = \frac{\sigma_1}{\sigma_0}. \tag{1.7}$$

Here, σ_0 and σ_1 are the conductivities on respective side of the threshold and ε measures the distance from the threshold. The scaling character may be expressed in the following general forms [4]:

$$\sigma(\varepsilon, x) = \sigma_0 \lambda \psi(|\varepsilon| \lambda^{-s}, x\lambda^{-t}) \qquad (\varepsilon > 0),$$
$$\sigma(\varepsilon, x) = \sigma_1 \lambda \psi(|\varepsilon| \lambda^{u}, x\lambda^{-v}) \qquad (\varepsilon < 0). \tag{1.8}$$

Here s, t, u, and v are scaling exponents. By examining $\sigma(\varepsilon, x)$ in the limit $x \to 0$, and choosing λ such that

$$\begin{aligned} \lambda &= \varepsilon^{1/s} & (\varepsilon > 0), \\ \lambda &= (-\varepsilon)^{-1/u} & (\varepsilon < 0), \end{aligned} \tag{1.9}$$

one can recover the power-law behavior of the conductivity.

6.2. Scaling theory of Anderson localization

Anderson's pioneering work [3] on electron localization did not attract wide attention immediately. However twenty years later, that is, in 1978, Mott [5] suggested that when the concentration of impurities reaches a certain level, localized electron waves start overlapping with each other and form a continuous impurity band. The bottom of this band is still localized so that there exists a sharp boundary between the localized and extended states. Hence, the mobility of electrons is finite if the electron energy is above a critical value ε_c, and is zero below. When the energy is decreased toward the critical value, the conductivity will decrease until it reaches a minimum value $\sigma_{\min}$, below which it drops to zero, as shown by the dashed lines in Fig. 6.2. Mott's argument appeared plausible, but the scaling theory of localization developed by Abrahams et al. [6] in the following year showed that it was incorrect. The conductivity should decrease continuously to zero, as shown schematically by the solid curve in Fig. 6.2. This scaling theory has allowed significant progress concerning Anderson localization.

In the scaling theory, the conductance G of a given system is considered as a function of its size L:

$$G_L = G(L). \tag{2.1}$$

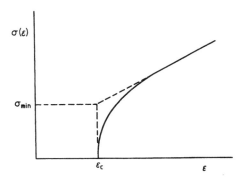

FIG. 6.2. Conductivity as a function of electron energy.

If the system is macroscopic, the conductance should be proportional to the cross-section and inversely proportional to the length L. In d dimensions, the cross-section is L^{d-1}, so that

$$G(L) = \sigma L^{d-2}. \tag{2.2}$$

σ is the conductivity, which is independent of L.

While Eq. (2.2) represents what we expect of a macroscopic systems, let us now consider changing the length scale. For an arbitrary length L', the conductance is assumed to obey the scaling form

$$G(L') = f(G(L), L'/L), \tag{2.3}$$

where f is a universal function independent of the microscopic structure of the system. A logarithmic differentiation of G with respect to

$$x = L'/L$$

at $x = 1$ yields

$$\frac{d \ln G}{d \ln L} = \beta(G), \tag{2.4}$$

where $\beta(G)$ is a universal function.

In the ordinary metallic region where Eq. (2.2) holds,

$$\beta(G) = d - 2. \tag{2.5}$$

As the disorder increases, the conductance will decrease and eventually will vanish. For large systems, its approach to zero may be assumed to follow

$$G_L \sim \exp(-\alpha L). \tag{2.6}$$

This form yields

$$\beta(G) = \ln G. \tag{2.7}$$

The expressions (2.5) and (2.7) correspond to two extreme cases. In the scaling theory, it is assumed that there exists a single function $\beta(G)$ whose limiting expressions are given by Eqs. (2.5) and (2.7). The resulting $\beta(G)$ is schematically illustrated in Fig. 6.3. For $d = 3$, $\beta(G)$ is negative below a certain value of G and is positive above. For $d = 2$ and 1, $\beta(G)$ stays below the abscissa. This means that the system is localized always.

Let us now examine the behavior of G near the point at which

$$\beta(G) = 0. \tag{2.8}$$

No L dependency of G is expected. Hence, the point is called a *fixed point*.

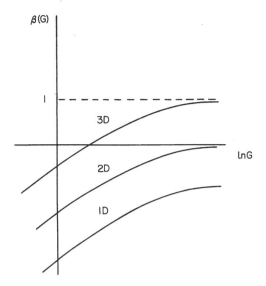

FIG. 6.3. Schematic variation of $\beta(G)$.

On the conductive side near the fixed point, a linear variation of $\beta(G)$ may be assumed:

$$\beta(G) = a[G(L) - G_c], \tag{2.9}$$

where G_c is the conductance at the fixed point, and a is the G-derivative of $\beta(G)$ at G_c.

In order to incorporate the asymptotic property that

$$\beta(G) \to 1 \qquad (G \to \infty). \tag{2.10}$$

Let us adopt a simple form:

$$\beta(G) = \frac{a[G(L) - G_c)}{1 + a[G(L) - G_c]}. \tag{2.11}$$

Although this form specifies a special approach to the asymptotic value, the final result we are going to obtain shortly is independent of this particular form.

Let us use Eq. (2.11) in Eq. (2.4) and integrate the result from a macroscopic length L_0 to a much shorter length L. The result of integration is

$$\left(\frac{G(L) - G_c}{G(L_0) - G_c} \frac{G(L_0)}{G(L)} \right)^{\nu} \frac{G(L)}{G(L_0)} = \frac{L}{L_0}, \tag{2.12}$$

where

$$v = \frac{1}{aG_c}. \tag{2.13}$$

For the macroscopic length L_0, one can use Eq. (2.2) so that the conductivity is given by

$$\sigma = \frac{G(L_0)}{L_0}. \tag{2.14}$$

We choose L such that $G(L)$ is sufficient close to G_c. Since $G(L_0)$ is much larger than G_c, we arrive at

$$\sigma = \sigma_c \left(1 - \frac{G_c}{G}\right)^v, \tag{2.15}$$

where σ_c is a certain constant. Because v is positive, this formula shows that the conductivity vanishes as G approaches G_c. Furthermore, under the assumption that the conductance is proportional to electron energy, one can rewrite Eq. (2.15) in the form

$$\sigma = \sigma_c \left(1 - \frac{\varepsilon_c}{\varepsilon}\right)^v. \tag{2.16}$$

The constant ε_c corresponding to G_c represents a mobility edge. That is, at the mobility edge the conductivity is expected to vanish rather than to approach a constant.

The above consideration is made for absolute zero. One and two dimensions are special in the sense that electrons are always localized in the macroscopic limit. Correspondingly, there appears no point at which $\beta(G)$ vanishes, as shown in Fig. 6.3. In two dimensions, the conductivity has the dimension of e^2/h. Hence, in consideration of Fig. 6.3, we assume

$$\beta(G) = -\frac{e^2 \lambda}{2\pi^2 \hbar} \frac{2}{G}. \tag{2.17}$$

Here, λ is a certain constant and $\beta(G)$ is assumed to approach zero as $1/G$. Equation (2.4) then yields

$$G(L) = G(L_0) - 2G_a \ln(L/L_0). \tag{2.18}$$

$G(L_0)$ is the value of G at a certain macroscopic length L_0, and G_a is defined by

$$G_a = \frac{e^2 \lambda}{2\pi^2 \hbar}. \tag{2.19}$$

Note that

$$\frac{e^2}{2\pi^2\hbar} = 1.2 \times 10^{-5}\ \Omega^{-1}.$$ (2.20)

The inverse G variation in Eq. (2.17) is only an approximate assumption, but for a certain length L less than L_0 the conductance is expected to vary logarithmically with length L.

The logarithmic variation of the conductance represents a precursor to localization and therefore is identified as representing *weak localization*. Such a logarithmic variation has actually been observed.

For finite temperatures electrons can diffuse about, maintaining phase coherence until inelastic scattering takes place. Hence, the elastic mean free path that determines conductivity is expected to be proportional to $(D\tau_i)^{1/2}$, where τ_i is the inelastic scattering time and D is the diffusion constant. The scattering time is infinite at absolute zero and decreases as temperature increases so that one may adopt

$$\tau_i \sim T^{-p},$$ (2.21)

where p is a constant of order 1. Hence, as temperature decreases, the conductivity is expected to deviate from the Drude conductivity logarithmically. The deviation is given by

$$\Delta\sigma(T) = \frac{e^2}{2\pi^2\hbar}\ \lambda p \ln(T/T_0)$$ (2.22)

below a characteristic temperature T_0. This logarithmic variation is well exhibited in the data on copper films shown in Fig. 6.4, obtained by

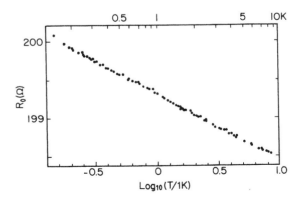

FIG. 6.4.   Resistance of a copper film. (From Kobayashi et al. [7])

Kobayashi et al. [7]. There are several other data supporting the logarithmic variation.

6.3. Weak localization

According to linear response theory the conductivity is expressed in terms of a current–current correlation function. If electron energy at the Fermi level $\sigma_{xx} = \sigma$ is given by

$$\sigma = \frac{i\hbar^2 e^2}{V} \langle \text{Tr}[v_x \delta(\varepsilon - H) v_x \delta(\varepsilon - H)] \rangle, \tag{3.1}$$

where v_x is the velocity component in the direction x of an external electric field, $\langle \cdots \rangle$ represents an average over impurity configurations, and H is the Hamiltonian of a single electron:

$$H = H_0 + V. \tag{3.2}$$

H_0 may include many electrons with Coulomb interaction, but avoiding theoretical complications let us assume that it is just the kinetic energy of a single electron. V is the impurity potential, which is assumed to be short-ranged and of the form

$$V(r) = \sum_i V_0 \delta(\mathbf{r} - \mathbf{R}_i). \tag{3.3}$$

The Dirac delta functions in Eq. (3.1) representing energy conservation can be expressed as

$$\delta(\varepsilon - H) = \frac{i}{2\pi} \left[\frac{1}{\varepsilon - H + i0} - \frac{1}{\varepsilon - H - i0} \right]. \tag{3.4}$$

Here, $i0$ in the denominator means a small imaginary parameter to be brought to zero after its use. The right-hand side of Eq. (3.2) can be expressed in terms of the resolvents $G^\pm(\varepsilon)$ defined by

$$G^\pm(\varepsilon) = (\varepsilon - H \pm i0)^{-1}. \tag{3.5}$$

To proceed, we note the identity that holds for two arbitrary operators A and B:

$$\frac{1}{A + B} = \frac{1}{A} - \frac{1}{A} B \frac{1}{A + B}.$$

By iteration, we can expand $1/[\varepsilon - H]$ in powers of V such that

$$\frac{1}{\varepsilon - H} = \sum_{n=0}^{\infty} \frac{1}{\varepsilon - H_0}\left(V\frac{1}{\varepsilon - H_0}\right)^n. \tag{3.6}$$

The trace in Eq. (3.1) can conveniently be taken in momentum space. Accordingly, we introduce the matrix element

$$G(\mathbf{k}, \mathbf{k}') = \langle\mathbf{k}| \frac{1}{\varepsilon - H} |\mathbf{k}'\rangle, \tag{3.7}$$

which is a one-electron Green's function. We also define

$$G_0(\mathbf{k}, \mathbf{k}') = \langle\mathbf{k}| \frac{1}{\varepsilon - H_0} |\mathbf{k}'\rangle = (2\pi)^2\delta(\mathbf{k} - \mathbf{k}')G_0(\mathbf{k}). \tag{3.8}$$

By the above expansion we obtain for two dimensions:

$$G(\mathbf{k}, \mathbf{k}') = (2\pi)^2\delta(\mathbf{k} - \mathbf{k}')G_0(\mathbf{k}) + G_0(\mathbf{k})V_{\mathbf{k}-\mathbf{k}'}G_0(\mathbf{k}')$$
$$+ \sum_{s=1}^{\infty} \frac{1}{(2\pi)^{2s}} \int d\mathbf{k}_1\, d\mathbf{k}_2 \cdots d\mathbf{k}_s G_0(\mathbf{k})V_{\mathbf{k}-\mathbf{k}_1}G_0(\mathbf{k}_1)V_{\mathbf{k}_1-\mathbf{k}_2}$$
$$\cdots G_0(\mathbf{k}_s)V_{\mathbf{k}_s-\mathbf{k}'}G_0(\mathbf{k}'). \tag{3.9}$$

Formula (3.1) can be expressed by an impurity average of a product of Green's functions. In the first approximation we can employ a method that is equivalent to effective medium theory. In this method we replace G by $\langle G\rangle$:

$$G \to \langle G\rangle,$$

and then use

$$\langle(G^+ - G^-)(G^+ - G^-)\rangle \to (\langle G^+\rangle - \langle G^-\rangle)(\langle G^+\rangle - \langle G^-\rangle).$$

That is, the average of the product of the Green's functions is replaced by the product of the averaged Green's functions. Furthermore, we assume that the averaged ideal Green's function has the following form:

$$\langle G_0(\mathbf{k}, \mathbf{k}')\rangle = (2\pi)^2 \frac{\delta(\mathbf{k} - \mathbf{k}')}{\varepsilon - \varepsilon_0 - \Sigma}. \tag{3.10}$$

Here ε_0 is the kinetic energy and Σ, called the self-energy, is generally complex:

$$\Sigma = \Delta - i\Gamma. \tag{3.11}$$

The real part Δ represents an energy shift and the imaginary part Γ represents level broadening. Equation (3.10) implies that the electron moves as in the ideal case although the energy levels are broadened and shifted by the impurities. One can then show that σ is given by the familiar form

$$\sigma_0 = \frac{ne^2\tau}{m}, \tag{3.12}$$

where

$$\tau = \frac{\hbar}{2\Gamma}. \tag{3.13}$$

It is clear that the above derivation neglects the quantum interference between the two electron propagations represented by the Green's functions. If the conductivity is high, the approximation may be acceptable, but as the impurity concentration increases, the interference of electron waves going around the impurities should become more and more important. Hence, in order to approach the regime of weak localization it is necessary to improve the Drude type form given by Eq. (3.12).

For this purpose, it is helpful to resort to a diagrammatic representation. Let us use a solid line to represent the propagation from $\mathbf{k}'$ to $\mathbf{k}$ represented by $G_0(\mathbf{k}, \mathbf{k}')$ and one dashed line for one impurity potential V_k. Since G is expressed by Eq. (3.9), which is an infinite series, its product with another G in Eq. (3.1) will include too many terms to evaluate. However, so-called *fan diagrams* can be summed to infinite order, yielding a significant contribution. Figure 6.5(a) illustrates such a diagram in which the product $G_0(\mathbf{k}, \mathbf{k}')G_0(\mathbf{k}', \mathbf{k})$ forms a loop. Note that it can be redrawn as in (b), which is a ladder diagram. The most dominant contribution from such diagrams to the conductivity is given by

$$\Delta\sigma = \frac{\hbar}{\pi}\left(\frac{e\hbar}{m}\right)^2 \int \frac{d\mathbf{k}\,d\mathbf{k}'}{(2\pi)^2} k_x k_x' G_0^+(\mathbf{k})G_0^-(\mathbf{k})G_0^+(\mathbf{k}')G_0^-(\mathbf{k}')\Lambda(\mathbf{k} + \mathbf{k}'), \tag{3.14}$$

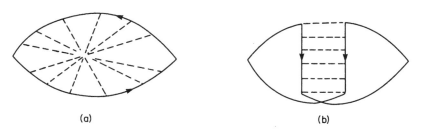

(a)   (b)

FIG. 6.5. Fan diagram and an equivalent ladder diagram.

where

$$\Lambda(q) = \frac{n_i^2 V_0^2}{1 - n_i V_0^2 \lambda(q)}. \qquad (3.15)$$

Here, n_i is the impurity concentration, V_0 is the strength of the impurity potential, and

$$\lambda(q) = \frac{1}{(2\pi)^2} \int d\mathbf{k}\, G_0^+(\mathbf{k}) G_0^-(\mathbf{q} - \mathbf{k}). \qquad (3.16)$$

Note that the result (3.15) obtained by a summation over the ladder diagrams is characterized by a single wavenumber $\mathbf{q}$. This feature makes the ladder diagram contribution significant. Moreover, the right-hand side of Eq. (3.14) contributes most significantly when $q \to 0$. Hence, $\lambda(q)$ may be obtained by expanding $G_0(\mathbf{q} - \mathbf{k})$ for small $\mathbf{q}$. This results in

$$\lambda(q) = \lambda(0)[1 - \tau D q^2], \qquad (3.17)$$

where D is the diffusion coefficient defined by

$$D = \frac{\tau}{2} \left(\frac{\hbar k_F}{m} \right)^2. \qquad (3.18)$$

Hence, the quantum correction is

$$\Delta\sigma = -\frac{2e^2}{\pi\hbar} \frac{1}{(2\pi)^2} \int \frac{d\mathbf{q}}{q^2}. \qquad (3.19)$$

The integral is divergent in infinite space. For a finite system with a characteristic length L it is appropriate to limit q such that

$$\frac{1}{L} < q < \frac{1}{L_0},$$

where

$$L_0 = \frac{\hbar k_F \tau}{m}. \qquad (3.20)$$

Adding σ_0 we arrive at

$$\sigma(L) = \sigma_0 - 2 \left(\frac{e^2}{2\pi^2\hbar} \right) \ln \frac{L}{L_0}. \qquad (3.21)$$

This is in agreement with the result of the scaling theory and shows that the logarithmic variation is caused by quantum interference.

6.4. Negative magnetoresistance

The term "negative magnetoresistance" sounds odd, but it is used for a decrease in magnetoresistance, that is, an increase in magnetoconductivity, due to weak localization. In this section we shall discuss weak localization in a magnetic field mainly in two dimensions. A given system is two-dimensional if the characteristic length $(D\tau_i)^{1/2}$ is longer than its thickness, where D is the diffusion constant and τ_i is the inelastic scattering time. The role played by this and some other characteristic lengths will be discussed.

One can expect that an external magnetic field gives rise to a large effect on conductivity because electron motion is affected. In particular, a perpendicular magnetic field affects 2D electrons strongly, confining the electrons to cyclotron orbits. The magnetic length

$$l = \left(\frac{c\hbar}{eH}\right)^{1/2} \tag{4.1}$$

characterizes the cyclotron orbits. It is the radius of the lowest cyclotron orbit, its explicit magnitude being given by

$$l = 25.66H^{-1/2} \quad \text{(nm)}$$

when H is in Tesla.

Since the cyclotron orbits shrink as H increases, weak localization appears to be enhanced by a magnetic field. That is, it looks as if the magnetic field reduces magnetoconductivity. However, this is not the case in disordered systems in a weak magnetic field. In fact, in the regime of weak localization, the electrons are scattered by impurities very often before completing their cyclotron orbits. The primary field effect is to destroy the time-reversal symmetry in electron motion. Hence, for instance, H acts against the cancellation of two electron waves with opposite wavevectors $\mathbf{k}$ and $-\mathbf{k}$.

The magnetic length l must be compared with the size L of a given system. If l is longer than L, the conductance is expected to depend on the latter and is given by Eq. (3.21). If it is shorter, it is appropriate to replace L in that equation by l so that

$$\sigma(H) = \sigma_0 - 2\frac{e^2}{2\pi^2\hbar}\ln\frac{l}{L_0}. \tag{4.2}$$

Since l is inversely proportional to the square root of H, we find the following change in conductivity due to H:

$$\Delta\sigma(H) = \frac{e^2}{2\pi^2\hbar}\ln H. \tag{4.3}$$

That is, in the weak localization regime the conductivity change increases with magnetic field, representing negative magnetoresistance.

The above derivation of Eq. (4.3) is intuitive, so that it is desirable to take a more quantitative approach. According to Hikami et al. and Al'tshuler et al. [8] spin–orbit coupling and magnetic scattering due to impurity spins give rise to relaxation times in addition to that one due to the ordinary impurity scattering. However, for weak fields and if the inelastic scattering time τ_i is shorter than the other relaxation times, the change ΔH in magnetoconductivity is given in a relatively simple form:

$$\Delta\sigma(H) = \alpha g_v \frac{e^2}{2\pi^2\hbar}\left[\psi\left(\frac{1}{2} + \frac{1}{a\tau_i}\right) - \psi\left(\frac{1}{2} + \frac{1}{a\tau}\right) + \ln\frac{\tau_i}{\tau}\right], \qquad (4.4)$$

where α is a parameter that is of order 1 if spin–orbit and magnetic scattering are weak, g_v is the valley degeneracy factor for applications to silicon inversion layers, and

$$a^{-1} = \frac{c\hbar}{4DeH}, \qquad (4.5)$$

in which

$$D = \frac{\varepsilon_F \tau}{m}. \qquad (4.6)$$

$\psi(z)$ is the digamma function, which varies for $z \to \infty$ such that

$$\psi(z) = \ln z - \frac{1}{2z} - \frac{1}{12z^2} + O(1/z^4). \qquad (4.7)$$

This function originates from a summation of the type

$$\sum_{n=0}^{N_M} \frac{1}{(n + \frac{1}{2}) + 1/(a\tau_i)}$$

over the Landau levels, which appears in taking the trace for the magneto-conductivity. Since this series is divergent a cut-off N_M has been introduced. This cut-off parameter is given approximately by $1/(a\tau)$.

The parameter $1/a$ has the dimension of time and is proportional to the cyclotron frequency and to the ratio of the Fermi energy to the imaginary part of the self-energy. That is, it represents an effective cyclotron frequency for a broadened Landau level. If broadening of Landau levels due to impurities is small, both $1/a$ and τ are large. If $1/a$ is larger than τ but smaller than τ_i, that is, for a relatively weak magnetic field such that

$$(a\tau)^{-1} > 1 > (a\tau_i)^{-1},$$

Eq. (4.4) is reduced effectively to Eq. (4.3).

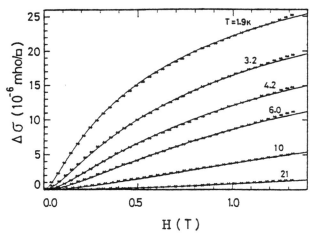

FIG. 6.6. Magnetoconductivity change $\Delta\sigma(H)$ in an n-channel [100] silicon inversion layer. (From Kawaji and Kawaguchi [9])

Figure 6.6 shows the field dependence of the magnetoconductivity of a silicon [100] inversion layer. The data are due to Kawaji and Kawaguchi [9]. The curves represent Eq. (4.4) in which τ_i and α are adjustable to fit the data.

If $1/a$ is much less than τ, the magnetic field causes the following changes approximately:

$$\Delta\sigma(H) = \begin{cases} \dfrac{\alpha e^2}{48\pi^2\hbar}\left(\dfrac{4DeH\tau_i}{c\hbar}\right)^2 & (a\tau_i \ll 1); \\[4mm] \dfrac{\alpha e^2}{2\pi^2\hbar}\ln\left(\dfrac{4DeH\tau_i}{c\hbar}\right) & (a\tau_i \gg 1). \end{cases} \qquad (4.8)$$

Here, g_v has been assumed to be 1. Note that $\Delta\sigma(H)$ is now proportional to H^2 and this proportionality is expected for a relatively strong magnetic field.

It is important to recognize that the logarithmic variation with H is characterized by a universal constant e^2/h. A similar logarithmic variation appears in the case of the Kondo effect, but the variation is not characterized by such a universal constant alone. Moreover, since the variation is independent of the direction of magnetic field, it can be distinguished from the case of weak localization.

In order to explain in detail experimental data on the magnetoconductivities of different samples at various temperatures, effects due to spin–orbit coupling or intervalley scattering must be taken into consideration [8]. The corresponding theoretical formulas have been developed out, and in general they have been found to be in good agreement with recent experimental results.

The case of three dimensions has also been solved [10]. The magneto-conductivity change is found to be proportional to $H^{1/2}$ for strong fields. Its limiting expression is

$$\Delta\sigma(H) = 0.605 \frac{e^2}{2\pi^2 \hbar l}$$

$$= 290 H^{1/2} \quad \text{S/mT}^{1/2}, \tag{4.9}$$

where S is siemens (i.e., Ω^{-1}) and T is Tesla.

The magnetoconductivity is proportional to H^2 for weak fields. Combining these two limiting cases, a semiempirical expression

$$\Delta\sigma(H) = aH^{1/2} + bH^2 \tag{4.10}$$

may be used, with adjustable parameters a and b. This type of variation has been observed.

6.5. Interference effects

Since the theoretical results in the previous section are generally in agreement with experiment, a question arises whether or not electron–electron interaction can be completely ignored. In the metallic regime interaction effects can be expected to be small, but in the regime of weak localization the effects must be assessed. When a correction to the Drude conductivity σ_0 is studied, it turns out that there are important interaction effects.

Before discussing these effects let us first reexamine the main results for weak localization in the previous sections. For two dimensions,

$$\sigma_0 = \frac{ne^2 \tau}{m}$$

is proportional to $\varepsilon_F \tau$. Thus, corrections to σ_0 could start with terms of order $\hbar/(\varepsilon_F \tau)$, which is dimensionless. In fact,

$$\frac{\Delta\sigma}{\sigma_0} = \frac{e^2/2\pi^2 \hbar}{e^2 \varepsilon_F \tau/\pi\hbar^2}$$

$$= \frac{\hbar}{2\pi\varepsilon_F \tau}. \tag{5.1}$$

However, the quantity

$$D = \frac{\varepsilon_F \tau}{m}$$

is the diffusion coefficient of an electron. This constant introduces a new length scale given by

$$L_D = \left(\frac{\hbar D}{kT}\right)^{1/2}. \tag{5.2}$$

If this length is longer than the localization length L, it is unimportant. In the opposite limit, L_D must replace L. This replacement is consistent with the transition from Eq. (2.18) to Eq. (2.22). At the same time, we can conjecture λ in these equations to be

$$\lambda = \frac{\hbar}{2\pi\varepsilon_F\tau}. \tag{5.3}$$

Thus, a theoretical improvement can be made by considering corrections of order $(1/\lambda)$ or higher. By doing so, such an expression as Eq. (2.17) for $\beta(G)$ will be modified into a new form that behaves properly for a wide range of conductance, as conjectured in Fig. 6.3. As it is, Eq. (2.17) can be assumed only for large G.

It turns out that electron interaction modifies the relaxation time τ and the density of states at the Fermi energy as follows [11]:

$$\tau^{-1} \to \tau^{-1}\left[1 + \lambda s \ln\left(\frac{\hbar}{4\pi\tau kT}\right)\right]; \tag{5.4}$$

$$\frac{m}{2\pi\hbar^2} \to \frac{m}{2\pi\hbar^2} s \ln\left(\frac{\hbar}{4\pi\tau kT}\right). \tag{5.5}$$

Here, s is an interaction parameter consisting of several terms that can be represented by diagrams for electron self-energy. Note that the logarithm of a new dimensionless parameter $\hbar/(4\pi\tau kT)$ appears in the above equations. This parameter can be converted into a length-parameter as in Eq. (5.2). One can say then that effects of Coulomb interaction are important when this length becomes small in comparison with L_i or L.

Due to electron interaction the temperature-dependent correction $\Delta\sigma(T)$ is modified. In the range where a logarithmic variation is observed it can be expressed as

$$\Delta\sigma(T) = \frac{e^2}{2\pi^2\hbar} \alpha_T \ln T. \tag{5.6}$$

Here,

$$\alpha_T = p + 4 - 3\left(\frac{2 + F}{F}\right)\ln\left(1 + \frac{F}{2}\right), \tag{5.7}$$

where p is the exponent for the temperature variation of τ_i, and

$$F = \frac{1}{2\pi} \int_0^\pi d\theta \, \frac{\kappa}{\kappa + 2k_F \sin(\theta/2)}, \qquad (5.8)$$

where κ is an inverse screening length, and k_F is the Fermi wavenumber. For silicon [100] inversion layers, $\kappa = 1.87 \times 10^7$ cm and $k_F = 1.77n^{1/2}$ may be used with the average dielectric constant of semiconductor and insulator. For electron density $n = 10^{12}$ cm^{-2}, $\kappa > k_F$ so that F is less than 1. To first order in F, α_T is given by

$$\alpha_T = p + 1 - \tfrac{3}{4}F. \qquad (5.9)$$

For $F \leq 1$, α_T is positive. Hence, near absolute zero,

$$\Delta\sigma(T) < 0. \qquad (5.10)$$

This means a positive additional resistivity.

The magnetoconductivity change $\Delta\sigma(H)$ of a 2D system is also affected by Coulomb interaction. An expression that corresponds to Eq. (5.6) is

$$\Delta\sigma(H) = \frac{e^2}{2\pi^2\hbar} \alpha_H \ln H, \qquad (5.11)$$

where

$$\alpha_H = 1 - \frac{F}{4}. \qquad (5.12)$$

Since $F \leq 1$, we expect $\alpha_H > 0$ and a positive $\Delta\sigma(H)$.

Because quantum interference produces similar logarithmic dependences, the interaction effects can be tested experimentally only through detailed analyses that determine either the parameter α_T or α_H. It is helpful for such analyses to use a more precise expression for $\Delta\sigma(H)$ than the above. By modifying Eq. (4.4) as indicated by Eq. (5.4), and for $kT\tau/\hbar \gg 1$ and $a\tau_i \gg 1$, one finds that $\Delta\sigma(H)$ is given by

$$\Delta\sigma(H) = \frac{e^2}{2\pi^2\hbar} \alpha_H \left[\psi\left(\frac{1}{2}\right) - \psi\left(\frac{1}{2} + \frac{\hbar}{kT\tau}\right) \right]. \qquad (5.13)$$

Note that this $\Delta\sigma(H)$ is independent of H and is dependent on $\hbar/(kT\tau)$, which represents an interaction effect. The corresponding resistivity change $\Delta\rho(H)$ is proportional to H^2 because it is proportional to $(\omega_c\tau)^2\Delta\sigma(H)$. It has been found experimentally [12] that this is indeed the case, with a proportionality constant $\alpha_H = 0.54$ that is close to the theoretical expectation 0.7 for the sample.

When spin–orbit coupling is strong, the parameters α_T and α_H are given by

$$\alpha_T = 1 - \tfrac{3}{4}F. \tag{5.14}$$

$$\alpha_H = -(1 + F)/2. \tag{5.15}$$

Thus, α_T can be positive even if the corresponding α_H is negative. The positive $\Delta\sigma(T)$ and negative $\Delta\sigma(H)$ observed simultaneously in gold and palladium films [13] may be due to strong spin–orbit coupling.

REFERENCES

1. N. F. Mott, *J. Noncryst. Solids* **1**, 1 (1968).
2. A. B. Fowler, A. Hartstein, and R. A. Webb, *Phys. Rev. Lett.* **48**, 196 (1982); *Physica* **117, 118B**, 661 (1983).
3. I. Webman, J. Jortner, and M. H. Cohen, *Phys. Rev. B* **16**, 2593 (1977).
4. P. W. Anderson, *Phys. Rev.* **109**, 1492 (1958).
5. N. F. Mott, *Rev. Mod. Phys.* **50**, 203 (1978).
6. E. Abrahams, P. W. Anderson, D. C. Licciardello, and T. V. Ramakrishnan, *Phys. Rev. Lett.* **42**, 673 (1979). P. W. Anderson, E. Abrahams, and T. V. Ramakrishnan, *Phys. Rev. Lett.* **43**, 1718 (1979).
7. S. Kobayashi, F. Komori, Y. Ootsuka, and W. Sasaki, *J. Phys. Soc. Jpn* **49**, 1635 (1980).
8. B. L. Al'tshuler, D. E. Khmelnitskii, D. E. Larkin, and P. A. Lee, *Phys. Rev. B* **22**, 5142 (1980). S. Hikami, A. I. Larkin, and Y. Nagaoka, *Prog. Theor. Phys.* **63**, 707 (1980). S. Kawabata, *J. Phys. Soc. Jpn* **53**, 3540 (1984).
9. S. Kawaji and Y. Kawaguchi, in *Application of High Magnetic Fields in Semiconductor Physics*, ed. by G. Landwehr (Springer-Verlag, New York, 1983), p. 53.
10. A. Kawabata, *Solid State Commun.* **34**, 432 (1980).
11. A. M. Finkel'shtein, *Zh. Eksp. Teo. Fiz.* **84**, 168 (1983) [*Sov. Phys.—JETP* **57**, 97 (1983)]. B. L. Al'tshuler and A. G. Aronov, *Solid State Commun.* **46**, 429 (1983). H. Fukuyama, Y. Isawa, and H. Yasuhara, *J. Phys. Soc. Jpn* **52**, 16 (1983).
12. K. K. Choi, D. C. Tsui, and S. C. Palmateer, *Phys. Rev. B* **33**, 8216 (1986).
13. W. G. McGinnis, M. J. Burns, R. W. Simon, G. Deutscher, and P. Chaikin, *Physica B & C* **107**, 5 (1981). T. Kawaguchi and Y. Fukumori, *J. Phys. Soc. Jpn* **51**, 703 (1982).

7

MAGNETISM

Magnetism is one of the oldest and most interesting subjects in condensed matter physics. Since the time of Gilbert who published "De Magnete" in 1600, many important discoveries have been made in the area of magnetism.

7.1. de Haas–van Alphen effect

In a magnetic field, free electrons show paramagnetism due to spin and diamagnetism due to orbital motion. At low temperatures and fields, the corresponding susceptibilities are constant. However, in 1930 de Haas and van Alphen discovered that the magnetic susceptibility of bismuth was not constant but oscillatory as a function of magnetic field. Subsequently, as the same oscillatory behavior was found in many other metals, the name *de Hass–van Alphen effect* emerged.

This effect is observed at low temperatures and in strong magnetic fields that satisfy

$$\varepsilon_F \gg \hbar\omega_c > kT.$$

The first inequality means that the electron system is quantum-mechanically degenerate even though, as required by the second inequality, the magnetic field is sufficiently strong. At 1 K, the field strength must be of order 1 T. On the other hand, as we shall see, the observability of de Haas–van Alphen (dHvA) oscillations is determined by

$$\frac{\Delta H}{H} \sim \frac{\hbar\omega_c}{\varepsilon_F} \sim 10^{-4}.$$

That is, for the observability of oscillations, fluctuations ΔH is an applied magnetic field should be small and the electron density should not be too high because the period depends on the ratio $\hbar\omega_c/\varepsilon_F$.

On the theoretical side, in his famous work on diamagnetism published in 1930, Landau noticed that the case of a finite magnetic field was not simple. However, he did not obtain any explicit result. In 1952, nearly twenty years after its discovery, the effect was finally explained by Onsager as a Fermi-surface phenomenon and also by Lifshitz and Kosevich. The latter authors derived an explicit formula for the oscillating susceptibility. Since then, the period of the dHvA oscillations has been used very effectively to determine the Fermi surface of metals.

The de Haas–van Alphen effect is due to quantization of electron energy levels in the direction perpendicular to an exernal magnetic field. As such, it is essentially a two-dimensional phenomenon. When the electron density is changed, the Fermi energy will move from one Landau level to another. Since each level has a natural width of eH/ch per unit area, the diamagnetic response of a given level increases initially as its electron population is increased but drops to zero when the level is filled. The same process repeats itself as the Fermi energy reaches the next Landau level, causing oscillations. Changing the magnetic field causes the same effect because the electron population in a Landau level depends on the ratio

$$v = \frac{n}{eH/ch}.$$

What happens in actual systems in which various electron scatterings take place? In contrast to the ideal case the existence of impurities broadens the Landau levels. Phenomenological consideration of such a broadening results in an exponential decrease of the amplitude of the dHvA oscillations. This reduction factor, known as the Dingle factor, has been determined experimentally. Thus, amplitude analyses of the dHvA oscillations provide useful information concerning electron scattering time.

7.1.1. *Three dimensions*

In order to see quantitatively how the dHvA oscillations take place in an interacting electron gas in three dimensions, first note that the Landau levels and the corresponding eigenfunctions of an electron are given by Eqs. (2.5) and (2.3) of Chapter 3. In terms of these we define a classical propagator in β and r space by

$$K^{\text{cl}}(\mathbf{r}_2 \beta_2; \mathbf{r}_1 \beta_1) = \sum_{n\sigma} \{\exp[-\varepsilon_n(\beta_2 - \beta_1)]\psi_{n\sigma}(\mathbf{r}_2)\psi_{n\sigma}^*(\mathbf{r}_1)\}. \tag{1.1}$$

Here, σ represents spin and $\beta = 1/kT$. Note that the partition function is obtained by taking the trace of this propagator. Using a contour integral we obtain for ideal fermions the grand partition function given by

$$\ln \Xi_0 = V \frac{\pi}{2\pi i} \int_{c-i\infty}^{c+i\infty} \frac{ds}{s \sin \pi s} e^{s/\delta} \operatorname{Tr} K^{\text{cl}}(\mathbf{r}, s\beta; \mathbf{r}, 0), \tag{1.2}$$

where V is the total value and

$$\delta = \frac{kT}{\varepsilon_F},$$

ε_F being the Fermi energy.

From the grand partition function the magnetic susceptibility is obtained as follows:

$$\chi = \frac{1}{\beta H V}\left(\frac{\partial \ln \Xi}{\partial H}\right)_Z, \tag{1.3}$$

where Z is the fugacity. For the present noninteracting case the susceptibility is given by

$$\chi = \frac{k_F}{2\pi^2}\left(\frac{e^2}{c^2}\right)\left[\left(\frac{g}{2}\right)^2 - \frac{1}{3} + \frac{\pi\delta}{\gamma^{3/2}}\sum_{s=1}^{\infty}(-)^{s+1}\frac{\cos(g\pi s/2)\sin[(\pi s v) - \pi/4]}{s^{1/2}\sinh(\pi^2 s/\alpha)}\right]. \tag{1.4}$$

Here units are such that $\hbar = 1$ and $2m = 1$, m being the electron mass and g the Landé factor. In three dimensions the susceptibility has no dimension. Its basic unit is given by

$$\chi_0 = \left(\frac{k_F}{2\pi^2}\right)\left(\frac{e}{c}\right)^2 = \left(\frac{k_F}{2\pi^2}\right)\mu_B^2 = 2g(0)\mu_B^2, \tag{1.5}$$

where $g(0)$ is the density of states per spin at the Fermi surface, and the parameters are defined by

$$\gamma = \frac{\omega_c}{2\mu}; \qquad v = \frac{2\varepsilon_F}{\omega_c} = \frac{1}{\gamma_0}; \qquad \alpha = \frac{\beta\omega_c}{2} = \frac{\mu_B H}{kT}. \tag{1.6}$$

μ is the chemical potential, which may deviate from the ideal Fermi energy. γ_0 is a natural dimensionless parameter that can be small. It is the ratio of the field energy to the Fermi energy. α is the ratio of the field energy to the thermal energy and is of order 1 for the dHvA effect. Formula (1.4) includes both the paramagnetic and diamagnetic susceptibilities with the ideal values given by

$$\chi_0^p = \chi_0(g/2)^2; \qquad \chi_0^d = -(\chi_0/3), \tag{1.7}$$

where n is the electron density and $\beta = 1/kT$. Note that for $g = 2$ the magnitude of the diamagnetic susceptibility is one-third of the paramagnetic susceptibility.

The oscillating part of the susceptibility depends on $1/\gamma_0 = \varepsilon_F/(\mu_B H)$. That is, the period of oscillation can be used to determine the Fermi energy.

Note that electron spin enters into the ideal susceptibility through the cosine function of the Landé factor g. In particular, for $g = 2$ the spin produces only a sign factor. The constant phase $\pi/4$ is characteristic of three dimensions so that an analysis of this phase reveals the dimensionaity of a given system. For two dimensions there is no such constant phase.

The contribution from the first-order exchange graphs to the grand partition function is given by [1]

$$\ln(\Xi_x/\Xi_0) = \frac{\beta}{2} \int \int d\mathbf{r}_1 \, d\mathbf{r}_2 \phi(|\mathbf{r}_2 - \mathbf{r}_1|)\tfrac{1}{2}K(\mathbf{r}_1, \beta; \mathbf{r}_2, 0)K(\mathbf{r}_2, \beta; \mathbf{r}_1, 0), \quad (1.8)$$

where $\phi(r)$ is the Coulomb potential and

$$K(\mathbf{r}_2\beta, \mathbf{r}_1 0) = \frac{\pi}{2\pi i} \int_{c-i\infty}^{c+i\infty} dt \, \frac{e^{t/\delta}}{\sin \pi t} K^{cl}(\mathbf{r}_2, t\beta; \mathbf{r}_1, 0). \quad (1.9)$$

The ideal and first-order exchange contributions can be combined to give the oscillating part of the susceptibility in the form

$$\chi = \frac{kT}{H^{3/2}} \sum_{s=1} \frac{(-)^{s+1} \cos(g\pi s/2) \sin(\pi s v - \pi/4)}{s^{1/2} \sinh(\pi^2 s/\alpha)} \exp(-s/\omega_c \tau_e). \quad (1.10)$$

Here τ_e, representing a lifetime due to electron–electron interaction, is defined by

$$\tau_e^{-1} = \left(\frac{2}{\pi}\right)^{1/2} \frac{p_F I_s}{a_0},$$

in which a_0 is the Bohr radius and I_s is an integral:

$$I_s = \int d\mathbf{r} \, \frac{1}{r} \frac{\sin(z^2/4\pi v s)}{r^{3/2}} J_{3/2}(r) J_0[(x^2 + y^2)^{1/2}].$$

Here, a new dimensionless variable $\mathbf{r} = (x, y, z)$ is used for $p_F r$. Formula (1.10) is based on the assumption that $1/(\omega_c \tau)$ is small. It shows that the exchange interaction decreases the amplitude of the dHvA oscillations.

7.1.2. Two dimensions

We have considered so far 3D electrons. In two dimensions the susceptibility has dimension. Its basic units is

$$\chi_0 = \frac{1}{2\pi} \left(\frac{e^2}{c^2}\right) = \frac{4\pi m}{h^2} \mu_B^2$$

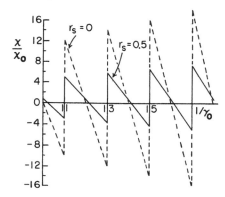

FIG. 7.1.   dHvA oscillations of a 2D electron gas at absolute zero. Solid lines are for ideal case; broken lines are for interacting case. The abscissa is the ratio of the Fermi energy to the field energy.

in contrast to Eq. (1.5). The first equality is in units in which $\hbar = 1$ and $2m = 1$.

When 2D electron systems are subject to a perpendicular magnetic field, the electron energy is completely quantized due to the lack of motion parallel to the field. This feature makes the 2D systems differ from the 3D case. The period of oscillations is still determined by $1/\gamma_0 = \mu_B H/\varepsilon_F = \nu$, but the oscillations are sawtoothlike at absolute zero because the Fermi energy must jump from one Landau level to the next when the electron density is changed.

The susceptibility of an interacting 2D electron system with $g = 2$ is illustrated in Fig. 7.1 as a function of the filling factor $\nu = 1/\gamma_0$ of a Landau level [2]. The ordinate is the ratio χ/χ_0, where χ_0 is given by $e^2/(2\pi c^2)$. The heavy lines correspond to the ideal case and the thin lines to the interacting case of $r_s = 0.5$, showing that the amplitude is reduced significantly.

While magnetic field effects are strong in two dimensions, Coulomb effects are also strong, and therefore their interplay becomes interesting. Since the dHvA effect is characteristic of an ideal Fermi gas, one can expect generally that the amplitude is suppressed at high densities by Coulomb interaction. The reduction of the dHvA amplitude at $1/\gamma_0 = 11$ is shown in Fig. 7.2 as a function of r_s. Here the ordinate represents the jump of the amplitude at $1/\gamma_0 = 11$ for the case $g = 2$. As r_s increases, the relative amplitude decreases and then at $r_s \sim 0.7$, where the ground-state energy of the electron system becomes negative, the phase of dHvA oscillations changes. The general behavior of the ground-state energy curve plotted against r_s has been known theoretically for many years, but no direct experimental test has been made of the point where the energy becomes negative. The phase change of the dHvA oscillations provides a possibility for this test.

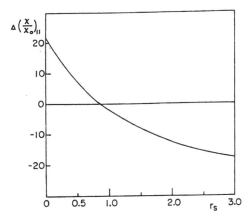

FIG. 7.2. Variation of the dHvA amplitude at $1/\gamma_0 = \nu = 11$ as a function of r_s. The abscissa is the Landau level filling factor. (From Isihara and Shiwa [3])

The dHvA effect has been observed in 2D systems, although the experiments are somewhat difficult. This has been achieved by increasing the surface area, because the total magnetic response is proportional to the surface area.

In two dimensions, the complete conversion between the field energy $\mu_B H$ and electron's kinetic energy causes very strong temperature oscillations under an adiabatic variation of magnetic field. This so-called magnetothermal effect is expressed by

$$dT = -\frac{(\partial S/\partial H)_{T,n}}{(\partial S/\partial T)_{H,n}} dH, \qquad (1.11)$$

where S is the entropy. Note here that the denominator is the electronic specific heat. One can show that in the presence of a magnetic field the linear specific heat is canceled out if the Landau levels are not broadened. Instead, the electronic specific heat can become exponentially small. In the presence of level broadening the linear specific heat is restored, but the specific heat oscillates, causing large temperature fluctuations.

When the Landau levels are broadened and at finite temperatures the abrupt changes at absolute zero are somewhat eased, but at low temperatures sharp changes can still be expected. Landau level broadening can be expressed by a scattering time τ. The time can in turn be related to electron's mobility. For silicon inversion layers we can use

$$\frac{\hbar}{2\tau} = 0.3047 \left(\frac{0.19 \, m_0}{m}\right) \left(\frac{10^4}{\mu(\text{cm}^2\text{V}^{-1}\text{s}^{-1})}\right) \quad \text{meV}, \qquad (1.12)$$

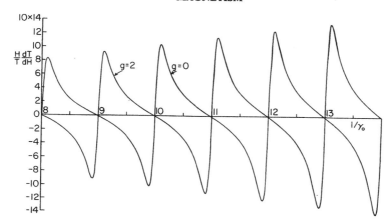

FIG. 7.3. Magnetothermal oscillations of a 2D electron system with elliptically broadened Landau levels at absolute zero.

where m_0 and m are the bare and effective masses respectively. For convenience this energy may be compared with the cyclotron energy:

$$\hbar\omega_c = 0.6093H(0.19\,m_0/m)\quad\text{meV}. \tag{1.13}$$

For 2D electrons in strong magnetic field an elliptic density of states is often used. This can be expressed by

$$g(\varepsilon) = \frac{A\alpha}{\pi\beta}\sum_0^\infty \frac{2\tau}{\pi}\,[1 - \tau^2(\varepsilon - \varepsilon_n)^2]^{1/2}, \tag{1.14}$$

where A is the surface area, and units are such that $\hbar = 1$ and $2m = 1$, m being the electron mass. According to Eq. (1.14), each Landau level is broadened about its center by an amount $\pm\hbar/\tau$ if $\hbar$ is explicitly displayed. Instead of τ, it is convenient to use a dimensionless broadening parameter Γ given by

$$\Gamma = \frac{\pi\hbar}{\tau a^2}, \tag{1.15}$$

where $a^2 = \hbar\omega_c/2$.

Figure 7.3 illustrates the magnetothermal oscillations [3] as represented by a dimensionless quantity given by

$$\frac{H\,dT}{T\,dH}.$$

The abscissa is $1/\gamma_0 = \varepsilon_F/\mu_B H = \nu$, which is the filling factor of a Landau level for two dimensions. The curves correspond to $g = 0$ and $g = 2$ for $\Gamma = 0.3$ at absolute zero. Note that the nodes occur precisely at integral values of the abscissa.

The amplitude of these oscillations can be a few hundred times larger than the 3D counterpart. Assuming that the electron energy loss is completely absorbed by a silicon MOSFET with a total heat capacity of 1×10^{-11} cal/K at 0.1 K, the temperature change of the system can be of order 0.03 K.

7.2. Spin glasses

While in all ordinary magnetic systems spins are ordered, a family of structurally random and yet magnetically interacting systems called *spin glasses* has been found and has attracted strong attention since 1972. In general, these systems show a cusp-type susceptibility anomaly at a temperature T_f where somehow spins are "frozen" in random directions. Typically, spin glasses can be obtained by alloying a magnetic element with a nonmagnetic host. An example is the CuMn alloy with 0.9% manganese. In a more modern method, a crystalline intermetallic compound is made amorphous by a sputtering or quench-in method as in the case of α-GdAl$_2$ or α-MnSi.

The freezing temperature is clearly observable as a cusp in the real part χ' of the ac susceptibility plotted against temperature at low frequencies. Above T_f, χ' follows Curie's law. Below T_f, the magnetic response to a step-field is characterized by a large instantaneous magnetic response and slow relaxation. The imaginary part χ'' of the susceptibility remains zero for $T \gg T_f$, and increases as the temperature approaches T_f, where it shows an inflection point. Generally, the height of the susceptibility cusp decreases as frequency increases, while the freezing temperature moves slightly higher.

These behaviors can be seen in Fig. 7.4 [4]. Here the real part of the ac susceptibility of CuMn is plotted against temperature. The inset shows in an expanded scale the frequency dependency of the cusp. Even though there is no macroscopic magnetism below T_f, a finite-time spin correlation $\langle S_i(t)S_i(0)\rangle$ exists. Moreover, below T_f metastable properties that depend on the rate of cooling, remanence, and hysteresis are observable. As the temperature approaches T_f from above, the spin correlation length ξ is observed to grow. Clearly a certain cooperative state has been reached.

Yet the magnetic specific heat shows only a very broad and smooth maximum around a temperature that is some 20% higher than T_f, as shown in Fig. 7.5 [5]. Below T_f, the specific heat is approximately linear, although deviations take place. The entropy evaluated from the specific heat is high above T_f and low below. Apparently, there are large spin fluctuations above T_f, while below it there are local spin correlations. However, a more complex picture of spin correlations is necessary.

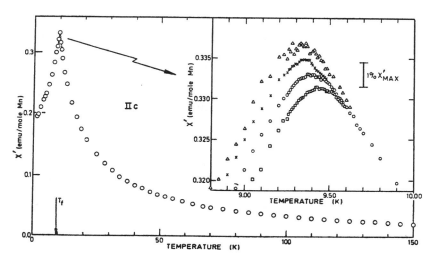

FIG. 7.4. Magnetic susceptibility of CuMn. Inset shows its frequency dependency. (From Mulder et al. [4]).

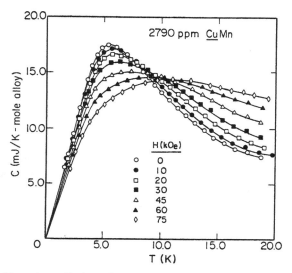

FIG. 7.5. Magnetic specific heat of manganese. (From Brodale et al. [5]).

183

Inelastic neutron scattering experiments indicate the existence of fluctuating ferromagnetic clusters, while quasielastic neutron diffuse scattering yields antiferromagnetic short-range order for first neighbors and ferromagnetic order for second neighbors. Neutron-polarization analysis measurements [6] on Cu–Mn single crystals show the presence of a long-period spin modulation with a concentration-dependent period. This modulation may correspond to spin density waves, but long-range order does not develop even though the modulation occurs over fairly large volumes in real space. Coexistent with these modulated regions are smaller regions where the spin correlations are determined by the atomic short-range order with ferromagnetic moments. The interactions between these smaller ferromagnetic and larger modulated regions seem to be the source for the spin-glass behavior.

Theoretically, a lattice with spins S_i in a field H may be represented by the Hamiltonian:

$$H = -2 \sum_{(ij)} J_{ij} S_i S_j - H \sum_i S_i. \tag{2.1}$$

Here J is the exchange integral and the sum is over the nearest-neighboring spin pairs. A positive J corresponds to ferromagnetism and a negative J to antiferromagnetism. In particular, if $S_i = S_z = 1/2$ or $-1/2$, the Hamiltonian represents the *Ising model*, where z specifies the z-component. If $S_z = 0$, it represents the *XY model*.

In the Edwards and Anderson model [7] the coupling constants J_{ij} are randomly distributed in accordance with

$$P(J_{ij}) = \frac{1}{(2\pi \bar{J}^2)^{1/2}} \exp(-(J_{ij} - J_o)^2 / 2\bar{J}^2). \tag{2.2}$$

The order parameter is given by

$$q(T) = \langle \langle S_i \rangle_T^2 \rangle_J, \tag{2.3}$$

where $\langle \cdots \rangle_T$ is a thermal average and $\langle \cdots \rangle_J$ is the configurational average in terms of the above distribution. Note that for spin glasses we expect that locally there exists $\langle S_i \rangle_T$ but its configurational average $\langle \langle S_i \rangle_T \rangle_J = 0$. A nonzero $q(T)$ results in a cusp-type singularity given by

$$\chi(T) = \frac{Ng^2 \mu_B^2 S(S+1)}{3kT} [1 - q(T)]. \tag{2.4}$$

S is the total spin at a site and g is the Landé factor.

This model has been studied well. In particular, its infinite-range solution has been given [8]. For finite ranges Monte Carlo simulations have resulted in a fairly well-defined susceptibility peak and a broad specific heat maximum [9].

When magnetic atoms such as manganese are disolved randomly in a nonmagnetic metal such as copper, the magnetic interaction between the manganese atoms is expected to depend on their distances. Hence, the corresponding spin glass may show a variety of properties. Although the order parameter q alone is not sufficient to describe complex behaviors of spin glasses at low temperatures, its use is convenient because it is determined by the square average of spins. Hence, its field conjugate variable is H^2. In the vicinity of the spin-glass transition point, the following scaling form for the nonlinear magnetization has been used [10]:

$$\frac{M}{H} \sim \varepsilon^\beta f\left(\frac{H^2}{\varepsilon^{\beta+\gamma}}\right), \tag{2.5}$$

where $\varepsilon = (T - T_f)/T_f$.

For $T > T_f$, $\lim_{x\to 0} f(x) = 0$ and

$$\frac{M}{H} \sim \frac{H^2}{\varepsilon^\gamma}, \tag{2.6}$$

with the order parameter response to magnetic field given by

$$\frac{dq}{dH^2} \sim \frac{(M/H^2)}{H^2} \sim \varepsilon^{-\gamma}. \tag{2.7}$$

The value of γ determined from the third derivative of the magnetization of AgMn is

$$\gamma = 2.2 \pm 0.2. \tag{2.8}$$

At T_f, $f(x) \sim x^{1/\delta}$, where

$$\delta = 1 + \frac{\gamma}{\beta}. \tag{2.9}$$

The temperature variation of the field derivative of $\Delta M/H$ is

$$p(T) = \frac{d(\Delta M/H)}{dH}$$

$$\sim \varepsilon^{(\beta-\gamma)/2}. \tag{2.10}$$

As shown in Fig. 7.6, M/H scales as H^ψ where

$$\psi = \tfrac{1}{2}(\gamma - \beta) = 0.6 \pm 0.05. \tag{2.11}$$

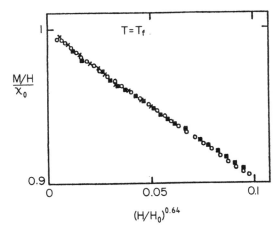

FIG. 7.6. Scaling behavior of the nonlinear susceptibility of AgMn at T_f. (From Bouchiat et al. [10])

Hence,

$$\beta = 1 \pm 0.01. \tag{2.12}$$

The exponent δ, determined from study of the low-field dependence of $\Delta M/H$ at T_f gives

$$\frac{M}{H} \sim H^{2/\delta}, \tag{2.13}$$

with

$$\delta = 3.1 \pm 0.2. \tag{2.14}$$

However, different values of the critical exponents have also been reported. Perhaps anisotropy, sample treatments, and other factors affect the critical behavior.

We remark that real or momentum-space renormalization group approaches have been developed for short-range cases with different scaling hypotheses and/or approximations [11]. For instance, one may introduce a scale-dependent coupling parameter $J(L)$ that varies for absolute zero as

$$J(L) \sim JL^y. \tag{2.15}$$

For an Ising ferromagnet the domain-interface energy has this form with $y = d - 1$, where d is the dimension. For a Heisenberg ferromagnet, $y = d - 2$. Numerical evaluations of an average over block spin couplings with random boundary conditions yield $y = 0.19$ for three dimensions [12].

For a finite temperature, the above coupling may be modified as

$$J(L) = (L/\xi)^y. \tag{2.16}$$

where ξ is the correlation length, which in turn varies as

$$\xi = \xi_0 \left(1 - \frac{T}{T_f} \right)^{-\nu} \tag{2.17}$$

and ν seems to be of order 3.

Despite these and many other theoretical efforts, actual spin glasses show complex properties beyond theoretical expectations. Indications are that there are even different low-temperature "ordered" states with barriers, and therefore several order parameters may be needed instead of one. Note that complex phenomena such as reentrance into spin-glass states, sequences of paramagnetic–ferromagnetic phases, and aging have been observed. Also, spin-glass behaviors are observed in high-T_c superconductors, although their connection with superconductivity remains to be explained. Hence there is still a long way to go to understanding of spin glass behaviors completely. Nevertheless, many interesting models and concepts have emerged from their study.

As an idealized model for spin glasses a triangular spin lattice with antiferromagnetic interaction is especially interesting because there is spin *frustration* to align three spins in a unit cell antiferromagnetically. This causes spins to fluctuate even though the structure itself is well-defined in distinction to spin glasses. Classically, the 120° configuration as in Fig. 7.7 is possible. However, Anderson [13] introduced the *resonating valence bond* (*RVB*) *state* for a spin-$\frac{1}{2}$ Heisenberg antiferromagnet. As will be discussed in Section 8.5, he has proposed a RVB state for high-T_c superconductors and a comment on this state is given below.

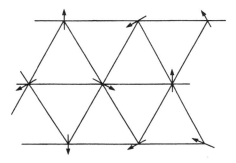

FIG. 7.7. 120° configuration of classical spins on a triangular lattice.

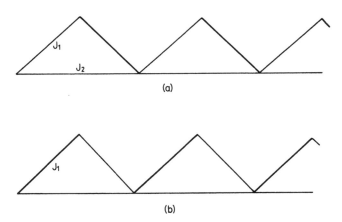

FIG. 7.8.   Equivalent singlet bond configurations in a 1D triangular lattice.

Note that a pair of nearest-neighboring spin pairs can be either in the spin triplet state or in the singlet state. The former is symbolically represented by

$$\uparrow\uparrow, \quad (\uparrow\downarrow + \downarrow\uparrow)/2, \quad \downarrow\downarrow,$$

while the latter is

$$(\uparrow\downarrow - \downarrow\uparrow)/\sqrt{2}$$

The eigenvalues of these states are $1/4$ and $-3/4$ respectively.

Now consider spins on a 1D triangular lattice with exchange integrals J_1 and J_2 as in Fig. 7.8, where (a) and (b) with heavy singlet bonds are equivalent singlet bond states. Anderson showed that the lowest energy state occurs when these two spin configurations are mixed equally. That is, a spin singlet state not fixed in space but resonating between equivalent configurations is energetically favorable. We might say that the system is a *spin liquid*. As one can anticipate, the number of equivalent states increases enormously as the dimension increases. However, numerical and other calculations indicate that in general the RVB state is indeed energetically viable [14].

7.3.   Itinerant electron ferromagnetism

In typical ferromagnetic iron or nickel the 3d electrons are responsible for ferromagnetism. The 3d orbit is not widely spread so that the electrons are localized at each atom. The first quantum theory of ferromagnetism, due to Heisenberg, considers the exchange interaction between the localized electrons in neighboring sites as the source of ferromagnetism. Although the Heisenberg model indicates quantum mechanically the origin of the Weiss molecular field, the exchange integral for atomic distances turns out to be too large and the Curie point, which is directly related to the exchange integral, comes out too high in comparison with experiment. Moreover, the saturation magnetic moment in iron or nickel is not an integral multiple of the Bohr magneton. In fact, it is $7.8\mu_B$ in iron and $0.6\mu_B$ in nickel. Such nonintegral values suggest that some of the d electrons move out of the shell and become itinerant. Therefore, the Heisenberg model is not considered applicable to transition metals such as iron, cobalt, or nickel, but rather a basic model for the magnetism of insulator compounds.

Stoner developed a theory of itinerant electron ferromagnetism. In this theory, a constant exchange energy J is assumed for a pair of Bloch electrons with parallel spins. The number $N_\uparrow$ (or $N_\downarrow$) of electrons with up-spins (or down-spins) is given by

$$N_\uparrow = \int_0^{\varepsilon_F - \Delta} g(\varepsilon)\, d\varepsilon; \qquad N_\downarrow = \int_0^{\varepsilon_F + \Delta} g(\varepsilon)\, d\varepsilon, \tag{3.1}$$

where $g(\varepsilon)$ is the density of states and where a constant energy shift 2Δ between the two spin-states is assumed. The total number of electrons is given by

$$N = N_\uparrow + N_\downarrow,$$

which is constant. The exchange energy can be expressed as

$$E_x = -\tfrac{1}{2}JN_\uparrow(N_\uparrow - 1) - \tfrac{1}{2}JN_\downarrow(N_\downarrow - 1) \tag{3.2}$$

The difference in the numbers of spin-up and spin-down electrons is given by

$$M = N_\downarrow - N_\uparrow$$
$$= 2g(0)\Delta[1 - b\Delta2 + \cdots], \tag{3.3}$$

where

$$b = \frac{1}{2}\left(\frac{g'^2}{g^2} - \frac{g''}{3g}\right), \tag{3.4}$$

and where $g = g(\varepsilon_F)$, g' and g'' are the density of states and its derivatives at the Fermi surface in the absence of Δ. The presence of M or Δ shifts the exchange energy such that

$$\delta E_x = -\tfrac{1}{4}JM^2. \tag{3.5}$$

The kinetic energy can be expressed as

$$E_0 = \int_0^{\varepsilon_F - \Delta} \varepsilon g(\varepsilon)\, d\varepsilon + \int_0^{\varepsilon_F + \Delta} \varepsilon g(\varepsilon)\, d\varepsilon. \tag{3.6}$$

By using a Taylor expansion, we find that the kinetic energy shift is given by

$$\delta E_0 = g\Delta^2[1 - \tfrac{3}{2}b\Delta^2 + \cdots].$$

$$= \frac{1}{4g} M^2 + \frac{1}{32g^3} bM^4 + \cdots. \qquad (3.7)$$

By comparing Eqs. (3.7) and (3.5) we find that the presence of M can decrease the total energy if the following *Stoner condition* is satisfied:

$$1 < Jg(\varepsilon_F). \qquad (3.8)$$

On the other hand, if $1 > Jg(\varepsilon_F)$ the system is paramagnetic with a susceptibility given by

$$\chi = \frac{2g(\varepsilon_F)}{1 - Jg(\varepsilon_F)} \mu_B^2. \qquad (3.9)$$

The numberator is Pauli's susceptibility:

$$\chi_0 = 2g(\varepsilon_F)\mu_B^2. \qquad (3.10)$$

The susceptibility in Eq. (3.9) enhances the Pauli susceptibility due to the exchange energy contribution in the denominator.

For a finite temperature, an expression similar to Eq. (3.8) with a temperature-dependent density of states $g(T)$ can be derived. The Curie temperature is obtained from

$$1 = Jg(T_c). \qquad (3.11)$$

A more explicit form is obtained if $g(T)$ is expanded in a Taylor series to order T^2. The susceptibility is then expressed in the form:

$$\chi = \frac{\chi_0}{a(T)(T - T_c)} \qquad (3.12)$$

Here,

$$a(T) = \frac{\pi^2 k^2}{6} (T + T_c) \left(\frac{g'^2 - gg''}{g^2} \right). \qquad (3.13)$$

The Curie temperature T_c is determined by

$$T_c = \left(1 - \frac{1}{Jg} \right) \frac{1}{a(0)}. \qquad (3.14)$$

The theoretical expression (3.13) is close to the Curie–Weiss form, particularly near T_c, but deviations occur as T increases. This is an important point because the Curie–Weiss law is observed in ferromagnets. The theoretical value of T_c determined from the spontaneous magnetization at 0 K is always too high. Moreover, the temperature dependence of the magnetization does not reproduce well the experimental variation.

The Heisenberg model with localized magnetic moments and Stoner model with itinerant electrons are two extremes. It has been found that the d electrons in insulators are localized but those in the transition metals are not as evidenced by the existence of Fermi surfaces. Since both models have drawbacks, immediate questions arise:

(a) Concerning the *itinerant electron model*. Is the assumption of a constant exchange energy realistic or can it be improved? Can electron–electron interaction be completely neglected? Is the Stoner condition (3.8) sound?

(b) Concerning the *localized moment model*. Is the magnitude of the localized magnetic moments really constant or can it fluctuate? If some electrons move out of the d-shell, how will the sign of the exchange interaction between the neighboring magnetic moments change?

The constant exchange energy is of course not realistic because the band structure is expected to affect J and because electrons interact with each other. The Stoner condition (3.8) can be expressed in terms of a constant susceptibility χ_0 such that $\chi_0 J / [2\mu_B^2] > 1$, but in terms of a wavenumber-dependent susceptibility χ_q a similar condition can be derived for the appearance of spin-density waves:

$$\frac{J\chi_q}{[2\mu_B^2]} > 1. \tag{3.15}$$

Hence, if $\chi_0 > \chi_q$, paramagnetic states with fluctuating magnetic moments become more stable than Stoner's paramagnetic state. This may be expected for a certain range of small q values.

Theoretically, one can guess that the exchange energy J is not constant but depends on Coulomb interaction, which in turn is a function of electron density. Theoretical studies show that the sign of the exchange interaction between the nearest-neighbor local moments does depend on the occupied fraction of the d-orbitals at each site in such a way that it is antiferromagnetic when each d-shell is nearly half-filled, while it is ferromagnetic if it is nearly full or nearly empty.

Near absolute zero Stoner's theory results in saturation magnetization given by

$$M = M_0(1 - cT^2), \tag{3.16}$$

where c is a constant that depends on J and $g(0)$. This quadratic dependence is qualitatively followed, particularly by data at moderately low temperatures. At low temperatures much lower T_c spin waves are excited, modifying the temperature variation such that

$$M = M_0(1 - aT^{3/2}).$$

(3.17)

This dependence is followed by experiment better than the T^2 dependence of Stoner's case, particularly at low temperatures. However, the theoretical value of a does not quite agree with experiment.

At high temperatures spin fluctuations may take place. These fluctuations can be considered as a major reason why Stoner's theory fails at high temperatures. We shall examine various magnetic systems from this point of view in the next section.

7.4. Spin orientation and fluctuation

In the Heisenberg localized theory the Curie–Weiss law

$$\chi = \frac{C}{T - T_c}$$

(4.1)

is characterized by the Curie constant C given by

$$C = Ng^2\mu_B \frac{S(S + 1)}{3k},$$

(4.2)

where S is the total spin. For $g = 2$, $2S$ is the total number of effective electrons per site. Hence, an effective number of localized electrons

$$q_c = gS$$

can be deduced experimentally from an observed Curie constant. Its ratio with q_s, which is the number of electrons deduced from the saturation magnetization, should be unity if all the electrons are localized. While this is found to be the case for $CrBr_3$, EuO, Gd, and others, a number of magnetic materials shows deviations. The ratio is 1.05 for Fe, 1.34 for Co, and 1.46 for Ni, but even larger deviations occur as shown in Fig. 7.9 [15]. Here the ratio is plotted against Curie temperature T_c. The light line corresponds to $q_c/q_s = 1$, and the heavy curve is a freehand curve. In particular, in the limit $T_c \to 0$ and $q_s \to 0$, which is the weak ferromagnetic limit, the heavy curve indicates divergence. Materials like Pd or $HfZn_2$ can be considered close to this limit.

The very wide variation of the ratio q_c/q_s indicates that spins are not always localized and the magnitude of each spin may be fluctuating. How

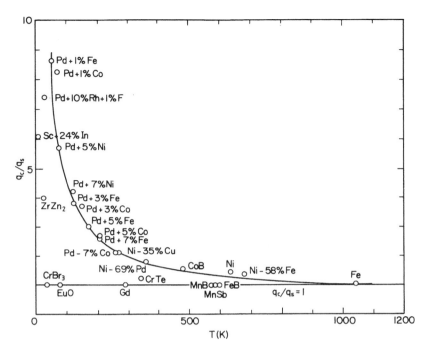

FIG. 7.9. Ratio q_c/q_s as function of Curie temperature. q_c is the number of electrons deduced from the Curie–Weiss constant, and q_s is the number deduced from saturation magnetization. (From Rhodes and Wolfarth [15])

does this come about? Why are there many different spin configurations? In trying to answer these questions it is convenient to use a momentum-space representation.

For spins S_j at positions r_j ($j = 1, 2, \ldots$) we define a Fourier transform

$$S_q = \frac{1}{N_0} \sum_j S_j \exp(-i q \cdot r_j). \tag{4.3}$$

This vector is used to describe spin waves, N_0 being the total number of spins. We assume that the interaction energy of neighboring spins is $-2J_{ij}S_i \cdot S_j$. This assumption is not always correct, but under this assumption the exchange energy can be expressed by

$$
\begin{aligned}
E_x &= -\sum_i \sum_i J_{ij}S_i \cdot S_j \\
&= -N_0 \sum_q J(q)S_q \cdot S_{-q}.
\end{aligned} \tag{4.4}
$$

Note that

$$\sum_{\mathbf{q}} \mathbf{S_q \cdot S_{-q}} = \frac{1}{N_0} \sum_i \mathbf{S}_i^2 = \langle S^2 \rangle \tag{4.5}$$

is the square average spin.

Let us consider the spin configuration in which all the spins have the same magnitude:

$$S_i^2 = S^2. \tag{4.6}$$

Such a configuration can minimize E_x for $\mathbf{q} = \mathbf{q}_0$ where $J(q)$ is a maximum. At Q the above condition becomes

$$\sum_{\mathbf{k}} \mathbf{S_k \cdot S_{-k}} + \sum_{\mathbf{q}_0 \neq 0} \sum_{\mathbf{k}} \mathbf{S_k \cdot S_{q_0-k}} \exp[i\mathbf{q}_0 \cdot \mathbf{r}_i] = S^2. \tag{4.7}$$

Since the right-hand side is independent of $\mathbf{r}_i$ and in view of Eq. (4.5), for $\mathbf{q}_0 = 0$ we set

$$\sum_{\mathbf{k}} \mathbf{S_k \cdot S_{-k}} = S^2, \tag{4.8}$$

and for $\mathbf{q}_0 \neq 0$

$$\sum_{\mathbf{k}} \mathbf{S_k \cdot S_{q_0-k}} = 0. \tag{4.9}$$

These conditions are satisfied if $|\mathbf{S_k}| = 0$ for all $\mathbf{k} \neq \mathbf{q}_0$ and if $\mathbf{S_{q_0}}$ is of the following form independently of $\mathbf{q}_0$:

$$\mathbf{S_{q_0}} = a(\mathbf{i} + i\mathbf{j})S, \tag{4.10}$$

where $\mathbf{i}$ and $\mathbf{j}$ are unit vectors that are mutually perpendicular, i is the imaginary unit, and $|a| = 1/2$. The above equation shows that in real space $\mathbf{S}_i$ in the plane spanned by $\mathbf{i}$ and $\mathbf{j}$ rotates as $\mathbf{r}_i$ varies. The period of rotation is $2\pi/q_0$. If the pitch of this oscillation is 0 and if the exchange integral is of the right sign, the corresponding spin configuration represents ferromagnetism. If its direction alternates 180° in adjacent lattice sites, the spin configuration corresponds to antiferromagnetism. In between, the spins may rotate helically.

The spin configurations can also be described in momentum space. The wavevector $\mathbf{q}_0$ of the helical wave is the one to maximize $J(q)$. Such a vector can be 0 (ferromagnetism) or $\pi\mathbf{G}$ (antiferromagnetism). where $\mathbf{G}$ is the reciprocal lattice vector.

We can conclude that even if the magnitude of each spin is constant, its projection on to a certain plane may vary, and a helical spin configuration can be expected. In extreme cases, this configuration may correspond to ferromagnetism or antiferromagnetism. A real helical spin configuration was first observed in MnF_2, and then in Tb, Dy, Ho, and other rare-earth metals.

Although the above simple consideration is used to describe various magnetic material, it is based on the assumed form (4.4) of spin interaction. Hence, for more specific descriptions the energy form itself must be examined. Also, it is more general if not only the direction of spins but also the magnitude is considered to vary.

In this respect, we note that temporal spin fluctuations can be described by a correlation function defined by

$$S(q, \omega) = \sum_{jk} \int_{-\infty}^{\infty} \exp[i\omega t - i\mathbf{q} \cdot (\mathbf{r}_j - \mathbf{r}_k)] \langle \delta \mathbf{S}_j(t + t') \cdot \delta \mathbf{S}_k(t') \rangle \, dt, \quad (4.11)$$

where the time-dependent spin deviation $\delta \mathbf{S}_j$ from its average is given in the Heisenberg representation by

$$\delta \mathbf{S}_j(t) = e^{itH/\hbar} \mathbf{S}_j e^{-itH/\hbar} - \langle \mathbf{S}_j \rangle. \quad (4.12)$$

Such a spin correlation function can be proved effectively by neutron scattering experiment.

Let us now examine magnetic materials in terms of spin fluctuations. In increasing order of spin localization and fluctuation in real space, magnetic materials may conveniently be classified as follows [16]:

(a) Nonlocal spin systems (spin fluctuations localized in q-space)). Weakly ferromagnetic metals: $ZrZn_2$, Sc_3In, Ni_3Al, $Fe_{0.5}Co_{0.5}Si$.

(b) Intermediate group: $MnSi$, α-Mn, $CeFe_2$, Cr, CrB_2, Fe_3Pt, γ-Mn, MnP, CoS_2, Ni, Co, Fe.

(c) Localized spin systems (local spin density fluctuations with a large constant amplitude): Magnetic insulator compounds; 4f-metals, Heusler alloys.

The list shows that magnetic materials vary widely from group (a) to group (c). In (a), spin fluctuations are small and extended in real space and localized in q-space. That is, the fluctuations occur only for small q and ω. Weak ferromagnetism occurs in systems with a small number of electrons relative to the number of the lattice sites. Such systems are characterized by small saturation magnetization and low T_c. In the case of $ZrZn_2$, the magnetization is $0.12\mu_B$ per atom and T_c is 25 K. The saturation magnetization is of order $0.1\mu_B$ per atom. According to Moriya and Kawabata [16], the spontaneous magnetization follows:

$$M(T) = M_0 \left(1 - \left(\frac{T}{T_c} \right)^{4/3} \right)^{1/2}. \quad (4.13)$$

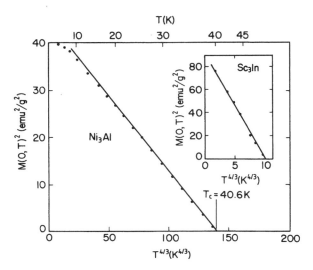

FIG. 7.10. Temperature variation of the magnetization of Ni_3Al and Sc_3In (insert). (From Masuda [17])

This dependency has been confirmed by experiment [17] as shown in Fig. 7.10. It contrasts with the $T^{3/2}$ variation of ferromagnetism.

The spin–spin correlation function can be defined for an arbitrary temperature. When spin fluctuations depend on temperature, various magnetic phases can be expected. In particular, *Laves phase compounds* MFe_2, where $M = Sc$, Ti, and so on, with hexagonal structure are known to show magnetic phases with different types and degrees of spin fluctuation and localization. For example, in $Sc_{1-x}Ti_xFe_{1.95}$, a ferromagnetic phase and an antiferromagnetic phase appear for $0.9 > x \geqq 0.7$ and below around 250 K in different crystal directions, as illustrated in Fig. 7.11. Two itinerant ferromagnetic magnetic states I and II with different saturation magnetic moments exist for $x < 0.7$. The phase boundary between these two ferromagnetic phases is not well defined. At high temperatures two paramagnetic phases I and II appear, but their boundary is not completely known. These two paramagnetic phases have different magnitudes of spin fluctuations. For $x > 0.9$ and at low temperatures, this material is antiferromagnetic.

Group (c) has large spin density fluctuations that are local in coordinate space. Typically, Heusler alloys consist of 10–25% aluminium, 18–26% manganese, and the rest copper. Cu_2MnAl belongs to this type; there are other compounds such as Pd_2MnSn and Ni_2MnSn. In all these Heusler alloys, manganese atoms are well separated from each other and have well-defined local moments and the Rhodes–Wohlfahrth ratio q_c/q_s is nearly 1.

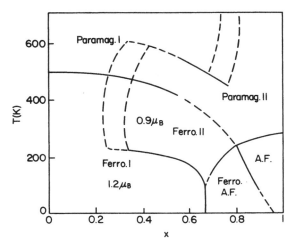

FIG. 7.11. Phase diagram of $Sc_{1-x}Ti_xFe_2$. (From Nishihara and Yamaguchi [18])

The spin fluctuations may be described by an effective Hamiltonian given by

$$H = -\sum_{i<j} U_{ij}(\langle S^2 \rangle)(\mathbf{S}_i \cdot \mathbf{S}_j) + \tfrac{1}{2}N_0 L(\langle S^2 \rangle)$$

$$= -\sum_{\mathbf{q}} U_{\mathbf{q}}(\langle S^2 \rangle)|S_{\mathbf{q}}|^2 + \tfrac{1}{2}N_0 L(\langle S^2 \rangle), \qquad (4.14)$$

where U_{ij} is the strength of the interaction of spins $\mathbf{S}_i$ and $\mathbf{S}_j$, which is assumed to depend on a square average local moment $\langle S^2 \rangle$ defined by Eq. (4.5). The second term represents effects of higher-order correlations. Its functional form $L(x)$ is unknown but x is assumed to be simply $\langle S^2 \rangle$.

The kinetic energy is of the form

$$E_0 = \sum_{\mathbf{q}} \left\{ \frac{1}{2\chi_{\mathbf{q}}} |S_{\mathbf{q}}|^2 + \cdots \right\}. \qquad (4.15)$$

In Eqs. (4.14) and (4.15) only the first-order terms in $|S_{\mathbf{q}}|^2$ are given. While the higher-order terms are not given, the free energy is illustrated schematically in Fig. 7.12 as a function of $|S_{\mathbf{q}}|^2$. Three different cases are shown here. If the kinetic energy is larger than the absolute magnitude of the exchange energy, we may have a curve like (a) without minimum. If this occurs for all q, the system is nonmagnetic. In curve (b) there is a minimum, but it is shallow and broad and appears near the origin. Hence, spin fluctuations may be expected but not localized in real space. Instead, they are localized in q-space. Curve (c) has a deep and relatively narrow minimum at a fairly large q. It may be considered to correspond to the localized spin systems.

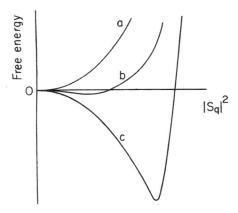

FIG. 7.12. Schematic free energy plotted against $|S_q|^2$.

7.5. Kondo effect

In general, the electric resistance of metals decreases in proportion to T^5 at low temperatures. However, a resistance minimum is observed in highly conductive metals such as copper, silver, and gold containing a very small amount of impurities such as chromium, manganese, and iron. Since these impurities have magnetic moments the cause for the upturn of their resistivity may be sought in the interaction mechanism of the conduction electrons with these impurities. Their typical interaction is the so-called *s–d interaction* between s and d electrons. In the case of manganese in copper the 3d electrons in manganese have an exchange s–d interaction with the 4s conduction electrons of copper.

In order to evaluate electrical resistivity it is necessary to determine the interaction between localized d electrons and itinerant s electrons. Such an interaction was considered by Zener [19] in 1951 for ferromagnetism in 3d metals. Although the Zener model is now considered not to be applicable to ferromagnetism in 3d metals, the s–d interaction is the key to understanding the resistance minimum. The s–d interaction was considered later by Kasuya also [19]. Once this interaction is established, the matrix element for electron scattering can be evaluated, which in turn leads to electrical resistance.

We have pointed out in Section 2.5 that the Anderson Hamiltonian can be used to derive an explicit form for such an interaction. Using this Hamiltonian, let us consider the case in which only one electron is in the d-level. This case is realized when in Eq. (5.8) of Section 2.5 the d-level is below the Fermi energy while $\varepsilon + U$ is above it:

$$\varepsilon_F \gg \varepsilon; \qquad \varepsilon + U \gg \varepsilon_F. \tag{5.1}$$

We then consider V_{sd}, which is the strength of the s–d interaction, as a small perturbation when the d electron is at the origin of a Bravais lattice. By using a canonical transformation [20], one can eliminate the first-order term in V_{sd} from the Hamiltonian. To second order in V_{sd}, and in an approximation in which products such as $V_{k'd}V_{dk}$ are replaced by a k-independent constant $|V|^2$, the s–d exchange energy can be expressed by an effective Hamiltonian:

$$H_{ex} = -2J\boldsymbol{\sigma}\cdot\mathbf{S} \tag{5.2}$$

Here the effective coupling constant J is given by

$$J = -N_0|V|^2\left(\frac{1}{U + \varepsilon - \varepsilon(\mathbf{k})} - \frac{1}{\varepsilon - \varepsilon(\mathbf{k}')}\right). \tag{5.3}$$

It corresponds to the second-order matrix element of the spin-flip process in which initially opposite spins of the magnetic impurity atom and the conduction electron are exchanged. N_0 is the total number of lattice sites, $\mathbf{S}$ and $\boldsymbol{\sigma}$ are the spins of the d electron and a conduction electron at the origin respectively. Note that the effective coupling constant J is negative because of the assumption in (5.1). Hence, the s–d interaction is antiferromagnetic.

The perturbation Hamiltonian can be expressed by summing the above type of exchange interactions that take place at various impurity sites. In addition, there can be ordinary potential scattering without spin flipping. By summing all of these, the perturbation Hamiltonian can be written down as

$$H' = \sum \exp[i(\mathbf{k}' - \mathbf{k})\cdot\mathbf{R}_n][V\delta_{ss'} - J\boldsymbol{\sigma}_{ss'}\cdot\mathbf{S}_n]a^\dagger_{\mathbf{k}s}a_{\mathbf{k}'s'}, \tag{5.4}$$

where the sum is over all impurity positions, $\mathbf{k}$, $\mathbf{k}'$, s, and s', and V is the strength of a δ-function-type potential assumed for the potential scattering.

The transition probability for a conduction electron to change from state $(\mathbf{k}\sigma)$ to state $(\mathbf{k}'\sigma')$ is determined by H'. If the initial and final states are represented by 1 and 2 respectively, perturbation theory gives the transition probability to third order in H' as

$$W(\mathbf{k}'\sigma'; \mathbf{k}\sigma) = \frac{2\pi}{\hbar}\left(H'_{12}H'_{21} + \sum_i{}'\frac{H'_{21}H'_{1i}H'_{i2} + \text{c.c.}}{E_1 - E_2}\right)\delta(E_1 - E_2). \tag{5.5}$$

Here i designates intermediate states, and E_1 and E_2 are the total energies of the initial and final states.

The transition from $\mathbf{k}$ to $\mathbf{k}'$ may or may not involve spin flipping. Considering both cases and all the intermediate states, Kondo [21] arrived at

$$W(\mathbf{k}', \mathbf{k}) = \frac{2\pi}{\hbar} \frac{J^2 S(S+1)c}{N_0} [1 + 4J\phi(\varepsilon_\mathbf{k})]\delta(\varepsilon_\mathbf{k} - \varepsilon_{\mathbf{k}'})$$

$$= w(\varepsilon_\mathbf{k})\delta(\varepsilon_\mathbf{k} - \varepsilon_{\mathbf{k}'}), \qquad (5.6)$$

where, for convenience, we have used the notation $\varepsilon_\mathbf{k}$ instead of $\varepsilon(k)$; c is the concentration of the magnetic impurities; and

$$\phi(\varepsilon_\mathbf{k}) = \frac{1}{N_0} \sum_\mathbf{q} \frac{f(\varepsilon_\mathbf{q})}{\varepsilon_\mathbf{q} - \varepsilon_\mathbf{k}}. \qquad (5.7)$$

The first term in Eq. (5.6) is due to ordinary potential scattering, and the second term comes from the third-order term in Eq. (5.5).

The relaxation time τ_k associated with the above transition is given for low temperatures by

$$\tau_k^{-1} = w(\varepsilon_\mathbf{k})g(\varepsilon_F), \qquad (5.8)$$

where $g(\varepsilon_F)$ is the density of states at the Fermi surface. The conductivity is then given by

$$\sigma = -\frac{e^2}{4\pi^3} \int \tau_k v_x^2 f'(\varepsilon_\mathbf{k}) \, d\mathbf{k}, \qquad (5.9)$$

where v_x is the velocity component in the direction of an external electric field, and f' is the derivative of the Fermi distribution function.

Because of the denominator of $\phi(\varepsilon_\mathbf{k})$, the integration over $\mathbf{k}$ in Eq. (5.9) is divergent at absolute zero, but for a finite temperature it results in a logarithmic term. The resistivity is given by $1/\sigma$. With some trivial approximations Kondo arrived at the resistivity due to the magnetic impurities:

$$\Delta\rho = \frac{3\pi\Omega}{\hbar e^2 v_F^2 N_0} cJ^2 S(S+1) \left(1 + 4J \frac{g(\varepsilon_F)}{N_0} \ln T\right). \qquad (5.10)$$

where Ω is the total volume.

The total resistivity includes a residual resistivity ρ_0 and a term that varies as T^5 at low temperature. Hence, it may be expressed as

$$\rho = \rho_0 + aT^5 + \Delta\rho; \qquad (5.11)$$

$$\Delta\rho = \Delta\rho_1 - bc|J| \ln T, \qquad (5.12)$$

where $\Delta\rho_1$ is the temperature-independent term of $\Delta\rho$ and bc is the coefficient of the logarithmic term of $\Delta\rho$. Note here that the constant b is positive since J is negative. The resistance minimum occurs at

$$T_K = (bc|J|/5a)^{1/5}. \qquad (5.13)$$

This temperature is called the *Kondo temperature*. It is proportional to $c^{1/5}$, in agreement with experiment. The temperature variation of $\Delta\rho$ also agrees with experiment.

REFERENCES

1. A. Isihara, J. Tsai, and M. Wadati, *Phys. Rev. A* **3**, 990 (1971). J. T. Tsai, M. Wadati, and A. Isihara, *Phys. Rev. A* **4**, 1219 (1971).
2. Y. Shiwa and A. Isihara, *Phys. Rev. B* **27**, 4743 (1983).
3. A. Isihara and Y. Shiwa, *J. Phys. C* **18**, 4703 (1985); *Solid State Commun.* **48**, 1081 (1983).
4. C. A. M. Mulder, A. J. van Duyneveldt, and J. A. Mydosh, *Phys. Rev. B* **23**, 1384 (1981).
5. G. E. Brodale, R. A. Fisher, W. E. Fogle, N. E. Philips, and J. van Curen, *J. Mag. Mag. Mater.* **31–34**, 1331 (1983).
6. J. W. Cable, S. E. Werner, G. P. Felcher, and N. Wakabayashi, *Phys. Rev. Lett.* **49**, 829 (1982).
7. S. F. Edwards and P. W. Anderson, *J. Phys. F* **5**, 965 (1975), **6**, 1927 (1976).
8. D. Sherrington and S. Kirpatrick, *Phys. Rev. Lett.* **32**, 1792 (1975).
9. K. Binder and K. Schröder, *Phys. Rev. B* **14**, 2142 (1976). W. Kinzel, *Phys. Rev. B* **19**, 4595 (1979)
10. H. Bouchiat, N. de Courtenay, P. Monod, M. Ocio, and P. Refregier, *Jpn. J. App. Phys. Suppl.* **26**, 1951 (1987).
11. K. Binder and A. P. Young, *Rev. Mod. Phys.* **58**, 901 (1986).
12. M. A. Moore, in *Cooperative Dynamics in Complex Physical Systems*, ed. by H. Takayama (Springer-Verlag, New York, 1988), p. 134.
13. P. W. Anderson, *Mater. Res. Bull.* **8**, 153 (1983).
14. T. Oguchi, H. Nishimori, and Y. Taguchi, *J. Phys. Soc. Jpn* **55**, 323 (1986).
15. P. Rhodes and E. P. Wohlfarth, *Proc. Roy. Soc. London* **273**, 247 (1963).
16. T. Moriya, in *Spin Fluctuations in Itinerant Electron Magnetism*, (Springer-Verlag, New York, 1985). T. Moriya and A. Kawabata, *J. Phys. Soc. Jpn* **34**, 639 (1973); **35**, 669 (1973).
17. Y. Masuda, *J. Mag. & Mag. Mater.* **31–34**, 259 (1983).
18. Y. Nishihara and Y. Yamaguchi, *J. Phys. Soc. Jpn* **55**, 920 (1986).
19. C. Zener, *Phys. Rev.* **81**, 440 (1951). T. Kasuya, *Prog. Theor. Phys.* **16**, 45 (1956).
20. M. Bailyn, *Adv. in Phys.* **15**, 179 (1966). J. R. Schrieffer and P. A. Wolf, *Phys. Rev.* **149**, 491 (1966).
21. J. Kondo, *Prog. Theor. Phys.* **32**, 1 (1964).

8

SUPERCONDUCTIVITY

The fascinating phenomenon of superconductivity was discovered by Kamerlingh Onnes in 1911 when mercury was cooled to around 4 K. That was a long time ago, but the basic mechanism for superconductivity started unfolding only in the 1950s. Finally, in 1957 Bardeen, Cooper, and Schrieffer [1] gave a satisfactory explanation for this phenomenon. However, the discovery of an unusually high-T_c superconductor by Bednorz and Müller [2] in 1986 has caused a complete change of the general view on superconductivity because it is no longer limited only to very low temperatures. It is even more fascinating than previously thought, and its industrial application is now conceivable.

8.1. Low-T_c superconductivity

Ordinary superconductivity, as it is called in distinction to high-T_c superconductivity, appears as a sudden drop in resistivity at a low but finite temperature. In the case of mercury this occurs at 4.15 K. Since its discovery constant effort has been made to find a new material with a high critical temperature. However, progress was slow. It took nearly sixty years to reach the 20 K range in niobium compounds. That is, a common feature of ordinary superconductivity is that it occurs at low temperatures.

There are several other important properties that helped to understand the mechanism for superconductivity. Among them are the following:

(1) Magnetic field effect. Superconductivity can be destroyed by a magnetic field. The critical magnetic field H_c follows approximately

$$H_c(T) = H_0\left(1 - \left(\frac{T}{T_c}\right)^2\right). \tag{1.1}$$

The critical field H_c can be determined well for the so-called Type I or "soft" superconductors since the transition is sharp. In Type II or "hard" superconductors the transition between the normal and superconducting states is not sharp but is spread over a finite range of magnetic fields, as shown in Fig. 8.1. These superconductors have a negative surface energy, and the critical fields are generally large. In Type I materials such as Al, Sn, In, and

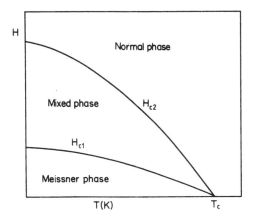

H

Normal phase

Mixed phase H_{c2}

H_{c1}

Meissner phase

T(K) T_c

FIG. 8.1.   Schematic phase diagram of Type II superconductors.

Pb, H_c is of order a few hundredths of a Tesla. In Type II superconductors there are two critical fields, the upper critical field H_{c2}, which separates the normal phase from superconducting phases, and the lower critical field H_{c1}, which separates what is called the superconducting mixed phase from the Meissner phase, which is the same as the superconducting phase of Type I. In Nb_3Sn which belongs to Type II the upper critical field is of order 10 T. Deviations from Eq. (1.1) occur positively or negatively: in lead, up to a few percent higher, and in tin a few percent lower than what is expected from the expression.

(2) *Meissner–Ochsenfeld effect.* Below T_c, Type I superconductors show perfect diamagnetism, expelling a magnetic field from their interior. The penetration depth λ of a magnetic field is of the order of 10^{-5}–10^{-6} cm. The mixed phase of Type II superconductors allows penetration of magnetic flux. Only below the lower critical field, H_{c1}, do they behave like Type I. The penetration depth λ is a function of temperature and drops sharply as the temperature decreases. In Type II superconductors, λ for 0 K is given by the London formula:

$$\lambda \sim \left(\frac{m^*c^2}{4\pi ne^2}\right)^{1/2},$$

where m^* is the effective mass and n is the superfluid carrier density.

(3) *Order of transition.* In the absence of a magnetic field, the phase transition is second-order. The electronic specific heat jumps to about three times the normal value just above the transition point. As temperature is decreased the electronic specific heat falls generally as $\exp(-2\Delta/kT)$, indicating the presence of an energy gap. The energy gap Δ depends on

temperature. That is, $\Delta = \Delta(T)$. The ratio $2\Delta(0)/kT_c$ is approximately 3.5. A magnetic field makes the transition first-order.

(4) *Isotope effect.* The transition point depends on the isotopic mass M as

$$T_c \sim M^{-1/2}. \tag{1.2}$$

This dependence gave a clue that superconductivity was related to lattice vibrations. The power is generally close to 1/2, but there are exceptions. For instance, in Ru, Mo, and Nb_3Sn it is almost zero.

(5) *Coherence length.* Experimentally, it is estimated that the boundary region between normal and superconducting parts of a specimen is determined by a length that is approximately ten times the penetration length. The rate at which the superconducting phase travels from a nucleus of super-conductivity suggests also that the order parameter for the appearance of superconductivity is characterized by a new length.

Such a length may be derived as follows: Superconducting electrons may have uncertainty of order $kT_c/(\hbar v_F)$ in wavenumber, and hence a semiclassical description of the particles is valid only within the length scale of order

$$\xi_0 = \frac{\hbar v_F}{kT_c}. \tag{1.3}$$

This is called the *coherence length.* This length is of order 10^4 Å in ordinary superconductors. This is small and indicates that detailed band structures are unimportant to superconductivity. When ξ_0 is compared with the penetration length, two cases arise:

$$\begin{array}{ll} \text{I:} & \xi_0 \ll \lambda; \\ \text{II:} & \xi_0 \gg \lambda. \end{array} \tag{1.4}$$

These two cases correspond respectively to Type I and Type II super-conductors.

8.2. BCS theory

Ordinary superconductivity was explained by Bardeen, Cooper, and Schrieffer in 1957. In this theory superconductivity occurs because two electrons bind each other by exchanging a phonon. The paired electrons obey Bose statistics, and thus can condense into a superfluid state. Since the phonon energy $\hbar\omega_k$ given by the zero-point lattice vibration is inversely proportional to the square root of the ion mass of a superconductor, this mechanism hinges upon the isotope effect.

In the ground state such a pair can be formed by electrons of opposite spins and wavenumbers. If one pair of states $(-\mathbf{k'}\downarrow, \mathbf{k'}\uparrow)$ is scattered into another pair of states $(\mathbf{k}\uparrow, -\mathbf{k}\downarrow)$, we can introduce an effective Hamiltonian:

$$H = \sum_{\mathbf{k}\sigma} \varepsilon_{\mathbf{k}} a^\dagger_{\mathbf{k}\sigma} a_{\mathbf{k}\sigma} + \sum_{\mathbf{k}\mathbf{k'}} V_{\mathbf{k}\mathbf{k'}} a^\dagger_{\mathbf{k}\uparrow} a^\dagger_{-\mathbf{k}\downarrow} a_{-\mathbf{k'}\downarrow} a_{\mathbf{k'}\uparrow}, \tag{2.1}$$

where the kinetic energy ε_k is measured from the Fermi energy. The second term assumes that such electron pairs act as units. Hence, new pair creation and destruction operators are introduced such that

$$b^\dagger_{\mathbf{k}} = a^\dagger_{\mathbf{k}\uparrow} a^\dagger_{-\mathbf{k}\downarrow}; \qquad b_{\mathbf{k}} = a_{-\mathbf{k}\downarrow} a_{\mathbf{k}\uparrow}. \tag{2.2}$$

If the number of such pairs is very large $b_{\mathbf{k}}$ may be replaced by a c-number which is equal to its average b^0_k given by

$$
\begin{aligned}
b^0_{\mathbf{k}} &= \langle b_{\mathbf{k}} \rangle \\
&= \mathrm{Tr}[e^{-\beta H} b_{\mathbf{k}}]/\mathrm{Tr}[e^{-\beta H}].
\end{aligned} \tag{2.3}
$$

This number is considered to determine the superconducting state. Accordingly, we define a gap parameter by

$$\Delta_{\mathbf{k}} = -\sum_{\mathbf{k'}} V_{\mathbf{k}\mathbf{k'}} b^0_{\mathbf{k'}}, \tag{2.4}$$

and proceed assuming that it is given. Later, it is determined in a self-consistent way such that b^0_k corresponds to the above average. The Hamiltonian becomes

$$H = \sum_{\mathbf{k}} \varepsilon_k a^\dagger_{\mathbf{k}\sigma} a_{\mathbf{k}\sigma} - \sum_{\mathbf{k}} (\Delta_{\mathbf{k}} a^\dagger_{\mathbf{k}\uparrow} a^\dagger_{-\mathbf{k}\uparrow} + \Delta^*_{\mathbf{k}} a_{-\mathbf{k}\downarrow} a_{\mathbf{k}\uparrow} - \Delta^*_{\mathbf{k}} b^0_{\mathbf{k}}). \tag{2.5}$$

The right-hand side of this Hamiltonian is now quadratic. Hence, it can be diagonalized by the canonical transformations:

$$a_{\mathbf{k}\uparrow} = u_{\mathbf{k}} c_{\mathbf{k}1} + v^*_{\mathbf{k}} c^\dagger_{\mathbf{k}2};$$
$$a^\dagger_{-\mathbf{k}\downarrow} = -v_{\mathbf{k}} c_{\mathbf{k}1} + u^*_{\mathbf{k}} c^\dagger_{\mathbf{k}2},$$

where the new operators $c_{\mathbf{k}1}$ and $c_{\mathbf{k}2}$ are Fermion annihilation operators. The parameters $u_{\mathbf{k}}$ and $v_{\mathbf{k}}$ are chosen such that the coefficients of the mixed terms such as $c^\dagger_{\mathbf{k}1} c^\dagger_{\mathbf{k}2}$ in the Hamiltonian vanish. This can be satisfied if

$$2\varepsilon_k u_{\mathbf{k}} v_{\mathbf{k}} + \Delta_k v^2_{\mathbf{k}} - \Delta^*_{\mathbf{k}} u^2_{\mathbf{k}} = 0. \tag{2.6}$$

Hence, we impose another condition:

$$|u_{\mathbf{k}}|^2 + |v_{\mathbf{k}}|^2 = 1. \tag{2.7}$$

These two conditions determine the two parameters.

The resulting energy is given by

$$E_{\mathbf{k}} = (\varepsilon_{\mathbf{k}}^2 + |\Delta_{\mathbf{k}}|^2)^{1/2}. \tag{2.8}$$

This is the energy associated with a pair-excitation. The gap parameter is determined by

$$\Delta_{\mathbf{k}} = -\sum_{\mathbf{k}'} V_{\mathbf{k}\mathbf{k}'}\Delta_{\mathbf{k}'}[1 - 2f(E_{\mathbf{k}'})]/2E_{\mathbf{k}'}, \tag{2.9}$$

where $f(E)$ is the Fermi distribution:

$$f(E) = \frac{1}{e^{\beta E} + 1}. \tag{2.10}$$

Equations (2.8) and (2.9) are coupled. The latter equation indicates that $\Delta_{\mathbf{k}}$ depends on $\beta = 1/kT$. This is reasonable because above the critical temperature T_c there is no superconductivity.

In order to determine the critical temperature we assume a constant gap for $|\varepsilon_k| < \hbar\omega$ and

$$\Delta_{\mathbf{k}} = \Delta, \tag{2.11}$$

and introduce a simplifying expression:

$$V_{\mathbf{k}\mathbf{k}'} = \begin{cases} -V, & |\varepsilon_{\mathbf{k}}| < \hbar\omega, \\ 0, & |\varepsilon_{\mathbf{k}}| > \hbar\omega. \end{cases} \tag{2.12}$$

In other words, the gap parameter and the interaction are assumed to be constant. The interaction is important only at $k = k_F$, where it is attractive, mediated by phonons.

For $\Delta \neq 0$, the condition (2.9) becomes

$$1 = V\sum_{\mathbf{k}} \frac{1 - 2f(E_{\mathbf{k}})}{2E_{\mathbf{k}}}. \tag{2.13}$$

Electron pairing takes place only if this condition is satisfied. Thus, this is the condition under which superconductivity occurs. Here, the sum over $\mathbf{k}$ is subject to the condition that $|\varepsilon_{\mathbf{k}}| < \hbar\omega$. The k-sum in Eq. (2.13) can be replaced by integration by using $g(\varepsilon_F)$, which is the density of states per spin at the Fermi level. In order to determine the critical temperature, we anticipate that Δ considered as a function of temperature vanishes there:

$$\Delta = 0. \tag{2.14}$$

Hence, neglecting Δ and using a variable $x = \varepsilon/kT_c$ we obtain

$$1 = g(\varepsilon_F)V \int_0^{\hbar\omega/kT_c} dx\, \frac{1 - 2f(x)}{x}. \tag{2.15}$$

Note that the condition $|\varepsilon_k| < \hbar\omega$ has been used. The right-hand side integral can be integrated by parts, and the upper limit of the integral thus obtained can be replaced by infinity. We can then make use of the known integral:

$$\ln(2\gamma/\pi) = 2 \int_0^\infty dx \ln x \, \frac{d}{dx}\left(\frac{1}{e^x + 1}\right), \tag{2.16}$$

where γ is Euler's constant. Note that the replacement of the upper limit $\hbar\omega/kT_c$ by infinity is valid only if $T < T_c$. Neglecting $2f(\hbar\omega)$ in comparison with 1 we arrive at

$$kT_c = 1.13\hbar\omega \exp\left(-\frac{1}{g(\varepsilon_F)V}\right). \tag{2.17}$$

This equation determines the critical temperature T_c. This is low because $g(\varepsilon_F)V$ is around 0.01–0.1. Moreover, it is proportional to $\hbar\omega$ so that it is proportional to $M^{-1/2}$.

The determination of the gap parameter requires the use of the energy spectrum given by Eq. (2.8) in Eq. (2.13). We can adopt similar steps as above but the integral cannot be obtained analytically. However, in the vicinity of T_c an explicit result is obtained:

$$\begin{aligned}\frac{\Delta(T)}{\Delta(0)} &= \gamma\left[\frac{8}{7\zeta(3)}\left(1 - \frac{T}{T_c}\right)\right]^{1/2} \\ &= 1.74\left(1 - \frac{T}{T_c}\right)^{1/2}, \end{aligned} \tag{2.18}$$

with $\gamma = 1.77$ and $\zeta(3) = 1.202$.

Figure 8.2 illustrates the BCS relation for $\Delta(T)$. There are several methods for experimentally determining the energy gap. One simple method is to measure the tunneling current between two films of a superconducting material separated by a thin oxide layer. The current flows only when the voltage exceeds the energy gap 2Δ. These experiments have been found to be in good agreement with the above BCS formula.

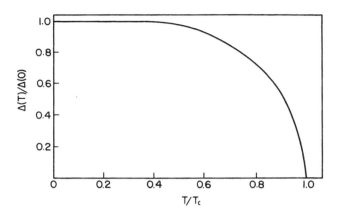

FIG. 8.2. Normalized energy gap $\Delta(T)/\Delta(0)$ as a function of temperature.

According to the BCS theory the coherence length is given by

$$\xi_0 = 0.18 \frac{\hbar v_F}{kT_c}. \tag{2.19}$$

This compares favorably with experiment: the numerical constant in aluminum is 0.15. The BCS theory gives a jump in the electronic specific heat at T_c in proportion to $d(\Delta^2)/dT$. Its form is

$$\Delta c = 1.43\gamma T_c. \tag{2.20}$$

The penetration depth is given by

$$\lambda(T) = \lambda_0[1 - (T/T_c)^4]^{-1/2}. \tag{2.21}$$

These results are generally in good agreement with experiment.

For a constant gap $\Delta > 0$, the excitation energy spectrum results in the BCS density of states relative to that of the normal states at ε_F as

$$\frac{N_s^{BCS}}{g(\varepsilon_F)} = \begin{cases} \dfrac{E}{[E^2 - \Delta^2]^{1/2}} & (E > \Delta) \\ 0 & (E < \Delta). \end{cases} \tag{2.22}$$

This expression represents the ratio of the density of quasiparticle states of the superconducting state with a gap Δ to that of the normal state. The quasiparticle states accumulate to $E = \Delta$, where the density diverges. Away from this point the ratio decreases and approaches 1 as $E \to \infty$.

8.3. Strong coupling

Although the BCS theory explains Type I superconductivity well, deviations occur, especially in the case of Type II. It was pointed out in Section 8.1 that the critical field $H_c(T)$ of lead deviated from the BCS parabolic variation as much as a few percent. There are other deviations as well. For instance, the BCS theory predicts

$$\frac{2\Delta(0)}{kT_c} = 3.52,$$

but the ratio is as large as 4.6 in mercury and 4.3 in lead. Note here that

$$E_{\text{exc}} = 2\Delta(0)$$

represents the minimum energy necessary to excite a pair of electrons from the ground state. The ratio of the superconducting to normal thermal conductivities of these superconductors has a limiting slope of order 10 at T_c in comparison with the value of order 1.5 expected from the BCS theory. Similar deviations in many other properties necessitate reexamination of the theory.

The BCS theory employs an effective interaction, energy transfers of order Debye frequency ω_D in phonon exchange, and other simplifications. It is a quasiparticle description with a constant effective interaction. However, in actuality the electron–phonon interaction causes a mass enhancement near the Fermi energy and a finite lifetime of a quasiparticle. With the excitation energy of the order of Debye frequency, the lifetime of a quasiparticle is short and its level width is of the order of the excitation energy. That is, its damping is very strong and a well-defined quasiparticle no longer exists. Hence, the quasiparticle picture becomes invalid. More detailed considerations of electron–electron interaction, frequency dependency in energy transfers, and other refinements are needed. The theory of strongly coupled superconductors was thus developed [3, 4].

The finite lifetime of a quasiparticle is expressed by the imaginary part of the electron self-energy, its real part being the gap $\Delta(\omega)$. As indicated in this notation, this gap parameter is not constant but is frequency-dependent. The determination of the self-energy is somewhat complicated, but Schrieffer [4] showed that the basic mechanism for renormalization and self-energy corrections can be understood simply by examining an addition of an electron to the ground state of a normal metal.

For an electron of momentum $\mathbf{p}$ such an addition is represented by the two virtual processes with the following initial momentum changes:

$$\mathbf{p} \rightarrow \mathbf{p}'; \qquad \mathbf{p}' \rightarrow \mathbf{p}.$$

The first is associated with the virtual emission and reabsorption of a phonon by an electron added to the ground state of a normal system, and the second with a similar process between the state $\mathbf{p}'$ within the Fermi sea and the state $\mathbf{p}$ above the Fermi sea. Note that the latter process blocks the addition of an electron to the same state $\mathbf{p}$. The self-energy $\Sigma(\mathbf{p}, \omega)$ depends on the momentum $\mathbf{p}$ of the electron and frequency $\omega_{\mathbf{p}-\mathbf{p}'}$ of the virtual phonon. It consists of the following two terms ($\hbar = 1$):

$$\Sigma(\mathbf{p}, \omega) = \sum_{\mathbf{p}'} |\tilde{h}_{\mathbf{pp}'}|^2 \left(\frac{f(p')}{\omega + \omega_{\mathbf{p}-\mathbf{p}'} - \varepsilon_{\mathbf{p}'} - i\delta} - \frac{1 - f(p')}{\varepsilon_{\mathbf{p}'} - \omega + \omega_{\mathbf{p}-\mathbf{p}'} - i\delta} \right), \quad (3.1)$$

where $f(p')$ is the Fermi distribution, $\tilde{h}_{\mathbf{pp}'}$ is a renormalized matrix element of electron–phonon coupling that will be given shortly, and $i\delta$ is a small imaginary parameter to be brought to zero. The electron energy $\varepsilon_{\mathbf{p}}$ is conveniently replaced by ω in the denominator, expressing the frequency dependence of Σ. Such a replacement can be made to order $(m/M)^{1/2}$, where m and M correspond respectively to the electron and ion masses. The electron energy is measured from the Fermi level:

$$\varepsilon_{\mathbf{p}} = \frac{p^2}{2m} - \mu$$

$$= \frac{p^2}{2m} - \frac{p_F^2}{2m}. \quad (3.2)$$

Before working on Eq. (3.1) we remark that the bare electron–phonon coupling associated with a momentum change $\mathbf{q} = \mathbf{p} - \mathbf{p}'$ is given by

$$h_{\mathbf{q}} = -i \left(\frac{N}{2M\Omega_q} \right)^{1/2} \mathbf{q} \cdot n_{\mathbf{q}} V(q), \quad (3.3)$$

where Ω_q is the phonon energy, $n_{\mathbf{q}}$ is a unit vector representing polarization, M is the ion mass, $V(q)$ is an effective electron-rigid ion interaction, and N is the number of ions. Strong electron–phonon coupling means that not only is the electron affected by the phonon but also the phonon is affected by the electron. Accordingly, the Green's functions of an electron and a phonon are coupled with each other.

The ideal Green's function of an electron is given by

$$G_0(\omega) = \frac{1}{\omega - \varepsilon_{\mathbf{p}} + i\delta_{\omega}}, \quad (3.4)$$

where $\delta_{\omega} = \delta \, \mathrm{sign}(\omega)$. In the presence of a screened Coulomb interaction between the electrons, the Green's function is modified such that

$$G(\omega) = \frac{1}{\omega - \varepsilon_p - \Sigma(\mathbf{p}, \omega)}. \tag{3.5}$$

The self-energy depends on processes that can conveniently be represented by diagrams. For the present Coulomb correction alone, we need only the mechanism in which the electron emits and reabsorbs momentum representing the screened Coulomb interaction. Note that the real part of the self-energy represents an energy gap, and the imaginary part yields the lifetime τ given by

$$\frac{1}{2\tau} = \text{Im } \Sigma(\mathbf{p}, \omega)$$

$$= -\pi \sum_{p'} |\tilde{h}_{\mathbf{pp'}}|^2 [1 - f(p')] \delta(\varepsilon_{p'} + \omega_{\mathbf{p}-\mathbf{p'}} - \omega). \tag{3.6}$$

The self-energy can be expanded at p_F and $\omega = 0$. For $p \sim p_F$,

$$p - p_F = \frac{\varepsilon_p}{v_F}.$$

Hence, the denominator of Eq. (3.5) becomes

$$\omega - \varepsilon_p - \left[\Sigma(p_F, 0) + \varepsilon_p \frac{1}{v_F} \frac{\partial \Sigma}{\partial p} + \omega \frac{\partial \Sigma}{\partial \omega} \right]$$

$$= \omega \left(1 - \frac{\partial \Sigma}{\partial \omega} \right) - \varepsilon_p \left(1 + \frac{1}{v_F} \frac{\partial \Sigma}{\partial p} \right) - \Sigma(p_F, 0). \tag{3.7}$$

One can introduce a renormalization parameter

$$Z = 1 - \frac{\partial \Sigma}{\partial \omega}, \tag{3.8}$$

and a renormalized quasiparticle energy

$$Z\varepsilon_p^* = \left(1 + \frac{1}{v_F} \frac{\partial \Sigma}{\partial \omega} \right) \varepsilon_p.$$

$$= \varepsilon_p(m/m^*), \tag{3.9}$$

where m^* is an effective mass. Equations (3.8) and (3.9) represent how renormalization due to electron–electron interaction takes place. In terms of the renormalization parameter Z and the effective mass m^*, and with the replacement $-\Sigma(p_F, 0) \to i\delta$, the Green's function is modified into

$$G(\omega) = \frac{1}{Z} \frac{1}{\omega - \varepsilon_p^* + i\delta}. \tag{3.10}$$

In the above approximation the renormalization factor Z is constant, but more elaborate calculations with frequency-dependent $Z(\omega)$ are possible.

As the electron is affected by the phonon, the latter is also affected by the former since the lattice vibration takes place in a medium of dielectric constant $\varepsilon(q)$ due to the cloud of electrons. As a result, the phonon frequency Ω_q is renormalized as follows:

$$\omega_q = \frac{\Omega_q}{[\varepsilon_1(q)]^{1/2}}, \tag{3.11}$$

where $\varepsilon_1(q)$ is the real part of the dielectric constant of electrons. The imaginary part of the dielectric function causes damping of lattice vibrations. Also, similarly to the Coulomb correction Z in the electron propagator, the phonon propagator will have a correction Z_p. These and some other corrections lead us to a renormalized electron–phonon coupling function:

$$\tilde{h}_q = -i\left(\frac{N}{2M\omega_q}\right)^{1/2} q \cdot n_q \frac{\lambda(q)}{\varepsilon(q)} \frac{V(q)}{Z}, \tag{3.12}$$

where $\varepsilon(q)$ is the dielectric constant, and $\lambda(q)$ is a vertex correction.

Let us now return to Eq. (3.1). The summation over p' can be replaced by an integration over $\varepsilon_{p'}$ and over the Fermi surface:

$$\sum_{p'} \to \int d\varepsilon_{p'} \frac{1}{(2\pi)^3} \int_{p_F} \frac{dS'}{v_F'},$$

where the first integral is over the value of $\varepsilon_{p'}$ and the last integral is over the Fermi surface on which $p' = p_F'$, v_F' being the Fermi velocity. Since $\omega_{p-p'}$ represents the phonon energy, we separate this energy and all other terms that include p' from those to do with the electron by introducing

$$\delta(\omega' - \omega_{p-p'}) \, d\omega'.$$

The resultant integral is similar in form to the phonon density of states $N_p(\omega')$ except for a factor $|\tilde{g}_{pp'}|^2$ because (without a sum over the polarization directions)

$$N_p(\omega') = \frac{1}{(2\pi)^3} \int d p' \, \delta(\omega' - \omega_{p-p'}). \tag{3.13}$$

This integral can be converted into that over the Fermi surface in a similar manner as above. We replace the dS' integral by its average:

$$\int dS' \rightarrow \frac{\int dS \int dS'}{\int dS},$$

where dS corresponds to the integration over the Fermi surface on which $p = p_F$. Thus, the phonon part is expressed by

$$\alpha^2(\omega')N_p(\omega') = \int dS\left[\frac{1}{(2\pi)^3}\int dS' \frac{1}{v_F'} |\tilde{h}_{\mathbf{pp'}}|^2 \delta(\omega' - \omega_{\mathbf{p-p'}})\right]\bigg/\int dS. \quad (3.14)$$

$\alpha^2(\omega')$ represents an effective electron–phonon coupling.

In terms of Eq. (3.14) the electron self-energy becomes

$$\Sigma(\omega) = \int_0^\infty d\varepsilon_{\mathbf{p'}} \int_0^\infty d\omega' \, \alpha^2(\omega')N_p(\omega')\left[\frac{1}{\varepsilon_{\mathbf{p'}} + \omega + \omega' - i\delta} - \frac{1}{\varepsilon_{\mathbf{p'}} - \omega + \omega' - i\delta}\right].$$

$$(3.15)$$

The terms in the square brackets can easily be renormalized in the presence of screened Coulomb interaction, following the scheme discussed earlier. Equation (3.15), which is based on a simple mechanism, represents the pertinent self-energy expression. The first integral over $\varepsilon_{\mathbf{p'}}$ can be converted into a frequency integral similarly to the second integral over ω'. The frequency integral may then be expressed in terms of the density of states associated with superconductivity that is characterized by an energy gap. The density of states relative to that in the normal state is given for an energy-dependent complex gap by

$$N_s(\omega) = g(\varepsilon_F) \, \mathrm{Re}\left(\frac{|\omega|}{[\omega^2 - \Delta^2(\omega)]^{1/2}}\right), \quad (3.16)$$

where $g(\varepsilon_F)$ is the density of states at the Fermi surface of the normal state. In the immediate vicinity of $\Delta(\omega) = \Delta = \text{const.}$, the density of states can be approximated by

$$N_s(\omega) = N_s^{\text{BCS}}\left[1 + \frac{1}{2}\left(\frac{\partial\Delta}{\partial\omega}\right)_\Delta\right] \quad (\omega \gtrsim \Delta). \quad (3.17)$$

This represents that the density of states is slightly enhanced over the BCS case. For lead the enhancement factor is 1.025.

We must note that the renormalization processes as discussed above are necessary not only for the superconducting state but also for the normal state. The existence of a gap Δ causes the renormalization parameter Z_s of the superconducting state to differ from the Z_n of the normal state. However, if the electron–phonon coupling is weak, they must be nearly the same:

$$Z_s \sim Z_n.$$

Thus, strong coupling is characterized by the difference $Z_s - Z_n$. According to a calculation of Bardeen and Stephen [5] based on an Einstein phonon model with a characteristic frequency ω_E,

$$\begin{aligned}
\text{Re}[Z_s(\omega) - Z_n(\omega)] &\sim \left(\frac{\Delta}{\omega_E}\right)^2 \ln\left(\frac{\omega_E}{\Delta}\right) & (\omega < \omega_E), \\
&\sim \left(\frac{\omega_E}{\omega}\right)^2 \left(\frac{\Delta}{\omega}\right)^2 \ln\left(\frac{\omega_E}{\Delta}\right), & (\omega > \omega_E).
\end{aligned} \tag{3.18}$$

Here, Δ is a constant gap independent of frequency at $T = 0$. For Type I superconductors,

$$(\Delta/\omega_E)^2 \lesssim 10^{-3}$$

so that the difference $\text{Re}[Z_s(\omega) - Z_n(\omega)]$ is approximately 0.1–1%. On the other hand, in Type II,

$$\Delta_0/\omega_E \sim 1/3.$$

Hence, the difference is in the range 10–15%.

The above considerations have been restricted to absolute zero, and therefore it is necessary to treat the case of finite temperatures. For instance, the critical field is given by [5]

$$\begin{aligned}
\frac{H_c^2(T)}{8\pi} = g(\varepsilon_F)\,\text{Re} &\int_0^\infty d\omega \tanh\left(\frac{\beta\omega}{2}\right) \\
&\times \left[[Z_s(\omega) - Z_n(\omega)]\left(\omega - (\omega^2 - \Delta^2(\omega))^{1/2} - \frac{\Delta^2(\omega)}{2[\omega^2 - \Delta^2(\omega)]^{1/2}}\right) \right. \\
&\quad \left. + \frac{[Z_s(\omega) - Z_n(\omega)]\Delta^2(\omega)}{2(\omega^2 - \Delta^2(\omega))^{1/2}} \right].
\end{aligned} \tag{3.19}$$

Here $g(0)$ is the density of states per spin at the Fermi surface. Note the appearance of the difference $Z_s - Z_n$ in the integrand. Numerical calculations show that this expression explains the deviation from the parabolic H_c–T curve.

However, for a close examination of small deviations from the weak coupling results it is necessary to treat electron–phonon interaction in an elaborate way, to consider the differences in the normal metal properties, and to solve the gap equations with sufficient accuracy. It is worth mentioning that the theoretical parameter $\alpha^2 N_p$ and the effective Coulomb potential of lead have been determined in comparison with electron tunneling data and the gap equations have been solved numerically with very good agreement with experiment [4].

8.4. High-T_c superconductivity

Since the discovery of the phenomenon of superconductivity, constant effort has been made to search for a new material with a higher transition point. Nevertheless, even after more than a half-century, the highest critical temperature was still in the region of 20 K. It appeared as if the T_c of 23.3 K in Nb_3Ge was a limit. However, in the 75th anniversary year of superconductivity, that is in 1986, Bednortz and Müller [2] discovered that LaBaCuO can be a superconductor at 35 K. This was a total surprise not only because of the high value of T_c but because the compound is a ceramic and is entirely different from all the previously known superconducting materials. The discovery triggered an exciting search for new materials in the new domain, causing a flood of reports on the subject, including new materials with T_c as high as 90 K. The number of new materials has reached approximately forty. Since there are now so many new materials, we discuss below only two representative families.

(1) *2–1–4 compounds.* Related to the first high-T_c superconductor is a family of compounds with the atomic structure $La_{2-x}M_xCuO_{4-y}$, where M is Ba, Sr, or Ca, x is of order 0.15, and y is nearly zero. The family is commonly called the 2–1–4 copper oxide in correspondence to the atomic composition ratio of the basic case in which $x = y = 0$. This family has T_c of order 40 K, and strontium appears to yield the highest.

At high temperatures, the 2–1–4 compounds have the body-centered tetragonal structure. In this structure, each copper atom is at the center of an elongated $Cu-O_6$ octahedron with four strongly bonded oxygens at a distance of 1.9 Å and two weakly bonded oxygens at a distance 2.4 Å. Along the c-axis, this octahedron structure separates the $Cu-O_2$ planes with two lanthanum and oxygen layers in between.

Figure 8.3 shows the structure in which Cu, O, and La or M atoms are represented respectively by black white and hatched circles. The $Cu-O_2$ planes are hatched for distinction. With this layered structure the compounds are highly anisotropic, and superconductivity is associated with the $Cu-O_2$ planes.

The compounds have the body-centered tetragonal structure at high temperatures and the orthorhombic structure at low temperatures. These two structures and also the superconducting phase depend sensitively on

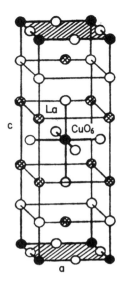

FIG. 8.3. Crystal structure of La_2CuO_{4-y}. White circles are oxygen atoms and black circles represent copper atoms, hatched circles represent lanthanum atoms.

oxygen doping. Figure 8.4 illustrates the phase diagram as a function of x in $La_{2-x}Sr_xCuO_{4-y}$. Below a certain temperature, the orthorhombic phase is metallic, and above insulating. There is a tiny antiferromagnetic phase, which is enhanced as y is increased. The graph shows the plane at $y = 0$. The antiferromagnetic phase is insulating.

The parent compound La_2CuO_{4-y} is not superconductive. In its ground state, the charges on La^{3+} and Cu^{2+} are balanced by O^{2-}. When doped with M, that is, in $La_{2-x}M_xCuO_{4-y}$, where M can be Sr, there are $x - 2y$ holes per cell. These holes are considered to go into O(2p) states and move about on each CuO_2 plane.

(2) *1–2–3 compounds.* This family has the general structure $RBa_2Cu_3O_{7-\delta}$, where R is Y, Eu, Gd, and so on. At the center of a unit cell, the copper atom is coordinated with four oxygen atoms, two at a distance 1.94 Å and the other two at 1.85 Å. Above and below the Cu–O_4 structure are two CuO_2 planes, although slightly buckled at the sites. In addition, copper at the center of each unit cell forms a Cu–O chain along the b-axis. Figure 8.5 shows the structure. The CuO_2 planes are hatched for clarity. Between these two planes are two 4Ba–1O planes. Above 500°C, the insulating tetragonal phase is stable.

The phase diagram of $YBa_2Cu_3O_{7-\delta}$ is shown in Fig. 8.6 as a function of the oxygen content parameter δ. Note that as δ decreases, the hole concentration increases; the hole concentration is given by $(1 - 2\delta)$ per cell.

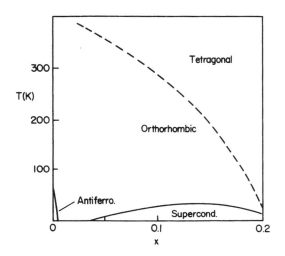

FIG. 8.4. Phase diagram of $La_{2-x}Sr_xCuO_{4-y}$.

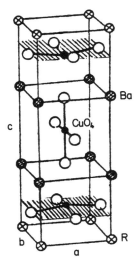

FIG. 8.5. Structure of $RBa_2Cu_3O_7$. Crossed circles at the corners of the unit cell of orthorhombic structure represent R, which can be ytterbium, europium, etc. White circles are oxygen atoms.

217

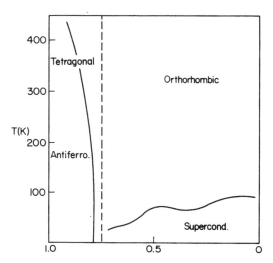

FIG. 8.6. Phase diagram of $YBa_2CuO_{7-\delta}$.

The critical temperature can be as high as 93 K for $\delta = 0$. The antiferromagnetic insulating phase appears when δ is above around 0.7. Below this value, the compounds are metallic.

There are other types of high-T_c superconductors, but let us now discuss their basic characteristic properties in comparison with the ordinary superconductors.

(3) *Characteristic properties.* One of their most distinguishing properties is that the coherence length is very small:

$$\xi_0 \sim \frac{0.2\hbar v_F}{kT_c}$$

$$\sim 10 \text{ Å}. \qquad (4.1)$$

It is approximately 1/10 to 1/100 of the ordinary superconductor's coherence length and is of order of the dimension of the unit cell. This smallness indicates that a local structure is important. In the case of $Y_1Ba_2Cu_3O_{7-\delta}$, ξ_0 has been estimated to be around 3 Å in the c-direction. This length is significantly smaller than the spacing of the Cu–O planes. Moreover, the coherence length is anisotropic.

The coherence length is temperature-dependent and diverges at the critical temperature. If this dependence is expressed by the Ginsberg–Landau form $\xi_0(1 - T/T_c)^{-1/2}$, which is equated to the spacing 8 Å between Cu–O planes of the 1–2–3 compounds, one obtains 74 K [6] for the crossover

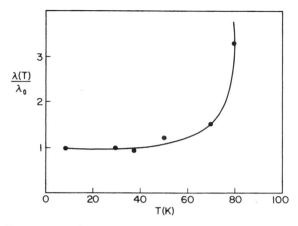

FIG. 8.7. Temperature variation of the penetration length λ.

temperature from 2D to 3D behavior. High-resolution torque magnetometry on an untwinned single crystal of $Y_1Ba_2Cu_3O_{7-\delta}$ with $T_c = 90.5$ K showed that the compounds are three-dimensional at 80 K and above, while being two-dimensional below [6].

The spin-spin correlation length in $La_{2-x}Sr_xCuO_4$ is proportional to $x^{-1/2}$ with a proportionality constant $3.8A$. Correspondingly, the magnetic susceptibility increases with x. It is of order

$$\chi \sim 3 \times 10^{-7} \text{ emu/g.}$$

The penetration depth λ follows the empirical relation

$$\lambda(T) = \lambda_0[1 - (T/T_c)^4]^{-1/2}. \tag{4.2}$$

The value of λ_0 is 1415 ± 15 Å in the ab plane of a single-crystal $YBa_2Cu_3O_{7-\delta}$ with $T_c = 82 \pm 0.3$ K. Figure 8.7 shows the data of Harshman et al. [7]. The curve represents the above relation (4.2).

The specific heat jumps at T_c. The jump in the 1–2–3 compounds is given by

$$\frac{\Delta c}{\gamma T_c} = 1.23\text{--}1.5. \tag{4.3}$$

These observed values of the jump constant compare favorably with the BCS value of 1.43. However, there are reports on the existence of a linear specific heat even in the superconducting state. Let the Sommerfeld constant γ be around $10 \, \text{mJ}/[(\text{molCu})\text{K}^2]$. For an isotropic 3D system, γ is given by

$$\gamma = k^2 \left(\frac{\pi}{3}\right)^{2/3} \frac{m^*}{\hbar^2} n^{1/3}, \tag{4.4}$$

where n and m^* are the normal state electron density and effective mass respectively. Since λ is extremely anisotropic, the effective mass should also be anisotropic. For a stack of 2D systems with layer spacing w, γ is

$$\gamma = \frac{\pi}{3} k^2 \frac{m^*}{\hbar^2} \frac{1}{w}. \tag{4.5}$$

For $w = 5.85 \, \text{Å}$, and $\gamma = 10 \, \text{mJ}/[(\text{molCu})\text{K}^2]$, Eq. (4.5) yields

$$\frac{m^*}{m} \sim 10. \tag{4.6}$$

If this effective mass is used in Eq. (4.4), the carrier density is estimated to be [7]

$$n \sim 1.4 \times 10^{22} \, \text{cm}^{-3}.$$

Concerning the linear specific heat, Phillips et al. [8] have found that Cu^{2+} magnetic moments in impurity phases such as $BaCuO_2$ and in the YBa_2O_7 lattice largely determine the specific heat constant and that these moments act as pair-breaking centers. There are apparently nonsuperconductive regions in the vicinity of these moments that are imbedded in perfectly superconducting areas.

According to Tsai et al. [8] the energy gap $\Delta(T)$ in the directions parallel and perpendicular to the CuO_2 planes of $YBa_2Cu_3O_{7-\delta}$ is given by

$$\frac{2\Delta(0)}{kT_c} = \begin{cases} 5.9 \pm 0.2 & \text{(parallel)}, \\ 3.6 \pm 0.2 & \text{(perpendicular)}. \end{cases} \tag{4.7}$$

Friedl et al. [8] have obtained a precise Raman scattering value of

$$\frac{2\Delta(0)}{kT_c} = 4.95 \pm 0.10$$

for $YBa_2Cu_3O_{7-\delta}$. These values are consistent with the BCS value of 3.53. The BCS gap equation is

$$2\Delta(T) = 3.53 T_c \left(1 - \frac{T}{T_c}\right)^{1/2}. \tag{4.8}$$

Angle-resolved photoemission studies reveal that the energy spectrum of outgoing electrons grows as the momentum associated with the angle approaches the Fermi surface calculated from conventional theory, but the lifetime broadening varies linearly with the energy difference from the Fermi energy rather than quadratically expected from Fermi liquid theory. The density of states just below the Fermi energy increases, similarly to the BCS case.

These observations indicates that the 1–2–3 compounds can be Type II superconductors. They are weakly linked materials in that islands of normal or weakly superconducting regions at Cu^{2+} magnetic moments are imbedded in superconductive regions [8]. These magnetic moments exist in impurity phases such as $BaCuO_2$ and also on the YBa_2CuO_7 lattice. Ordering occurs near 10 K in the former case and only below 1 K in the latter case. These magnetic moments cause pair-breaking, but T_c is nearly independent of the concentration of the magnetic moments on the 1–2–3 lattice after the specific heat change $\Delta c(T_c)$ has decreased by more than a factor of 2. The value of $\Delta c(T_c)$ seems to be a measure of the volume fraction of superconductive regions rather than an indicator of the effect of the concentration on the order parameter. The existence of these magnetic moments explains the low critical current, the Curie–Weiss behavior of the susceptibility, and low values of $\Delta c(T_c)$.

A triangular flux lattice exists in ordinary Type II superconductors in a certain range of magnetic fields and temperatures. Such a lattice consists of superconductive regions interspersed with a regular array of vortex lines. The upper critical field H_{c2} is associated with melting of the flux lattice. Similar melting has been observed in single crystals of $YBa_2Cu_3O_7$ with an applied field in the c direction. Figure 8.9 shows the phase diagram for flux-lattice melting determined by Gammel et al. [9]. When a magnetic field is perpendicular to the c-axis, the vortex mobility transition occurs at H_{c2} as in 3D superconductors. However, in a parallel field melting occurs 3.2 K below the critical point, where apparently a transition into a vortex-liquid state takes place as in the case of 2D superconducting films. In contrast to the 3D case, a 2D flux lattice does not have long-range positional order and can melt well below the critical point. Such early melting causes low critical currents, hampering industrial applications.

The upper critical field is extremely high in the 1–2–3 compounds so that its measurements are limited to the vicinity of T_c where a variation of the type

$$H_{c2} \sim (T_c - T)^{1.5} \qquad (4.9)$$

has been observed.

The isotope effect is nearly absent. That is, the exponent in

$$T_c \sim M^{-\alpha}$$

is very small. For instance, α in the 1–2–3 oxides is as small as [10]

$$\alpha = 0.019 \pm 0.005. \qquad (4.10)$$

8.5. Theoretical models

Both 1–2–3 and 2–1–4 compounds have an insulating antiferromagnetic phase below a certain temperature. The antiferromagnetic phase is due to the unpaired spins of copper electrons. Doping converts them into spin-liquids, metals, and then superconductors. The basic electronic structure and some theoretical models of high-T_c superconductors are addressed in this section.

The CuO_2 planes play an important role for superconductivity, even though there are copperless materials. In fact, the critical temperature is sensitive to the oxygen atoms in these planes. Each copper atom has ten electrons in the 3d shell, which consists of one $d(x^2 - y^2)$ orbital and one $d(z^2)$ orbital. The former has four lobes directed toward the four oxygen atoms in the same xy plane, while the latter has two lobes pointed to the two oxygen atoms above and below the plane and one circular orbital in the xy plane. The single 4s electron and one of the ten 3d electrons of copper hybridize with the oxygen 2p electrons to form La_2CuO_4, keeping the $d(x^2 - y^2)$ orbital partially empty while the $d(z^2)$ orbital is filled. The remaining nine electrons in the $d(x^2 - y^2)$ orbital invite oxygens in the same plane to come closer. On the other hand, the electrons in the filled $d(z^2)$ orbital expel the oxygens above and below the xy plane. These configurations are illustrated in Fig. 8.8 in which the $d(z^2)$ orbital is shaded.

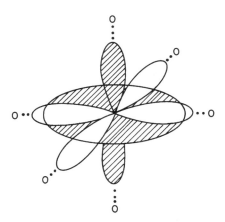

FIG. 8.8.   Copper d-orbitals. The four lobes of $d(x^2 - y^2)$ orbital are white and the $d(z^2)$ orbital is hatched. The locations of the neighbouring oxygen atoms are indicated. The top and bottom oxygens are at a greater distance than those on the horizontal plane.

Note that eight out of nine electrons in the Cu $d(x^2 - y^2)$ are paired, while one is unpaired. Thus, at each Cu site is a hole with a localized spin. Since the $d(x^2 - y^2)$ orbital is strongly coupled with the O(2p) orbital, one can talk about O(2p) or Cu(3d) holes.

The localized spin of the ninth, called d9, electron of copper causes antiferromagnetism. It is difficult for an unpaired spin to move about in an antiferromagnetic configuration due to energy costs. However, this configuration can easily be destroyed by doping or by some other disorder, particularly in two dimensions.

The replacement of La^{3+} by Sr^{2+} in $La_{2-x}Sr_xCuO_{4-y}$ creates $(x - 2y)$ holes per cell. The copper atoms appear to keep the same valence state, Cu^{2+}, even after doping. Hence, the holes seem to be on the oxygen sites, creating O^-. There are $(1 - 2\delta)$ holes per cell in the 1–2–3 compounds $RBa_2Cu_3O_{7-\delta}$ with R^{3+}. Accordingly, the 1–2–3 compounds can have more holes than the 2–1–4. Note that their critical points are also higher. Since the superconductive phase stretches beyond $\delta = 0.5$ some Cu^{2+} might be converted into Cu^+ as the hole concentration in the plane increases.

It becomes easier for the holes on copper (or oxygen) sites to move about once the antiferromagnetic regularity is destroyed. The high critical point indicates that a certain process involving high energy plays a role in pairing of holes. The destruction of the antiferromagnetic configuration by doping cannot be neglected in this respect, particularly because the resultant spin glass phase is not metallic but is insulating. The superconductive transition in the 2–1–4 compounds is preceded by an insulator–metal transition, but a direct transition from a spin glass state to a superconducting state without entering a metallic phase appears to take place in the 1–2–3 compounds near absolute zero.

The holes created by doping are primarily on the O^- sites in the $Cu-O_2$ planes. In consideration of their hopping from site to site, including copper, sites we express the Hamiltonian of a single $Cu-O_2$ plane as follows:

$$H = \sum_{i\sigma} \varepsilon_{ij} a_{i\sigma}^\dagger a_{j\sigma} + \frac{1}{2} \sum_{ij\sigma\sigma'} U_{ij} a_{i\sigma}^\dagger a_{i\sigma} a_{j\sigma'}^\dagger a_{j\sigma'}. \qquad (5.1)$$

The operator $a_{i\sigma}^\dagger$ creates a hole with spin σ in the $2p_x$ or $2p_y$ orbital at the copper site i. The hole is in the $3d(x^2 - y^2)$ orbital of copper. The diagonal energies will be either (ε_p, U_p) or (ε_d, U_d) for the $2p$ or $3d$ state respectively.

The choices:

$$\varepsilon_{ij} = t; \qquad U_{ij} = U$$

simplify the Hamiltonian. In addition, if

$$U_p = U_d = U; \qquad U_{pd} = 0,$$

the above Hamiltonian is reduced to a single-band Hubbard Hamiltonian:

$$H = t \sum_{(ij)} a_{i\sigma}^{\dagger} a_{j\sigma} + U \sum_{j} n_{j\uparrow} n_{j\downarrow}, \tag{5.2}$$

as in Eq. (5.7) of Chapter 2.

The same Hamiltonian can of course describe electron hopping. Its properties depend on the relative strength of t and U. The first term represents hopping between neighboring sites (ij), and the second term represents the interaction at the same site j. If this interaction is repulsive and large such that $U \gg t$, no two electrons can be on the same site. Hence, each lattice site is taken by only a single electron with a certain spin. As a consequence, the electrons can hardly move. Due to the large U, the band is split into two with a gap between. That is, a half-filled Hubbard model corresponds to an insulator with an energy gap between the lower occupied and upper unoccupied states. Thus, this Hamiltonian may be adopted for the insulating phase of high-T_c materials.

It is convenient to start with the above Hamiltonian, not distinguishing the copper and oxygen sites from each other. However, the single-band model is symmetric under a particle–hole transformations. Thus, removing holes from the $Cu-O_2$ planes is equivalent to adding them. This symmetry can be broken by a more elaborate *copper–oxygen model*. In this model, the removal of holes from the copper sites produces Cu^+. The energy of Cu^+ can be higher or lower than ε_d of Cu^{2+}. If it is higher, and if oxygen's ε_p is located between the two energies, any additional hole will go into the oxygen sites. Only in the opposite case, in which ε_d is higher than ε_p, can the holes go into the copper sites. Spectroscopic observations of excess holes on oxygen sites favor the copper–oxygen model. These excess holes are the charge carriers.

Let us discuss the effect of doping on the antiferromagnetic configuration that appears in both 2–1–4 and 1–2–3 compounds. Doping supplies additional oxygens and weakens magnetic coupling. Thus, spin flipping takes place, causing local spin-parallel configurations. This occurrence can be seen by examining the interaction of spins S_1 and S_2 on two neighboring Cu^{2+} with spin σ of an oxygen hole:

$$H = -J(S_1 + S_2) \cdot \sigma. \tag{5.3}$$

In order to minimize this energy, σ prefers to be parallel (antiparallel) to both S_1 and S_2 if $J > 0$ ($J < 0$). That is, regardless the sign of J, S_1 and S_2 are preferably parallel. Moreover, since the oxygen hole is presumably located closer to copper than the original Cu–O distance, the above energy would overcome the antiferromagnetic energy.

The local parallel-spin configurations created by doping stir up spin frustration so that the material becomes a quantum spin liquid. This liquid state is still insulating but may be considered as a parent state for superconductivity. Note that the ground state of a 1D Bethe lattice corresponds to a spin liquid. On the other hand, Raman scattering studies [11] have revealed that spin fluctuations in nonsuperconducting La_2CuO_4 are characterized by an extremely high exchange constant $J \sim 1100 \text{ cm}^{-1} = 137 \text{ meV}$. A similar magnitude $J \sim 950 \text{ cm}^{-1}$ has been found in $YBa_2Cu_3O_{7-\delta}$. Therefore, energies of order 1000 K may be involved for pairing. Increasing the oxygen concentration causes broadening and weakening of the spin-pair peak and dilution of the spin system in the planes. That is, spins are removed as the oxygen concentration is increased. This indicates that magnon exchange may not be responsible to pairing. In fact, there are perovskites such as $BaPbO_3$ that do not show any special magnetic properties but have T_c of order 30 K. It is also known that the excitations from the Bethe state are not spin waves but are quasifermions called *spinons*.

The existence of the O–Cu–O configuration before doping requires a close examination of energy changes due to excess oxygen atoms in relation to their motion in the Cu–O_2 planes. For instance, Emery and Reiter [12] solved a model in which an oxygen hole moves through a ferromagnetic copper spin background. This model suggests that pairing of these holes is mediated by enhanced superexchange coupling.

On the other hand, noting that a metal–insulator transition is close to the superconducting transition, Anderson [13] suggested that the insulating phase is an RVB (see Section 7.2). With sufficient doping, the magnetic singlet pairs in the insulating state become charged superconducting pairs. His model may be described in a simple way by starting with a half-filled Mott insulator in a simple square lattice. This system corresponds to a Heisenberg antiferromagnet and is represented by the Hamiltonian

$$H = J \sum_{(ij)} (\mathbf{S}_i \cdot \mathbf{S}_j - \tfrac{1}{4}). \tag{5.4}$$

In terms of Hubbard's t and U the exchange constant $J = 4t^2/U$. The spin operators can be rewritten in terms of the electron operators such that

$$H = -J \sum_{(ij)} b_{ij}^\dagger b_{ij} \tag{5.5}$$

with the local constraints $n_{i\uparrow} + n_{i\downarrow} = 1$. Here the singlet operators $b_{ij}^\dagger$ are defined by

$$b_{ij}^\dagger = \frac{1}{2^{1/2}} (a_{i\uparrow}^\dagger a_{j\downarrow}^\dagger - a_{i\downarrow}^\dagger a_{j\uparrow}^\dagger). \tag{5.6}$$

It is interesting to observe that the new Hamiltonian has the local gauge symmetry for $a_{i\sigma}^\dagger \rightarrow e^{i\theta_i} a_{i\sigma}^\dagger$. A similar gauge symmetry has been discussed for the fractional quantum Hall effect. The spins behaving as fermions are spinons. If an electron is removed by doping a hole, called a *holon*, is created. The holons do not carry spins but only charges. The effective Hamiltonian for a doped material can be expressed in terms of holon and spinon operators of the BCS case. At temperatures below $J \sim 1000$ K, the spinons do not hop. The dominant process is tunneling of a holon pair, which involves a virtual excitation of a spinon.

In addition to the above two models there are several other models. However, a convincing description at a finite value of doping is still lacking and the basic mechanism is yet to be disclosed. Nevertheless, the 2D character of high-T_c superconductivity suggests a certain connection with the FQHE. In fact, it has been shown that the ground state of the frustrated Heisenberg antiferromagnet in two dimensions and the fractional quantum Hall state for bosons are equivalent and the spinons and holons obey $\frac{1}{2}$-fractional statistics [14]. If a gas of such particles is described by

$$H = \sum_j^N \frac{p_j^2}{2m} \qquad (5.7)$$

any eigenstate may be expressed as

$$\Psi(z_1, z_2, \ldots, z_N) = \left[\prod_{j<k}^N \frac{(z_j - z_k)^\nu}{|z_j - z_k|^\nu} \right] \Phi(z_1, \ldots, z_N), \qquad (5.8)$$

where z_j is the position of the jth particle in two dimensions, $\nu = 1/2$, and Φ is a fermion wavefunction. When $H\Psi = E\Psi$, Φ satisfies $H'\Phi = E\Phi$ with

$$H' = \sum_j^N \frac{1}{2m} |\mathbf{p}_j + \mathbf{A}_j|^2, \qquad (5.9)$$

where

$$\mathbf{A}_j(r_j) = \nu \sum_{k \neq j} \frac{\hat{z} \times \mathbf{r}_{jk}}{|\mathbf{r}_{jk}|^2}. \qquad (5.10)$$

Therefore, each particle appears to carry a magnetic solenoid with it as it moves about in the system. On the other hand, the magnetic analog of the charge-$\frac{1}{3}$ quasiparticle of the fractional quantum Hall effect is a spin-$\frac{1}{2}$ excitation which may be considered as a spin-down electron in a spin liquid.

Hall experiments on the high-T_c compounds show that charge carriers are holes in the superconductive regime while they are electrons in the ordinary metallic regime and that T_c is roughly proportional to the oxygen concentration. This concentration is proportional to the Fermi energy in two dimensions. Also, it has been revealed [15] that the carriers are electrons in the superconductive domain while they are holes outside the domain in the case of a new high-T_c material $Nd_{2-x}Ce_xCuO_4$. Such a carrier conversion suggests that not only holes but also electrons play a role in high-T_c superconductivity.

Given the Type II character of the 1–2–3 compounds, what causes high T_c? This question may be answered if the process mediated by phonons in the semiconducting or insulating states is clarified. We would then see the difference from the ordinary metallic case. Self-consistent linearized-augmented-plane-wave calculations [16] show that the Cu 3d and O 2p energies are very close to each other if the energy bands of La_2CuO_4 are fitted to tight-binding parameters. This makes copper unique among transition metals so far as high-T_c materials are concerned. The modulation of Cu–O bonds by breathing-type oxygen vibrations appears to couple very strongly with the conduction band at the Fermi surface. Moreover, replacement of La^{3+} by Ba^{2+} in $La_{2-x}Ba_xCuO_4$ leads to significant changes in the copper and oxygen densities of states $g(\varepsilon_F)$ and soft phonon modes due to the vibration of the light oxygen atoms play an important role. However, since the isotope effect appears to be suppressed as T_c becomes higher, there may be a mechanism in addition to phonon-mediated pairing.

Figure 8.9 is a phase diagram for flux-lattice melting in $YBa_2Cu_3O_7$ single crystals with an applied field in the c-direction [9]. In this case, melting occurs 3.2 K below H_{c2}, where apparently a transition into a vortex-liquid state takes place similarly to the case of 2D superconducting films. In contrast to the 3D case, a 2D flux lattice does not have long-range positional order and can melt well below the critical point. Such early melting causes low critical currents, hampering industrial applications. Based on a mean field treatment, it has been shown that flux states are stable when the flux per plaquette is commensurate with the electron density and the dependence

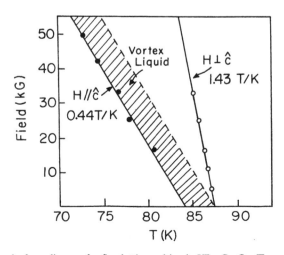

FIG. 8.9.   A phase diagram for flux-lattice melting in $YBa_2Cu_3O_7$. (From Gammel et al. [9])

of these states on the gauge degrees of freedom leads to superconductivity [17].

REFERENCES

1. J. Bardeen, L. N. Cooper, and J. R. Schrieffer, *Phys. Rev.* **108**, 1175 (1957).
2. J. R. Bednorz and K. A. Müller, *Z.Phys.B* **64**, 189 (1986). For general information concerning high-T_c superconductivity, see for instance: H. Ehrenreich and D. Turnbull (eds.) (Academic Press, New York, 1989), **42**. J. C. Phillips, *Physics of High T_c Superconductivity* (Academic Press, New York, 1989). J. W. Lynn (ed.), *High Temperature Superconductivity* (Springer-Verlag, New York, 1990). D. M. Ginsberg (ed.), *Physical Properties of High Temperature Superconductors* (World Scientific, Teaneck, NJ, 1989, 1990).
3. G. M. Eliashberg, *Zh. Eksp. Teor. Fiz.* **38**, 966 (1960) [*Sov. Phys.—JETP* **11**, 696 (1960)]; **43**, 1005 (1962) [**16**, 780 (1963)]. D. J. Scalapino, J. R. Schrieffer, and J. W. Wilkins, *Phys. Rev.* **148**, 263 (1966).
4. J. Robert Schrieffer, *Theory of Superconductivity* (W. A. Benjamin, Elmsford, NY, 1964). Douglas J. Scalapino, in *Superconductivity*, Vol. 1, ed. by R. D. Parks (Marcel Dekker, New York, 1969), p. 449. W. L. McMillan and J. M. Rowell, in *Superconductivity*, Vol. 1, ed. by R. D. Rowell (Marcel Dekker, New York, 1969), p. 561.
5. Y. Wada, *Phys. Rev. A* **135**, 1481 (1964). J. Bardeen and M. Stephen, *Phys. Rev. A* **136**, 1485 (1964). D. J. Scalapino, Y. Wada, and J. C. Swihart, *Phys. Rev. Lett.* **14**, 102 (1965).
6. D. E. Farrell, J. P. Rice, D. M. Ginsberg, and J. Z. Liu, *Phys. Rev. Lett.* **64**, 1573 (1990).
7. D. R. Harshman, L. F. Schneemeyer, J. V. Waszczak, G. Aeppli, R. J. Cava, B. Batlogg, L. W. Rupp, E. J. Ansaldo, and D. Li. Williams, *Phys. Rev. B* **39**, 851 (1989). L. Krusin-Elbaum, R. L. Greene, F. Holtzberg, A. P. Malozemoff, and Y. Yeshurun, *Phys. Rev. Lett.* **62**, 217 (1989).
8. J. S. Tsai et al., *Physica C* **163**, 1385 (1988). J. M. Imer et al., *Phys. Rev. Lett.*, **62**, 336 (1989). B. Friedl, C. Thomas, and M. Cardona, *Phys. Rev. Lett.* **65**, 915 (1990). N. E. Phillips et al., *Phys. Rev. Lett.* **65**, 357 (1990).
9. P. L. Gammel, L. F. Schneemeyer, J. V. Waszczak, and D. J. Bishop, *Phys. Rev. Lett.* **61**, 1666 (1988).
10. D. E. Morris, R. M. Kuroda, A. G. Markelz, J. H. Nickel, and H. Y. T. Wei, *Phys. Rev. B* **37**, 5936 (1988).
11. K. B. Lyons, P. A. Fleury, J. P. Remeike, and T. J. Negran, *Phys. Rev. B* **37**, 2353 (1988). K. B. Lyons, P. A. Fleury, J. P. Remeike, A. S. Cooper, and T. J. Negran, *Phys. Rev. B* **37**, 2353 (1988).
12. V. J. Emery and G. Reiter, *Phys. Rev. B* **38**, 4547 (1988).
13. W. Anderson, *Science* **235**, 1196 (1987). J. M. Wheatley, T. C. Hsu, and P. W. Anderson, *Phys. Rev. B* **37**, 5897 (1988).
14. V. Kalmeyer and R. B. Laughlin, *Phys. Rev. Lett.* **59**, 2095 (1987). R. B. Laughlin, *Phys. Rev. Lett.* **60**, 2677 (1988).
15. Y. Tokura, H. Takagi, and S. Uchida, *Nature* **337**, 345 (1989).
16. L. F. Mattheiss, *Phys. Rev. Lett.* **58**, 1028 (1987). W. E. Pickett, H. Krakauer, D. A. Papaconstantopoulos, and L. L. Boyer, *Phys. Rev. B* **35**, 7252 (1987).
17. P. Lederer, D. Poilblanc, and T. M. Rice, *Phys. Rev. Lett.* **63**, 1519 (1989).

9

LIQUID ^{4}He

Soon after his successful liquefaction of a helium gas in 1908 Kammerlingh-Onnes noticed that the liquid was strange because it started expanding as the temperature was reduced below around 2.2 K. Later discoveries of the discontinuity in the thermal expansion coefficient and the λ-type specific heat anomaly made it appropriate to distinguish the high-temperature side (He I) from the low-temperature side (He II) of the λ-point. In 1938 Kapitza observed that the viscosity measured by the capillary flow method dropped below the λ point by many orders of magnitude to an extremely small value. Thus, thirty years after its liquefaction, it became evident that the liquid makes a transition into a new superfluid phase. In the same year, London associated this transition with the condensation of an ideal Bose gas. While liquid ^{4}He is not a gas but a strongly interacting liquid, superfluid ^{4}He adsorbed on Vycor glass has been found more recently [1] to share many of the features of a low density three-dimensional gas.

9.1. Condensate fraction

In 1938, London [1] pointed out that if the total number of molecules of an ideal Bose gas is evaluated by replacing the summation of the Bose distributions over the momentum states by integration in a usual way, the contribution from the zero-momentum state is missed because the integration carries a weight factor $4\pi p^2$, where p is momentum. Therefore, he suggested separating out this particular term as follows:

$$N = \sum_p f(p) = \frac{z}{1-z} + \frac{V}{(2\pi)^3} \int f(p) 4\pi p^2 \, dp \tag{1.1}$$

$$= N_0 + N_1, \tag{1.2}$$

where V is the total volume, $f(p)$ is the ideal Bose distribution, and z is the fugacity, and N_0 and N_1 represent the first and second terms respectively. When z is close to 1, the first term becomes macroscopically large. This can occur because the molecules obeying Bose statistics prefer to be in the lowest momentum state.

The number N_0 is temperature-dependent. Its approximate form can be obtained by integrating the second term in Eq. (1.1) from 0 to ∞. Due to the factor $4\pi p^2$, the lower bound of integration can still be zero. The temperature variation is given by

$$N_0 = N\left[1 - \left(\frac{T}{T_0}\right)^{3/2}\right], \tag{1.3}$$

where T_0 is the condensation temperature and is determined by

$$n = 2.61\frac{(2\pi mkT_0)^{3/2}}{h^3}, \tag{1.4}$$

where n is the number density and m is the helium mass.

Note that N_0 starts increasing macroscopically at T_0. Since these molecules have zero momentum they do not contribute to kinetic properties, and thus sudden changes in the thermodynamic properties of the system can be expected at this point. Thus, an ideal Bose gas undergoes a phase transition called *Bose condensation* at T_0. The molecules in the zero-momentum state form the *condensate*. At absolute zero the condensate fraction of an ideal Bose gas is 1.

It turns out that the changes in the thermodynamic properties due to Bose condensation resemble those at the λ-point of liquid helium, but the changes are not as sharp; T_0 not only comes out higher than the λ-point but the specific heat does not diverge in a λ-shape but rather shows only a small kink. Thus, an ideal gas model is inadequate for liquid helium, and therefore a question arises concerning the condensate fraction.

The transition into the superfluid phase of liquid ^{4}He can still be associated with Bose condensation, but its fraction needs experimental determination. For this purpose neutron scattering can be used. The reason is simply that neutron scattering at a sufficiently high momentum transfer q is due to individual atoms. For ordinary materials the minimum value of q is prohibitively high, but for liquid ^{4}He, it is approximately 15 Å^{-1}. This value is still high but is experimentally accessible.

In the limit of large q, the dynamic structure factor $S(q, \omega)$ approaches that due to individual atoms given by [2]

$$S_I(q, \omega) = \sum_{\mathbf{p}} \delta\left(\omega - \frac{\hbar q^2}{2m} - \frac{\hbar\mathbf{q}\cdot\mathbf{p}}{m}\right)n(p) \qquad (q \to \infty), \tag{1.5}$$

where $n(p)$ is the momentum distribution function in the interacting system. This result is called the *impulse approximation*.

When Bose condensation takes place for $q = 0$, the momentum distribution function can be given by a sum of the contributions from the

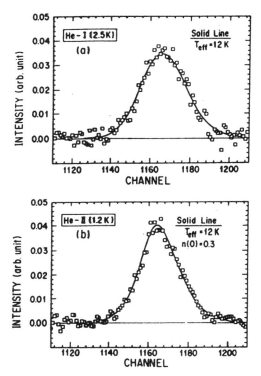

FIG. 9.1. Neutron scattering data at around 150 Å⁻¹ above and below T_λ. (From Ikeda and Watanabe [3])

condensed and noncondensed atoms such that

$$n(p) = n(0)\delta_p + [1 - n(0)]n_1(p). \tag{1.6}$$

In principle, the condensation at $T_\lambda = 2.19$ K is expected to affect the structure factor. However, it is difficult to observe this effect with momentum changes around $q = 15$ Å⁻¹ because of large broadening due to the final state of the scattered atoms. This final-state effect decreases slowly as q increases and finally becomes unimportant for $q > 100$ Å⁻¹ and is replaced by Doppler broadening. Pulsed-spallation neutron sources can give sharp and intense neutrons in this range.

Experimental time-of-flight spectra at $q \sim 150$ Å⁻¹ above and below the λ-point of liquid helium are shown in Fig. 9.1 [3]. The spectrum in (a) corresponding to 2.5 K is broader than that in (b) for 1.2 K. The solid curve is a theoretical one obtained for a Maxwellian distribution of noncondensate particles with an effective temperature of 12 K. A small variation of this effective temperature about this temperature indicates that this choice is the best. The solid curve in (b) is obtained for $n(0) = 0.3$ and for the same noncondensate particle distribution as in (a). Although this spectrum does not show a particular condensate line, a condensate fraction of around 0.1–0.3 seems to apply.

The condensate fraction has been estimated theoretically also. Penrose and Onsager [4] suggested the use of an off-diagonal element of the singlet density matrix. If $\Psi(\mathbf{r}_1, \mathbf{r}_2, \ldots, \mathbf{r}_N)$ is the ground-state wavefunction, this element is defined by

$$\rho_1(\mathbf{r}_1, \mathbf{r}_1') = \int \Psi^\dagger(\mathbf{r}_1, \mathbf{r}_2, \ldots, \mathbf{r}_n)\Psi(\mathbf{r}_1', \mathbf{r}_2, \ldots, \mathbf{r}_N) \, d\mathbf{r}_2 \cdots d\mathbf{r}_N. \quad (1.7)$$

For a fluid, $\rho_1(\mathbf{r}_1, \mathbf{r}_1')$ may be assumed to depend only on $|\mathbf{r}_1 - \mathbf{r}_1'| = r$. Then, $\rho_1(r)$ may be expanded in plane waves:

$$\rho_1(r) = \sum_{\mathbf{k}} \exp(i\mathbf{k} \cdot \mathbf{r}) n_k. \quad (1.8)$$

n_k represents the occupation number of state $\mathbf{k}$. In particular,

$$\lim_{r \to \infty} \rho_1(r) = n_0. \quad (1.9)$$

If n_0 is finite, there is off-diagonal long-range order. Bose–Einstein condensation can be represented by this long-range order.

On the other hand, the variation of $n_1(p)$ in Eq. (1.6) with temperature is found to be small above the λ-point T_λ. If this remains so even below T_λ, the difference between $n_p(T)$ and n_1 may be used to estimate $n_0(T)$ in high-resolution measurements of $S(q, \omega)$.

The condensate fraction $n_0(T)$ at a temperature T may be fitted to an empirical expression:

$$n_0(T) = n_0(0)\left[1 - \left(\frac{T}{T_\lambda}\right)^\alpha\right]. \quad (1.10)$$

Note here that T_λ instead of T_0 is used, and the exponent α is not equal to 3/2 as for ideal Bose gas.

In Fig. 9.2 experimental $n_0(T)$ values determined by various methods are plotted [5]. The solid curve represents a least-squares fit of Eq. (1.10) to the experimental values. For $T_\lambda = 2.17$ K, the parameters are found to be $n_0(0) = 13.9\%$ and $\alpha = 3.6$. An estimate based on surface tension yields similar results [5]. Theoretical estimates [4] range from around 8% to 13% so that a condensate fraction of 10% is an average.

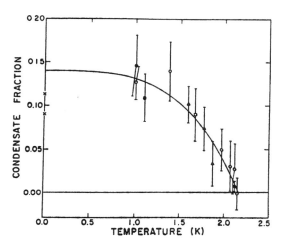

FIG. 9.2. Condensate fraction. The crosses are theoretical estimates. The solid curve is a least-squares fit of Eq. (1.10) to the experimentally determined values. (From Sears et al. [5])

The momentum distribution n_k is usually considered to be of the form of the ideal Bose distribution. For interacting systems, there can be modifications, but the use of an effective mass would enable retention of the same form. If a perturbation method can be used, one can show that to first order in the interaction potential $u(q)$ the limiting value of the singlet distribution function is given by

$$\lim_{r \to \infty} \rho_1(r) = n_0 + n_0^2 \kappa_T n\langle u(q)\rangle, \tag{1.11}$$

where for the ideal Bose distribution $f(q)$

$$n\langle u(q)\rangle = \frac{1}{(2\pi)^3} \int f(q)u(q) \, d\mathbf{q}. \tag{1.12}$$

$$u(q) = \int \phi(r)e^{i\mathbf{q}\cdot\mathbf{r}} \, d\mathbf{r}. \tag{1.13}$$

In correspondence to n_0, $f(0) = N_0$ represents the number of molecules in the zero-momentum state. Their contribution to the isothermal compressibility is denoted by κ_T. This term represents the fluctuation in the number of molecules in the condensed state. As such, it is related to the scattered intensity of electromagnetic waves in the incident direction. If the Bose distribution function is peaked sharply at zero momentum, $\langle u(q)\rangle = u(0)$. As we shall discuss shortly, this is the quantity that determines the velocity of sound in liquid helium.

9.2. Energy dispersion

Liquid helium is not an ideal gas but a strongly interacting Bose system. Therefore, it is understandable that London's theory does not quantitatively explain the properties of liquid helium. Landau [6] then developed a phenomenological theory assuming a single energy dispersion curve with two distinctive branches for the excitation energy He postulated that for low energy, only sound can be excited so that the spectrum is phononlike:

$$\varepsilon(q) = c\hbar q, \tag{2.1}$$

where c is the sound velocity and q is a wavenumber. For high energy, the excitation energy is particlelike near a certain wavenumber q_0:

$$\varepsilon(q) = \Delta + \hbar^2 \frac{(q - q_0)^2}{2\mu}, \tag{2.2}$$

where Δ is a minimum excitation energy or the energy gap and μ is an effective mass. These particlelike excitations are called *rotons*. Figure 9.3 illustrates a dispersion curve in which the solid circles represent the data obtained by Cowley and Woods [7]. The solid curve is theoretical. We shall discuss this, the thin straight line in the figure, and some other theoretical aspects of the elementary excitations in liquid helium later in this section and also in the next section.

With the phonon spectrum (2.1), the specific heat at low temperatures is expected to be proportional to T^3, as in experiment. The energy gap Δ results in an exponentially increasing specific heat as temperature increases toward the λ-point. The experimentally determined values for the roton

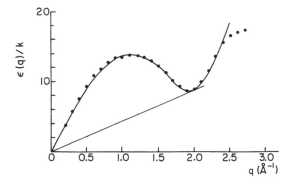

FIG. 9.3. Energy dispersion of liquid helium. Data points due to Cowley and Woods [7]; solid curve is theory [14].

parameters are

$$\Delta = 8.5 \text{ K}, \qquad q_0 = 1.9 \times 10^{-8} \text{ cm}^{-1}, \qquad \mu = 0.16 m_{\text{He}}.$$

Although it did not determine the λ-point, Landau's theory agreed well with experiment on liquid He II. Therefore, it became a theoretical task to derive microscopically the assumed energy spectrum. Bogoliubov [8] was the first to demonstrate the existence of a phonon spectrum in a Bose gas with weakly repulsive forces. Bogoliubov replaced the number density operator for zero momentum by a number under the assumption that the number of molecules in this state is very large. This and other theoretical subjects of liquid helium have been examined by the collective coordinate method and other methods [9].

These theories have shed light not only on the Bogoliubov theory but on the later Feynman theory [10] and also the role played by the interactions between elementary excitations at high energies. While Bogoliubov's theory did not reproduce the roton excitation, Feynman succeeded in deriving a Landau-type dispersion curve based on a variational method. His trial wavefunction consists of two factors [see Eq. (4.1) below]. One is the ground state in which overlapping of helium molecules does not occur. The other, representing density fluctuations for sound, is expressed by the Fourier transform of the density given by

$$n(r) = \sum_i \delta(\mathbf{r} - \mathbf{r}_i). \tag{2.3}$$

Assuming that the ground state is given, he obtained a formula for the excitation energy:

$$\varepsilon(q) = \frac{\hbar^2 q^2}{2mS(q)}, \tag{2.4}$$

where $S(q)$ is the static structure factor. Since $S(q)$ increases from the origin first linearly and then reaching a maximum, $\varepsilon(q)$ shows phonon and roton excitations.

Feynman's theory reproduces the energy dispersion curve qualitatively well, but it turns out that near the roton minimum the theoretical energy dispersion curve is too high in comparison with experimental data. As has been shown [11], a basic difficulty with a pairwise trial function such as adopted by Feynman stems from the fact that the radial distribution function evaluated by such a trial wavefunction does not vanish at the origin but rather reaches a positive value. This is of course in contradiction with our theoretical expectation.

On the other hand, Bogoliubov's theory is limited to a weakly interacting Bose gas. Since his model does not represent real liquid helium, a hard-sphere model has been investigated. For example, the ground-state energy per molecule is given by [12] [see Section A.5]:

$$\varepsilon_g = 4\pi a n \left[1 + \left(\frac{128}{15\pi^{1/2}} \right) (a^3 n)^{1/2} \right] \left(\frac{\hbar^2}{2m} \right)$$

for small $(a^3 n)$, where a is the hard-sphere diameter and n is the number density. The pair distribution function has also been evaluated [13]. For large distances it decreases as r^{-4}, in correspondence to a phonon excitation energy. The structure factor that can be used in Eq. (2.4) is given by

$$S(q) = q[q^2 + 16\pi a n]^{-1/2}.$$

A hard-sphere model is well defined but unfortunately it does not reproduce a roton spectrum. For this reason a soft-potential model [14] has been introduced. In this model a Fourier transform $u(q)$ of the interaction potential between helium molecules is assumed. If $u(q)$ is given the pair distribution function can be evaluated. In particular for collective couplings, the chain and chain-exchange diagrams are important particularly for small q. From the pair distribution function, the internal energy can be obtained. In contrast to the Bogoliubov and Feynman theories, which were developed for absolute zero, Isihara et al. have shown that the internal energy can be expressed in terms of a Bose distribution function of massless quasiparticles with energy $\varepsilon(q)$ in such a way that

$$U(T) = \frac{1}{(2\pi)^3} \int \varepsilon(q) f(\varepsilon(q)) g(\varepsilon(q)) \, d\mathbf{q}, \tag{2.5}$$

where $g(\varepsilon)$ is the density of states and $f(\varepsilon)$ is the quasiparticle distribution function given by

$$f(\varepsilon) = \frac{1}{\exp[\beta\varepsilon(q)] - 1}. \tag{2.6}$$

Moreover, the structure factor is expressed by [15]

$$S(q) = \frac{\hbar^2 q^2}{2m\varepsilon(q)} [1 + 2f(\varepsilon(q))], \tag{2.7}$$

where m is an effective mass, and

$$\varepsilon(q) = \frac{c\hbar q}{u(0)^{1/2}} \left(u(q) + \frac{\hbar^2 q^2}{4mn} \right)^{1/2}. \tag{2.8}$$

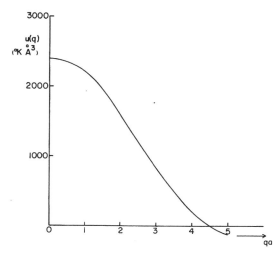

FIG. 9.4. Fourier transform $u(q)$ of a soft potential. (From Isihara et al. [14])

The excitation energy given by Eq. (2.8) agrees with what Bogoliubov and Zubarev [8] obtained. For a soft potential, $u(q) \to u(0) > 0$ as $q \to 0$, so that the sound velocity is given by

$$c = \left(\frac{nu(0)}{m}\right)^{1/2}. \tag{2.9}$$

For small q, $\varepsilon(q)$ represents phonon excitations.

Based on the above theory, explicit results have been obtained for a soft repulsive potential with a Lennard–Jones attractive tail. Namely,

$$\phi(r) = \begin{cases} V_0 & (r \leqq a) \\ \varepsilon^*[(a/r)^{12} - (a/r)^6] & (r \geqq a). \end{cases} \tag{2.10}$$

V_0 represents a finite repulsive core and a corresponds to a soft core. The Fourier transform $u(q)$ of a soft potential is illustrated in Fig. 9.4 [14]. For small q, $u(q)$ assumes a constant positive value $u(0)$. If this value is much larger than kT the potential is essentially a hard-sphere type. The finiteness may be considered to originate from a sum over ladder diagrams of two helium molecules with strong repulsive interaction. As q increases $u(q)$ becomes negative, reaches a shallow minimum, and finally decreases to zero as $q \to \infty$.

The solid curve in Fig. 9.3 represents a theoretical result obtained for the $u(q)$ of Fig. 9.4. The potential parameters are such that $V_0 = 30.6$ K,

$m^* = 1.71m$, $\varepsilon^* = 19.03$ Å, and the helium density is $2.18\,\text{Å}^{-3}$. A similar curve has been obtained for helium films also.

One can then show that for small momentum,

$$\varepsilon(q) = c\hbar q[1 + \delta_1 q^2 - \delta_2 q^3 + \cdots], \tag{2.11}$$

where δ_1 and δ_2 are positive functions of the potential parameters. For $V_0 = 35.65$ K, $\varepsilon^* = 30.1$ K, and $a = 2.73$ Å, it is found [14]

$$\delta_1 = 1.5\,\text{Å}^2; \qquad \delta_2 = 1.48\,\text{Å}^3.$$

When δ_1 is positive, the energy dispersion is called *anomalous* and its curve is concave-up. If the dispersion is convex-down, it is called *normal*. After many years of controversy, it has been established that in liquid helium, the dispersion is anomalous. In other liquids the dispersion is convex-down and is normal.

For small q, the structure factor can be expanded as [15]:

$$\frac{S(q)}{S(0)} = 1 + s_1 q^2 + s_2 q^3 + \cdots. \tag{2.12}$$

Here,

$$s_1 = \frac{1}{12}\left(\frac{c\hbar}{kT}\right)^2 - 2\delta_1. \tag{2.13}$$

$$s_2 = 2\delta_2. \tag{2.14}$$

Hence, at

$$T_I = \frac{c\hbar}{k(24\delta_1)^{1/2}} \tag{2.15}$$

an inversion of the slope of $S(q)$ plotted against q is expected to occur.

Figure 9.5 illustrates the variation of $S(q)$ for small q. The solid curves are theoretical [15], and the dots are the experimental data due to Svensson et al. [16]. The actual value of T_I is found to be 2.78 K, and the minimum of $S(q)$ occurs at

$$q_I = \frac{|s_1|}{3\delta_2}. \tag{2.16}$$

For 4.27 K and $|s_1| = 2.16\,\text{Å}^2$, $q_I = 0.49\,\text{Å}^{-1}$.

Figure 9.6 illustrates the structure factor of liquid helium. The solid curves are theoretical [15] and solid circles represent the experimental data of Sears et al. [5]. Since the ordinate is shifted for the four curves, it is somewhat difficult to observe but, below the λ-point, the higher the temperature the higher the first peak. This feature was first theoretically

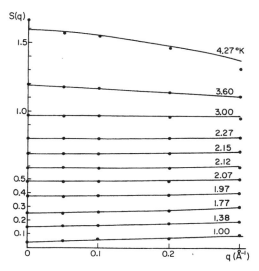

FIG. 9.5. Variation of $S(q)$ for small q. Curves are theory [15]; dots are data [16].

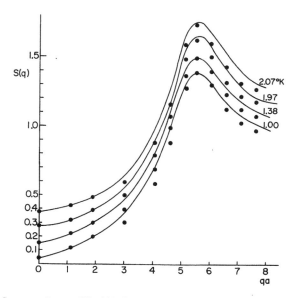

FIG. 9.6. Structure factor of liquid helium. Curves are theory [15]; dots are data [16].

shown by Isihara and Samulski [13]. They also obtained a temperature-dependent excitation energy.

How high would the dispersion curve increase for large momenta? One can think of 2Δ as a threshold energy because breaking of a roton into two can be expected. However, according to experiments by Graf et al. [17] near the maximum area, called "maxon region", the peak energy exceeds 2Δ. Note in this respect that the Landau spectrum corresponds to a single excitation and multiphonon excitations take place for momentum $q > 25$ nm^{-1}. The corresponding excitation spectrum has been determined by Cowley and Woods. Analyses of neutron scattering and more recent Raman scattering data at high energies have shown that rotons attract each other. Roton–roton interaction has been treated theoretically also [18] and the reason why a two-maxon peak had not been observed has been clarified.

The lifetimes of the excitations are determined by scattering between them and are inversely proportional to the concentration of prevailing excitations. Their determination is difficult at low temperatures because the lifetimes are long. However, Forbes and Wyatt [17] made direct time-of-flight measurements of roton–roton scattering at a temperature of around 0.1 K by using two crossing beams of ballistic rotons. One of the beams was weak and was the probe. The attenuation of the probe-pulse rotons was found to increase significantly with the wavevector near 2.15 Å^{-1}. That is, the scattering is strongly wavevector-dependent.

9.3. Superfluidity and critical velocity

The existence of a phonon spectrum is a sufficient condition for the occurrence of superfluidity. This can be understood as follows. Consider a macroscopic body with mass M that moves slowly with velocity $\mathbf{v}$ in the liquid at absolute zero. Its energy and momentum may be transferred into the liquid, and as a result, its velocity becomes $\mathbf{v}'$. The energy and momentum conservation laws state

$$\tfrac{1}{2}Mv^2 = \tfrac{1}{2}Mv'^2 + \hbar\omega; \qquad M\mathbf{v} = m\mathbf{v}' + \hbar\mathbf{q}.$$

Here, $\hbar\omega$ is the excitation energy of the fluid. These two equations result in the condition:

$$\mathbf{v}\cdot\mathbf{q} = \frac{\hbar q^2}{2M} + \omega.$$

The lowest velocity v_c satisfying this equation is

$$v_c = \min\left(\frac{\omega}{q}\right). \tag{3.1}$$

If ω is proportional to q^2, there is no finite value of v_c. If ω is proportional to q, that is, if the excitation is phononlike, the minimum velocity is given by the sound velocity:

$$v_c = c. \tag{3.2}$$

If the body moves at a velocity less than the sound velocity, which is 238 m/s, no energy transfer takes place so that the liquid exhibits superfluidity. This critical velocity corresponds to the linear part of the excitation spectrum at low energies. No phonon can be excited if the body moves below the sound velocity.

A roton excitation can take place at high energies. Correspondingly, there is a second solution of Eq. (3.1). If the excitation spectrum is Landau-like, $\min(\omega/q)$ can be given by a straight line that passes through the origin and is a tangent to the dispersion curve near the roton minimum, as shown by a thin straight line in Fig. 9.3. In this case,

$$v_c \sim \frac{\Delta}{q_0}. \tag{3.3}$$

This yields approximatly 58 m/s.

In both (3.2) and (3.3), c and Δ are temperature-dependent. Empirically, Δ is reduced at finite temperatures in accordance with

$$\frac{\Delta}{q} = 8.68 - 0.0084T^7.$$

Such a reduction reduces the critical velocity also. Moreover, there are some other factors that must be considered for finite temperatures. Therefore, it is safe if an experimental test is carried out at low temperatures.

One must be careful about the manner in which rotons are excited because excitations with only one roton are different from those involving two. These two cases can be distinguished experimentally on the basis, for instance, of the difference in the average drift velocity of an ion in the dissipative regime of liquid helium; it is proportional to $E^{2/3}$ in the former case and to $E^{1/3}$ for the latter, where E is an external electric field [19]. By plotting the drift velocity against E, the latter type of excitation is found to take place. By extrapolation to $E = 0$, the corresponding limiting drift velocity is found to be 46 m/s at 0.35 K [15]. This limiting velocity represents the critical velocity associated with pair-roton excitations and is found to be lower than for a single-roton excitation. The cause for the apparently easier pair-roton excitation in the ion experiment is unknown, but it may be related to the attractive interaction between rotons. However, their binding energy of -0.37 K is significantly smaller than $2\Delta \sim 17$ K.

9.4. Vortex motion

Strange hydrodynamical properties of liquid ^{4}He were first studied by the two-fluid model under the assumptions that the flow is irrotational and that the superfluid component flows without friction. However, it has been observed that a series of vortex lines threads the fluid in a rotating vessel. Also, the predicted critical velocity has been found to be too high. As we discuss below, the appearance of vortices in the superfluid is actually not limited to a rotating fluid. Instead, it is a characteristic of superflow.

Let us assume that the condensate in liquid helium is described by a single wavefunction given by

$$\psi(\mathbf{r}, t) = \psi_0(\mathbf{r}, t) \exp[i\phi(\mathbf{r}, t)], \tag{4.1}$$

where ψ_0 is a real function of position $\mathbf{r}$ and is normalized such that

$$\psi_0^2(\mathbf{r}, t) = n_s, \tag{4.2}$$

n_s being the number density of the superfluid component. $\phi(\mathbf{r}, t)$ is a phase factor and is also real.

Note that microscopically, ψ_0 and ϕ are symmetric functions of the positions $(\mathbf{r}_1, \mathbf{r}_2, \ldots, \mathbf{r}_N)$ of the helium molecules. Feynman assumed the above form of the wavefunction with ψ_0 representing the ground state and

$$\phi(\mathbf{r}) = \sum_i \mathbf{r}_i \cdot \mathbf{k}_i. \tag{4.3}$$

According to quantum mechanics, the current density of the superfluid at position $\mathbf{r}$ is given by

$$\mathbf{j} = \frac{\hbar}{2mi} (\psi^* \nabla \psi - \psi \nabla \psi^*)$$

$$= \frac{\hbar}{m} |\psi_0|^2 \nabla \phi. \tag{4.4}$$

Here, m is the helium mass. The superfluid velocity $\mathbf{v}_s$ can be expressed as

$$\mathbf{v}_s = \frac{\hbar}{m} \nabla \phi. \tag{4.5}$$

These equations show that ϕ plays the role of an order parameter for the superfluid.

The flow given by the gradient in Eq. (4.4) is irrotational. That is,

$$\text{rot } \mathbf{v}_s = 0. \tag{4.6}$$

The function $\phi(\mathbf{r})$ that generates this irrotational flow is single-valued. Microscopically as a function of the coordinates of molecules, $\phi(\mathbf{r}_1, \mathbf{r}_2, \ldots, \mathbf{r}_N)$ is a symmetric function of the coordinates. Under this condition, if a small displacement $d\mathbf{r}_i$ is introduced to $\mathbf{r}_i$, the phase will change by an amount

$$\sum_i [\nabla_i \phi] \cdot d\mathbf{r}_i.$$

If the displacements do not generate density fluctuations or any other gross changes in the system, and in compliance with Bose statistics, the phase change from one allowed configuration to another must be an integral multiple of 2π. The same is true for a single-valued function wavefunction because going around a closed loop does not change its value. As a result, a loop integral of the vector $\mathbf{v}_s$ should satisfy

$$\zeta = \oint \mathbf{v}_s \cdot d\mathbf{r} = j\frac{h}{m} \qquad (j = 0, \pm 1, \pm 2, \ldots). \tag{4.7}$$

This integral is the *circulation*, its contour being within the fluid. According to Eq. (4.7), the circulation of the irrotational superfluid flow is quantized in units of the quantum of circulation given by

$$\zeta_0 = \frac{h}{m} = 0.997 \times 10^{-3} \, \text{cm}^2/\text{s}. \tag{4.8}$$

The quantization of circulation, discussed by Onsager [20] and Feynman [20], has been confirmed by experiment since 1961. Figure 9.7

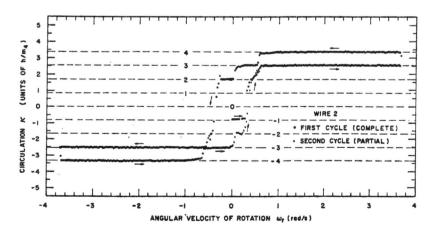

FIG. 9.7. Quantization of circulation. (From Karn et al. [21])

depicts the experimental data of Karn et al. [21]. The dashed lines are equally spaced, showing the quantum levels inferred from the data. The rotation cycle was begun by accelerating the apparatus from rest counterclockwise as seen from above. The arrows represent the sequence in which data were recorded. Four quantized levels were recorded in the case of wire 4, as can be seen in the graph.

In a simply connected space, a loop integral such as that in Eq. (4.7) can be converted into a surface integral of (rot $\mathbf{v}_s$) in accordance with Stokes' theorem. That is,

$$\oint \mathbf{v}_s \cdot d\mathbf{r} = \int (\text{rot } \mathbf{v}_s) \cdot d\mathbf{A}. \qquad (4.9)$$

Since rot $\mathbf{v}_s = 0$, the loop integral vanishes. The experimental fact that the circulation ζ is finite means that there is a vortex core in which

$$\text{rot } \mathbf{v}_s \neq 0.$$

Now imagine a superfluid flow circulating a cylinder in the plane perpendicular to the cylindrical axis. The superfluid flow is concentric with velocities varying only with distance r from the axis. Hence, Eq. (4.7) yields

$$|\mathbf{v}_s| = v_s = \frac{\zeta}{2\pi r}. \qquad (4.10)$$

The kinetic energy associated with a vortex line is given per unit length by

$$\begin{aligned}
\varepsilon_0(r) &= \int_a^r \frac{\rho_s v_s^2}{2} 2\pi r \, dr \\
&= \frac{\rho_s \zeta^2}{4\pi} \ln(r/a),
\end{aligned} \qquad (4.11)$$

where ρ_s is the mass density and a is a core radius.

In the situation in which a superfluid flows out of a tube of diameter d into an open space, a number of vortex lines are created. For speed v_s the number of vortex lines per centimeter is $v_s/(h/m)$, and the number rolling out per second is v_s times this number. The energy needed to create these vortex lines is

$$\frac{v_s^2}{(h/m)} \varepsilon_0(d).$$

On the other hand, the total kinetic energy of the fluid is $(\rho_s v_s^2/2)v_s d$ per second. Hence, if v_c is the velocity corresponding to the minimum energy needed to create the vortices,

$$\frac{\rho_s v_c^2}{2} v_c d = \frac{v_c^2}{(h/cm)} \varepsilon_0(d).$$

That is,

$$v_c = \frac{\hbar}{md} \ln(d/a), \tag{4.12}$$

with $h/m = \zeta_0$ for the minimum of ζ. This criterion is due to Feynman.

The above simple argument ignores the normal fluid and the creation of vortices and other details concerning the vortex motion and therefore it may deviate from experiment. For example, in experiments in which the normal component is clamped, the observed critical velocity follows [22]

$$v_c = cd^{-1/4}, \tag{4.13}$$

where $c \sim 1$ cm$^{5/4}$/s and d ranges from a few angstroms in unsaturated films to a few centimeters in a superfluid wind tunnel.

To describe superfluid flows Anderson [23] introduced the concept of *phase slippage*. The rate of this slippage is given by the Josephson frequency relation:

$$\frac{\hbar \partial(\phi_2 - \phi_1)}{\partial t} = \mu_1 - \mu_2 = \Delta\mu, \tag{4.14}$$

where the ϕ's and μ's are respectively the phases of the superfluid wave function and the chemical potentials, $\Delta\mu$ being the difference in the chemical potentials between two quiescent regions, 1 and 2. This equation may be viewed as a generalization of the Euler–Landau equation for an ideal fluid,

$$\frac{\partial \mathbf{v}_s}{\partial t} + \nabla\left(\frac{\mu}{m} + \frac{\mathbf{v}_s^2}{2}\right) = 0, \tag{4.15}$$

to the case in which quantized vorticity is created.

When the flow does not create vortices, Eqs. (4.14) and (4.15) are consistent with Eq. (4.5) in which $\mathbf{v}_s$ and ϕ are related. However, above a critical velocity v_c vortices may be created, and phase slippage takes place. For a phase slip of 2π the total work done on the stream moving at velocity v_c in a channel with cross-sectional area D is given by

$$\Delta E = \rho\zeta_0 D v_c. \tag{4.16}$$

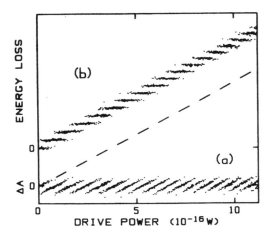

FIG. 9.8.   Phase slippages in liquid ^{4}He. (a) Peak amplitude change. (b) Energy loss in half a period. (From Avenel and Varoquaux [24])

This represents the kinetic energy loss, which is independent of the details of the mechanism for phase slippage.

Phase-slippage events have been observed by Avenel and Varoquaux [24]. They used a miniaturized low-frequency Helmholtz resonator of about 7 nm^3 internal volume immersed in a main superfluid bath. Highly reproducible isolated dissipation events were observed in the flow of pure superfluid ^{4}He through an aperture of submicron size at 10 mK, when the flow velocity exceeded a well-defined critical threshold. The system was driven on resonance by a sinusoidal voltage synchronized electronically to a SQUID (superconducting quantum interference device) output. A dc bias voltage was applied to linearize the excitation. Starting from the idle state, the peak amplitude of the oscillation was observed to grow linearly with time. At a critical level of oscillation this amplitude changed suddenly. As the drive level increased, ΔA followed the fishbone pattern shown in Fig. 9.8(a). This graph shows the peak amplitude-change from one half-period to another. The ratio of the amplitude change ΔA to the critical amplitude A_c is found to be constant and reproducible:

$$\frac{\Delta A}{A_c} = (0.0187 \pm 5)\%.$$

This sudden jump is associated with energy dissipation. It occurs at the maximum membrane velocity that corresponds to the critical velocity in the aperture. In Fig. 9.8, (b) represents the energy dissipation in a half-period. This graph corresponds to subtracting the dashed line from (a). That is, the energy loss is measured from that of the first fishbone pattern.

The critical volume flow is given by ΩA_c per unit hydraulic area, Ω being the angular frequency. With $\Omega/2\pi = 2.81$ Hz, and if the total hydraulic area is equal to the geometric area of the orifice, the energy change in Eq. (4.13) and the critical velocity are found to be given by

$$\Delta E = 1.01 \times 10^{-17} \text{ J}; \qquad v_c = 55 \text{ cm/s}. \tag{4.17}$$

These results show that phase slippages occur at a rate given by Anderson's relation and the critical velocity agrees with the value obtained from Feynman's criterion for $d = 0.3$ μm and $a = 1.3$ Å.

Despite these findings the slippage experiment involves some other aspects that require further study [25]. In actuality, the occurrence and breakdown of superflow are not completely understood. In contrast to the equilibrium case for which Landau's picture gives the basis, both Feynman's and Anderson's relations stem from simple energy-loss considerations without without detailed mechanisms. For many important experimental and theoretical works concerning superflow, the reader is referred to the review article by Langer and Reppy [26].

9.5. Sound in liquid helium

Liquid helium accommodates several sound modes depending on the conditions. In bulk helium, density fluctuations propagate as *first sound* and temperature fluctuations as *second sound*. The former occurs under adiabatic conditions with the normal and superfluid components moving in phase. The latter is observed when these two components are out of phase while the total density remains constant. In bulk helium, a mechanical wave (first sound) propagates approximately isothermally and a thermal wave (second sound) propagates approximately isobarically. Their velocities are related to each other at absolute zero by

$$c_2 = c_1/\sqrt{3}, \tag{5.1}$$

In thick films, first and second sound can still propagate. Their relation at absolute zero is

$$c_2 = c_1/\sqrt{2}. \tag{5.2}$$

As the film thickness decreases, the normal fluid component is clamped to the substrate more and more, and a surface wave of the superfluid component alone becomes observable. This is a longitudinal oscillatory motion of the superfluid component parallel to the substrate and is called *third sound*. When a superfluid is confined to narrow channels in packed powders or a superleak, the density oscillations can propagate as *fourth sound*. In the same geometry but under the pressure-released conditions, yet another mode, called *fifth sound*, is observed.

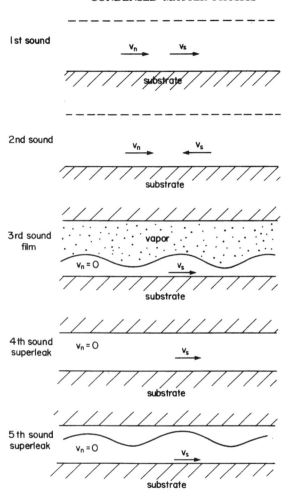

FIG. 9.9. Sound modes in liquid ^{4}He.

These sound modes are schematically illustrated in Fig. 9.9. In the propagation of third sound the motion of the superfluid component is mainly parallel to the wall and the normal component is stationary. Since the superfluid component carries no entropy, the accumulation of superfluid near a crest lowers the temperature there, whereas the increasing concentration of the normal component near a trough raises the temperature. Hence, strong temperature gradients are set up and a distillation of vapor from troughs to crests takes place. In fifth sound the average amplitude h must be much less than the mean free path of ^{4}He.

In a superleak the isothermal fourth sound is analogous to the first sound in bulk. Under adiabatic conditions, its velocity is given by

$$c_{4A}^2 = \frac{\rho_s}{\rho} c_1^2 + \frac{\rho_n}{\rho} c_2^2. \tag{5.3}$$

However, when second sound is suppressed only the first term can be used. In this case fourth sound depends directly on the superfluid component. The fifth sound velocity is related to c_2 by

$$c_5^2 = \frac{1}{\gamma} \frac{\rho_n}{\rho} c_2^2, \tag{5.4}$$

where γ is the specific-heat ratio.

All the sound velocities depend on temperature. They can be evaluated microscopically [27]. In units in which $\hbar = 1$ and $2m = 1$, the kinetic equation for the distribution function $f(\mathbf{q}, \mathbf{r}, t)$ of excitations is

$$\frac{\partial f}{\partial t} + [f, H] = 0, \tag{5.5}$$

where [   ] represents a Poisson bracket and

$$H = \varepsilon(q) + \mathbf{q} \cdot \mathbf{v}_s. \tag{5.6}$$

The equation of continuity is given by

$$\frac{\partial n}{\partial t} + \mathbf{V} \cdot \mathbf{j} = 0, \tag{5.7}$$

where the current density $\mathbf{j}$ is

$$\mathbf{j} = n\mathbf{v}_s + \frac{1}{(2\pi)^3} \int \mathbf{q} f \, d\mathbf{q}. \tag{5.8}$$

The equation of motion for the superfluid velocity $\mathbf{v}_s$ is

$$\frac{\partial \mathbf{v}_s}{\partial t} + \nabla \left(\mu + \frac{v_s^2}{2} \right) = 0, \tag{5.9}$$

where the chemical potential μ is given by

$$\mu = \mu_0 + \frac{\partial}{\partial n} \frac{1}{(2\pi)^3} \int \varepsilon f \, d\mathbf{q}. \tag{5.10}$$

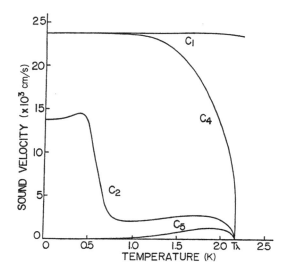

FIG. 9.10. Theoretical sound velocities as function of temperature. (From Oh et al. [27])

Here, μ_0 is the chemical potential at absolute zero. The normal fluid number density n_n is obtained from a consideration of the total momentum carried by the quasiparticles. With $\hbar$ and m restored, it is given by

$$n_n = \frac{1}{6\pi^2}\frac{\hbar^2}{2m}\int dq\, q^4\left(-\frac{\partial f_0(\varepsilon)}{\partial \varepsilon}\right) \quad \text{(bulk)}, \tag{5.11}$$

where $f_0(\varepsilon)$ is the equilibrium distribution. The total density consists of the normal and superfluid densities:

$$n = n_n + n_s. \tag{5.12}$$

The excitation energy spectrum and sound velocities can be evaluated based on a soft-potential model. Theoretical sound velocities for density $n = 2.18 \times 10^{-12}\,\text{Å}^{-3}$ are shown in Fig. 9.10 as functions of temperature [27]. The temperature dependence of the first sound velocity c_1 is relatively small, while that of second sound is more complex because it depends on the superfluid density. In contrast, c_4 and c_5 vary more smoothly toward zero at the λ-point.

Sound attenuation is an interesting aspect of sound propagation. The attenuation coefficient of first sound has a peak at around 1 K, and then diverges towards the λ-point. The effect of the anomalous energy dispersion on the 1 K peak has been discussed recently [28].

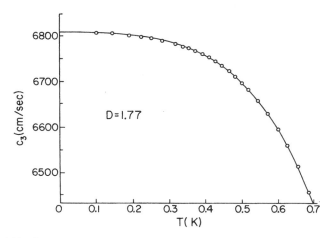

FIG. 9.11.   Temperature variation of third sound. Solid curve is theory [29]; points are data [30].

In films the superfluid component can accommodate density fluctuations while the normal fluid component is held on to the substrate. This type of propagation is called third sound. Its wavelength is longer than the film thickness. Its velocity is given by

$$c_3^2 = \frac{n_s \kappa}{m}, \tag{5.13}$$

where κ is the adiabatic eleastic constant, which is given by the second derivative of the energy with respect to the surface number density:

$$\kappa = \left(\frac{\partial^2 \varepsilon}{\partial n^2}\right)_s. \tag{5.14}$$

The normal fluid density in films is given by

$$n_n = \frac{1}{4\pi} \frac{\hbar^2}{2m} \int dq \, q^3 \left[-\frac{\partial f_0(\varepsilon)}{\partial \varepsilon} \right] \quad \text{(film)}, \tag{5.15}$$

in contrast to Eq. (5.11). The superfluid density n_s is then given by Eq. (5.12).

The velocity of third sound depends on the film coverages. The case of film coverage $D = 1.77$ in atomic layers is shown in Fig. 9.11 [29]. One atomic layer corresponds to 3.6 Å. The circles are data by Rutledge et al. [30], and the solid curve represents theoretical results obtained for a soft potential [29]. With this soft-potential model the temperature variation of the T^3 term of third sound can be explained in agreement with the data.

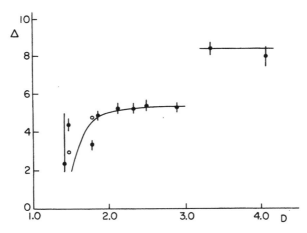

FIG. 9.12. Roton gap Δ as a function of film coverage D. Solid circles are data [30]; open circles are theory [29]. The solid curves represent an average variation.

The energy dispersion curve of a film is similar to that of bulk. However, the roton parameters can depend on film coverage D. The roton gap is given in Fig. 9.12. The solid circles represent the experimental values of Rutledge et al. The open circles represent theoretical values based on a soft-potential model. The curves smooth out the points. Figure 9.13 represents the roton momentum q_0. The symbols and curves are the same as in Fig. 9.12.

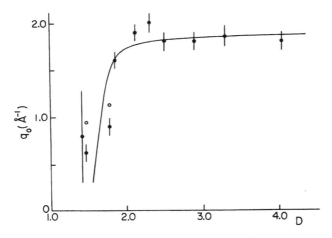

FIG. 9.13. Roton parameter q_0 as a function of film coverage D. The symbols and lines are the same as in Fig. 9.12.

In thin films with coverage D, the so-called *healing length* H must be introduced because helium layers immediately close to the substrate are immobilized by the van der Waals attraction. Hence, D is replaced by $D - D_0$, where an experimental value is

$$D_0 = 1.25 \pm 0.05.$$

The velocity of third sound decreases at first slowly and then more rapidly with increasing temperature. Since its square is proportional to the superfluid density, it is interesting to find how onset of superfluidity takes place in connection with the temperature variation of third sound. For bulk helium the onset of superfluidity is associated with the presence of long-range order. The existence of superfluidity in two dimensions suggests that long-range order is only a sufficient condition rather than a necessary condition.

According to the Kosterlitz–Thouless theory the 2D superfluid is characterized by bound vortex–antivortex pairs. At low temperatures isolated vortices cannot occur because their energy increases logarithmically with the size of the system. However, vortex pairs can have finite energy. These pairs do not destroy the topological long-range order of the system because the net vorticity is zero. At finite temperatures the phase of the order parameter varies with position but may be correlated from region to region. By a simple argument, Kosterlitz and Thouless derived a relation between the superfluid density at onset and the transition point. Subsequently Nelson and Kosterlitz [31] derived the following result:

$$\frac{\rho_s(T_0)}{T_0} = 3.52 \times 10^{-9} \text{ g/cm}^2 \text{ K}, \tag{5.16}$$

where the right-hand side is a universal constant. Experimentally, this relationship appears to be correct within 10% of accuracy. However, the Nelson–Kosterlitz theory does not specify the onset temperature itself.

Let us now try to derive a relation for T_0 and simultaneously the ratio $\rho_s(T_0)/T_0$ based on a simple consideration [32] of an anomalous dispersion of the excitation energy ε. We first note that in Eq. (5.11) the distribution function depends on ε/T so that the normal component density is a functional represented by $n_n[\varepsilon/T]$. If the distribution is of the form of a Bose distribution function of quasiparticles, we have

$$\frac{\partial n_n[\varepsilon/T]}{\partial \varepsilon} < 0. \tag{5.17}$$

We can always write $\varepsilon = \varepsilon_p + (\varepsilon - \varepsilon_p)$, where ε_p is the phonon energy. Since phonons are the lowest excitations, we assume that the onset temperature corresponds to the temperature at which

$$n = n_n[\varepsilon_p/T_0]. \tag{5.18}$$

Note that $n_n[\varepsilon_p/T_0]$ given by this equation may differ from the actual normal component density at T_0. Explicitly, this condition is expressed by

$$kT_0 = \left[\frac{8\pi nc^4 m}{3!\zeta(3)\hbar}\right]^{1/3} \hbar. \tag{5.19}$$

where $\zeta(s)$ is the Riemann zeta function. If a soft-potential model is used this relation becomes

$$kT_0 = n\left[\frac{4\pi(2\pi a^2 V_0)^2}{3!\zeta(3)2m\hbar}\right]^{1/3} \hbar. \tag{5.20}$$

In actual films consideration of a healing length requires use of a reduced density.

The superfluid component decreases with increasing temperature in accordance with

$$n_s[\varepsilon/T] = n - n_n[\varepsilon/T]. \tag{5.21}$$

At $T = T_0$,

$$n_s[\varepsilon/T_0] = n_n[\varepsilon_p/T_0] - n_n[\varepsilon/T_0]. \tag{5.22}$$

For the soft-potential model, this equation results in

$$n_s = kT_0 \frac{6!\zeta(5)}{[3!\zeta(3)]^{4/3}} \cdot \frac{2^{1/3}}{4\pi} \cdot \frac{1}{(a^2 V_0)^{4/3}} \left(\frac{\hbar^2}{2m}\right)^{1/3}. \tag{5.23}$$

For numerical results we note that the mass density corresponding to coverage D is given by $0.145 \times 3.6 \times 10^{-8} D$. With healing length $D_0 = 1.25$, and replacing D by $D - D_0$, we obtain

$$c^2 = c_0^2(D - D_0), \tag{5.24}$$

where

$$c_0^2 = 0.145 \times 3.6 \times 10^{-8} \pi(a^2 V_0)(m^*)^{-2} \left(\frac{cm}{s}\right)^2 D^{-1}. \tag{5.25}$$

Introducing such molecular parameters as $a = 2.65$ Å, $V_0 = 8.43$ K, and $m^* = 1.54m$ into Eq. (5.20), we arrive at

$$T_0 = 0.084(M - M_0) \quad \text{K}, \tag{5.26}$$

where M is the film thickness in $\mu mol/m^2$ and M_0 is the thickness of immobilized layers in the same unit. The superfluid density has been determined experimentally [33] using on third sound. According to experiment [34] $M_0 = 24.91$. For $M = 30$ we find

$$T_0 = 0.427 \text{ K}. \tag{5.27}$$

This agrees well with the experimental value of 0.43 K. Moreover, Eq. (5.18) yields

$$\frac{\rho_s(T_0)}{T_0} = 3.65 \times 10^{-9} \text{ g/cm}^2 \text{ K}. \tag{5.28}$$

This theoretical result is in accord with the experimental values of 3.42×10^{-9} g/cm^2 K and 3.86×10^{-9} g/cm^{-2} K for the glass and Grafoil substrates respectively [33].

While the numerical values are not much different, Eq. (5.28) has somewhat different physics from the Nelson–Kosterlitz relation in Eq. (5.16). The latter can be interpreted as the condition under which the pseudo-potential evaluated from the average separation between the superfluid atoms is equal to the thermal energy:

$$\frac{\hbar^2}{2m}\left(\frac{\rho_s}{m}\right) = kT_0. \tag{5.29}$$

On the other hand, Eq. (5.28) depends on the ratio of the potential energy to the pseudopotential $(\hbar^2/2m^*a^2)$ due to the hard core. Hence, it depends on the helium atoms as represented by the soft-potential model.

So far we have discussed only sound velocities. Sound attenuation is a topic of great importance. In fact, the anomalous character of the energy dispersion in liquid ^{4}He resulted from an analysis of the ultrasonic attenuation [35]. The attenuation coefficients of first and second sound can be evaluated explicitly as functions of temperature [36]. Near the λ-point dynamic scaling theory has successfully explained the critical attenuation of first sound, especially on the normal fluid side [37]. Scaling aspects of the critical attenuation in ^{3}He–^{4}He solutions will be discussed in Section 10.5.

REFERENCES

1. F. London, *Phys. Rev.* **54**, 947 (1938). J. D. Reppy, *Physica* **126B**, 335 (1984).
2. P. C. Hohenberg and P. M. Platzman, *Phys. Rev.* **152**, 198 (1966).
3. V. F. Sears, E. C. Svensson, E. C. Martel, and A. D. B. Woods, *Phys. Rev. Lett.* **49**, 279 (1982). H. A. Mook, *Phys. Rev. Lett.* **51**, 1454 (1983). S. Ikeda and N. Watanabe, *Physica A* **121**, 34 (1987).

4. O. Penrose and L. Onsager, *Phys. Rev.* **104**, 576 (1956). M. H. Kalos, M. A. Lee, P. A. Whitlock, and G. V. Chester, *Phys. Rev.* B **24**, 115 (1981). P. M. Lam and M. L. Rising, *Phys. Rev.* B **20**, 1960 (1979).

5. V. F. Sears, E. C. Svensson, P. Martel, and A. D. B. Woods, *Phys. Rev. Lett.* **49**, 279 (1982). H. A. Mook, *Phys. Rev. Lett.* **51**, 1454 (1983). L. J. Campbell, *Phys. Rev.* B **27**, 1913 (1983).

6. L. Landau, *J. Phys. USSR* **5**, 71 (1941), **11**, 91 (1947).

7. R. A. Cowley and A. D. B. Woods, *Can. J. Phys.* **49**, 177 (1971).

8. N. N. Bogoliubov, *J. Phys. USSR* **11**, 23 (1947). N. N. Bogoliubov and D. Zubarev, *Zh. Eksp. Teov. Fiz.* **28**, 129 (1955) [*Sov. Phys.* —*JETP* **1**, 83 (1955).]

9. D. Pines, in *Proc. Int. School of Physics "Enrico Fermi"* **21**, 147 (1963). D. Bohm and B. Salt, *Rev. Mod. Phys.* **39**, 894 (1967). T. Nishiyamna, *Prog. Theor. Phys.* **45**, 730 (1971). T. Kebukawa et al., *Prog. Theor. Phys.* **49**, 1802 (1973). F. Iwamoto, *Prog. Theor. Phys.* **44**, 1121 (1970).

10. R. P. Feynman, *Phys. Rev.* **94**, 312 (1954). R. P. Feynman and M. Cohen, *Phys. Rev.* **102**, 1189 (1956).

11. M. D. Girardeau, *J. Math. Phys.* **1**, 516 (1960); *Phys. Fluids* **5**, 1468 (1962).

12. T. D. Lee and C. N. Yang, *Phys. Rev.* **112**, 1419 (1958), T. D. Lee, K. Huang, and C. N. Yang, *Phys. Rev.* **106**, 1135 (1957).

13. A. Isihara and D. D. H. Yee, *Phys. Rev.* A **136**, 618 (1964). A. Isihara and T. Samulski, *Phys. Rev.* B **16**, 1969 (1977); *Physica* **86A**, 257 (1977). C. J. Nisteruk and A. Isihara, *Phys. Rev.* **154**, 1150 (1967). T. Morita and H. Hara, *Prog. Theor. Phys.* **41**, 60 (1969).

14. A. Isihara, S. T. Choh, W. H. Kang, and C. I. Um, *Physica* **100B**, 74 (1980).

15. A. Isihara, *Physica* **106B** & **C**, 161 (1981), **108**, 1385 (1981).

16. E. C. Svensson, V. F. Sears, A. D. B. Woods, and P. Martel, *Phys. Rev.* B **21**, 3638 (1981).

17. E. H. Graf, V. J. Minkiewicz, M. H. Bjerrum, and L. Passell, *Phys. Rev.* A **10**, 1748 (1974). O. W. Dietrich, E. H. Graf, C. H. Huang, and L. Passell, *Phys. Rev.* A **5**, 1377 (1972). A. C. Forbes and A. F. G. Wyatt, *Phys. Rev. Lett.* **64**, 1393 (1990).

18. F. Iwamoto, *Prog. Theor. Phys.* **44**, 1135 (1970). J. Ruvalds and A. Zawadowski, *Phys. Rev. Lett.* **25**, 333 (1970). K. Nagai, *Prog. Theor. Phys.* **49**, 46 (1973).

19. D. R. Allum, P. V. E. McClintock, A. Phillips, and R. M. Bowley, *Phil. Trans. Roy. Soc.* A **284**, 179 (1977). T. Ellis, C. I. Jewell, and P. V. E. McClintock, *Phys. Lett.* **78A**, 358 (1980).

20. L. Onsager, *Nuovo Cimento* Suppl. **6**, 249 (1949). R. P. Feynman, in *Progress in Low Temperature Physics I*, ed. by C. J. Gorter (North-Holland, Amsterdam, 1957), p. 17.

21. P. W. Karn, D. R. Starks, and W. Zimmermann, Jr., *Phys. Rev.* B **21**, 1797 (1980).

22. R. de Bruyn Ouboter, K. W. Taconis, and W. M. van Alphen, in *Progress in Low Temperature Physies V*, ed. by C. J. Gorter (North-Holland, Amsterdam, 1967), p. 44.

23. P. W. Anderson, *Rev. Mod. Phys.* **38**, 298 (1966).

24. O. Avenel and E. Varoquaux, *Phys. Rev. Lett.* **55**, 2704 (1985).

25. D. F. Brewer, *Phys. Rev. Lett.* **57**, 920 (1986). O. Avenel and E. Varoquaux, *Phys. Rev. Lett.* **57**, 921 (1986).

26. J. S. Langer and J. D. Reppy, in *Progress in Low Temperature Physies* 6, ed. by C. J. Gorter (North-Holland, Amsterdam, 1970), p. 1.

27. C. I. Um, W. H. Kahng, K. H. Yeon, S. T. Choh, and A. Isihara, *Phys. Rev. B* **29**, 5203 (1984). H. G. Oh, C. I. Um, W. H. Kahng, and A. Isihara, *Phys. Rev. B* **34**, 6151 (1986).
28. A. Isihara, C. I. Um, C. W. Chun, W. H. Kahng, and S. T. Choh, *Phys. Rev.* **37**, 7343 (1988). A. Isihara, *Phys. Rev.* **40**, 698 (1989).
29. A. Isihara, C. I. Um, W. H. Kahng, H. G. Oh, and S. T. Choh, *Phys. Rev. B* **28**, 2509 (1983).
30. J. E. Rutledge, W. L. McMillan, J. M. Mochel, and T. E. Washburn, *Phys. Rev. B* **18**, 2155 (1978).
31. D. R. Nelson and J. M. Kosterlitz, *Phys. Rev. Lett.* **39**, 1201 (1977).
32. A. Isihara, *Physica* **123B**, 207 (1984).
33. J. A. Roth, G. J. Jelatis, and J. D. Maynard, *Phys. Rev. B* **44**, 333 (1980). J. S. Brookds, F. M. Ellis, and R. B. Hallock, *Phys. Rev. Lett.* **40**, 240 (1978). I. Rudnick, *Phys. Rev. Lett.* **40**, 1454 (1978). R. Rosenbaum et al., *J. Low Temp. Phys.* **37**, 663 (1979). D. J. Bishop and J. D. Reppy, *Phys. Rev.* **167**, 1727 (1979).
34. D. J. Bishop, Thesis (Cornell University, 1978).
35. H. J. Maris and W. E. Massey, *Phys. Rev. Lett.* **25**, 220 (1970).
36. I. M. Khalatnikov, *Introduction to the Theory of Superfluidity* (Benjamin, New York, 1965). A. Isihara, *Phys. Rev. B* **40**, 69 (1989).
37. R. A. Ferrell, B. Mirhashem, and J. K. Bhattacharjee, *Phys. Rev. B* **35**, 4662 (1987).

10

LIQUID ^{3}He

Atmospheric ^{4}He gas contains usually a minute amount of isotopic ^{3}He atoms. Isotopic ^{3}He atoms were separated out in small amounts in the 1940s. Although it was thought earlier that liquefaction could not be done, a gas of ^{3}He was finally liquefied at Los Alamos in 1948. Subsequently, it became possible to obtain sufficient amounts of ^{3}He by making use of the nuclear reaction of neutrons in lithium and the β-decay of the reaction product tritium. Liquid ^{3}He attracted strong attention immediately because it is a Fermi liquid that is completely independent of electron systems and hence could be used as a new system to test Fermi liquid theory. It was soon discovered that this liquid has a linear specific heat around 30 mK, to everyone's delight. However, the story did not end with a mere confirmation of Fermi liquid theory. Since millikelvin temperatures have become accessible it has been found that the liquid is more complex and has fascinating properties. The discovery of superfluidity in the liquid is a triumph of condensed matter physics because superfluidity is considered to correspond to superconductivity of electrons.

10.1. Superfluidity

The linear specific heat and the increasing viscosity towards absolute zero provided convincing evidence of the role played by Fermi statistics in liquid ^{3}He. It was then conjectured that superfluidity took place by a mechanism similar to electron pairing in superconductivity. Thus, as low-temperature techniques were advanced by dilution refrigeration to the millikelvin range from around 1966, the race to discover superfluidity became intense. The pressure and NMR measurements performed by Osheroff et al. at Cornell in 1972 [1] indicated transitions into two new phases in the region of a few millikelvin. Around the same time, Wheatley and others [2] at La Jolla observed a reproducible pressure anomaly at constant volume that corresponded to one of Cornell's transition points. They also verified the other transition point. Subsequent fourth sound experiments at these laboratories clearly showed that in both phases the liquid was a superfluid.

A phase diagram of liquid ^{3}He is given in Fig. 10.1 [2]. This graph contains the data of several groups. The superfluid phases A and B are separated by the two T_{AB} lines. The white circles represent B → A transitions

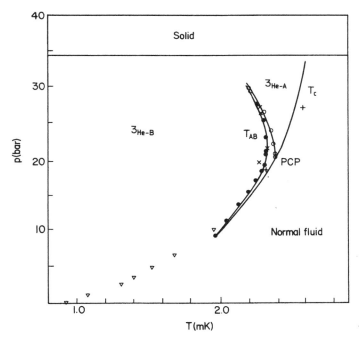

FIG. 10.1.   Phase diagram of ³He. (After Wheatley [2])

by heat flow in zero field obtained by Greytak et al. [2]. The point PCP is
the approximate location of the polycritical point. The solid circles are
magnetic thermodynamic A → B transitions in 378 G obtained by Paulson
et al. [2], and "exes" are those in $0.032T$ by Ahonen et al. [2]. The triangles
are the average of T_c and T_{AB}, where T_c is the second-order transition line.
The crosses represent second-order transitions observed by Ahonen et
al. [2].

In the normal liquid phase lowering the temperature or increasing the
pressure causes a second-order phase transition into the A phase. Further
decrease in temperature results in the B phase. The line T_{AB} divides the two
superfluid phases. In the presence of a small magnetic field, the T_c-line is
split into two, and in between another phase A_1 appears. The solid phase
occurs at high pressures.

Thus, the liquid is complex with five phases A, B, A_1, normal, and solid.
The A phase is sometimes called the ABM phase after Anderson, Brinkman,
and Morel [3]. The B phase is called the BW phase after Balian and
Werthamer [4]. The former is anisotropic and the latter is isotropic. The
orientation of the orbital state in the A phase has been probed by sound
attenuation. The B phase does not show such anisotropy.

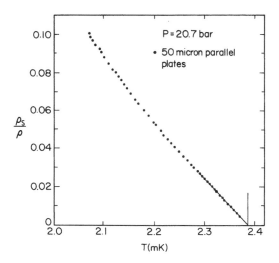

FIG. 10.2. Superfluid fraction in liquid ^{3}He. (From Kojima et al. [5])

The existence of superfluidity has been confirmed by fourth sound experiments because the velocity is proportional to ρ_s. The data due to Kojima et al. [5] in Fig. 10.2 show that the ratio ρ_s/ρ is proportional to $(1 - T/T_c)$ near T_c. This linear variation indicates that the transition is of BCS type.

In an ordinary diatomic molecule the state is spin singlet if the total angular momentum quantum number l is even and the total spin $S = 0$. It is spin triplet if l is odd and $S = 1$. Pairs in a spin singlet state are nonmagnetic, but in superfluid ^{3}He the susceptibility in the A phase is nearly same as that of the normal phase, and in the B phase it still exists although small. Hence, the two phases must be spin triplets and have odd l. In the phases A and A$_1$ liquid ^{3}He is about equally as magnetic as the normal liquid. When the temperature is reduced through T_{AB} into the B phase, the magnetism suddenly drops. In the B phase the magnetic susceptibility becomes independent of temperature as the temperature is reduced.

Hence, one can guess that opposite spin pairs form pairs in B. The corresponding spin wavefunction is a linear combination of

$$|\uparrow\uparrow\rangle; \quad |\downarrow\downarrow\rangle; \quad |\uparrow\downarrow\rangle + |\downarrow\uparrow\rangle.$$

One can say that in the B phase three superfluids are superimposed. On the other hand, in A only parallel spin pairs exist so that the wavefunction includes

$$\frac{1}{\sqrt{2}}\{|\uparrow\uparrow\rangle + |\downarrow\downarrow\rangle\}.$$

Hence, two interpenetrating superfluids with either $|\uparrow\uparrow\rangle$ or $|\downarrow\downarrow\rangle$ coexist. In A$_1$ only one of these two parallel spin pairs exists, because a magnetic field lines up the spins. The relative magnetization of the B phase is nearly universal when plotted against T/T_c, where T_c is the transition from A to normal liquid. This suggests that the B fluid has the same T_c as the A fluid.

The existence of the two interpenetrating superfluids in the A phase has been tested by experiment in a way analogous to a Josephson junction that couples two superconductors. If a voltage difference V is applied across a junction, the phase of the superconducting electron state on one side of the junction advances with respect to that on the other side at the rate $2\Delta\mu/h$, where $\Delta\mu = eV$ is the change in the chemical potential of a single electron advancing across the barrier. The factor 2 is introduced to represent tunneling of a pair of electrons.

In the superfluid A, $\uparrow\uparrow$ and $\downarrow\downarrow$ fluids are homogeneously coupled and not separated in space. However, if a magnetic field ΔH is suddenly applied to the liquid, the chemical potential of $\uparrow$ spins decreases by $\gamma\hbar\,\Delta H/2$ while that of $\downarrow$ spins increases by $\gamma\hbar\,\Delta H/2$. Thus, the chemical potential of a pair will increase by $2\gamma\hbar\,\Delta H$. The corresponding frequency is $2\delta\,\Delta H$. Thus, in units of $\gamma\,\Delta H$ the frequency can be expected to approach 2 near T_c.

In practice, the weak coupling of two interpenetrating superfluids causes oscillations in the magnetization when a small enough field change ΔH is applied. The corresponding frequency is called a *ringing frequency*. This frequency is proportional to $(1 - T/T_c)$. What is happening near T_c is a transfer of pairs from one superfluid to another, similar to the so-called Josephson plasma oscillation. Since the frequency $2\gamma\,\Delta H$ is temperature-independent, it becomes relatively large compared to the ringing frequency, which vanishes at T_c. Figure 10.3 shows [6] the ratio of the driven-mode ringing frequency f_R to $\gamma\,\Delta H/2\pi$ for both A and B superfluids in two geometries. Indeed, as $T \to T_c$, the ratio approaches 2.

As in the case of liquid ^{4}He shown in Fig. 9.8, phase slippage in superfluid ^{3}He-B has been observed. At low enough temperatures, periodic patterns in the shape of a staircase were observed by a miniature two-hole Helmholtz resonator with a submicron aperture. The observed individual phase slips at $0.69T_c$ are shown in Fig. 10.4 [7]. The main difference from the case of ^{4}He is the low value of the critical phase difference across the orifice. This phase difference decreases as the temperature becomes closer to T_c. The experimental results have been interpreted on the basis of the ac Josephson equation.

10.2. Spin dynamics

Superfluidity of liquid ^{3}He originates from pairing of two ^{3}He atoms as in the case of Cooper's state in superconductivity. A ^{3}He atom located at position $\mathbf{r}$ at time t will produce a polarization field at $\mathbf{r}'$ and t' in proportion to the susceptibility $\chi(\mathbf{r}' - \mathbf{r}, t' - t)$. In response to this field, a second atom

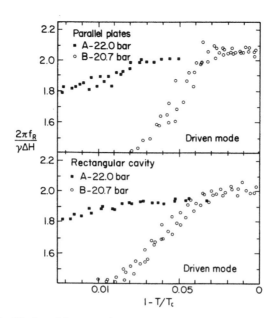

FIG. 10.3. Ringing of the magnetization. (From Wheatley [6])

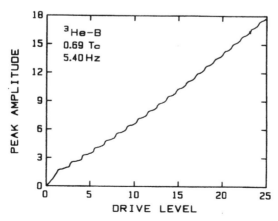

FIG. 10.4. Staircase pattern due to phase slippage in ^{3}He-B. (From Avenel and Varoquaux [7])

at $\mathbf{r}'$ and t' will return a polarization field to the first atom. Thus, the two atoms exchange virtual spin fluctuations and bind each other. An effective spin-dependent interaction between the two atoms may be expressed in the form

$$V(\mathbf{r}' - \mathbf{r}, t' - t) = -g^2 \chi(\mathbf{r}' - \mathbf{r}, t' - t) \boldsymbol{\sigma} \cdot \boldsymbol{\sigma}'. \qquad (2.1)$$

This interaction is attractive in the triplet state of pairs and is repulsive in the singlet state. Thus, this interaction favors odd angular momentum pairing.

The wavefunction $\psi(\mathbf{r} - \mathbf{r}'; \boldsymbol{\sigma}, \boldsymbol{\sigma}')$ of a pair of atoms can be Fourier-transformed into momentum space in which only the states at the Fermi surface are important for low temperatures. The transformed wavefunction will depend on a unit vector $\hat{n}$ on the Fermi surface so that its general form is $\tilde{\psi}(\hat{n}; \boldsymbol{\sigma}, \boldsymbol{\sigma}')$. We assume for a given $\hat{n}$ that the pair formation takes place in an eigenstate of spin projection zero. That is, if this eigenstate is represented by a vector $\mathbf{d}(\hat{n})$ in spin space and the total spin by $\mathbf{S}$,

$$\mathbf{d}(\hat{n}) \cdot \mathbf{S} \tilde{\psi}(\hat{n}; \boldsymbol{\sigma}, \boldsymbol{\sigma}') = 0. \qquad (2.2)$$

Note that $\mathbf{d}$ determines order in superfluid ³He.

Pair states $|\uparrow\uparrow\rangle$ and $|\downarrow\downarrow\rangle$ occur in the ABM. These are spin triplet states with odd angular momentum quantum number. A typical ABM state is represented by

$$d_x = d_A(n_x + in_y); \qquad d_y = d_z = 0. \qquad (2.3)$$

The corresponding orbital part is of the form $\sin \theta \, e^{i\phi}$ with reference to a particular axis.

On the other hand, in the isotropic BW state,

$$\mathbf{d} = d_B \hat{n}. \qquad (2.4)$$

Pairs are formed in this state with all three components of spin and orbital angular momentum combined in equal measure. Hence, the state represented by a linear combination of ³P₀, ³P₁, and ³P₂ has no net angular momentum because of the equal combination.

When two spins are involved, the BCS gap equation given by Eq. (2.9) in Section 8.2 must be generalized to a matrix equation:

$$\Delta(\mathbf{k}) = -\frac{1}{2} \sum_{\mathbf{k}'} V_{\mathbf{k}\mathbf{k}'} \frac{\Delta(\mathbf{k})}{E_{\mathbf{k}'}} \tanh(\tfrac{1}{2}\beta E_{\mathbf{k}'}), \qquad (2.5)$$

where the gap $\Delta(\mathbf{k}) = -\sum_{\mathbf{k}'} V_{\mathbf{k}\mathbf{k}'} \langle a_{\mathbf{k}\sigma} a_{-\mathbf{k}\sigma} \rangle$ has components determined by

$\sigma\sigma'$ and the quasiparticle energy E_k is given by

$$E_k^2 = \varepsilon_k^2 + \tfrac{1}{2}\mathrm{Tr}[\Delta^\dagger(k)\,\Delta(k)].$$ (2.6)

A term proportional to $\hat{k}\cdot\hat{k}'$ is most important in $V_{kk'}$ for p-wave pairing, where $\hat{\ }$ represents a unit vector. Hence, neglecting some other terms we write

$$V_{kk'} = -3V_{kk'}^*\,\hat{k}\cdot\hat{k}'.$$ (2.7)

$\Delta(k)$ is given by:

$$\Delta(k) = \Delta_A(\hat{k}_x + i\hat{k}_y)\begin{pmatrix} 1 & 0 \\ 0 & 1 \end{pmatrix} \qquad \text{(ABM)}, \tag{2.8}$$

$$\Delta(k) = \Delta_B\begin{pmatrix} -\hat{k}_x + i\hat{k}_y & \hat{k}_z \\ \hat{k}_z & \hat{k}_x + i\hat{k}_y \end{pmatrix} \qquad \text{(BW)}. \tag{2.9}$$

Equation (2.5) can be rewritten in the ordinary BCS form for $\Delta(k)$ and $V_{kk'}$ replaced by Δ and $-V_{kk}^*$ respectively. It has been shown that the BW state is the most stable one among the gap solutions for $V_{kk'}$ given by Eq. (2.7).

When the system deviates from equilibrium, interesting oscillations take place due to spin–orbit coupling, as shown by Leggett [8]. The spin dynamics developed by him and others has explained phenomena that are observed by NMR experiments in the liquid. In what follows we shall discuss only two of these phenomena: frequency shift and nonlinear oscillation.

Spin fluctuations taking place away from equilibrium cause changes in the orientation **d**, but a dipole energy produces a torque that tends to restore this vector back to the equilibrium direction. The dipole energy is commonly expressed as

$$H_D = \frac{1}{2}\gamma^2\hbar^2 \int\int d\mathbf{r}\,d\mathbf{r}' \left[\frac{\sigma(\mathbf{r})\cdot\sigma(\mathbf{r}')}{|\mathbf{r}-\mathbf{r}^-|^3} - \frac{3(\mathbf{r}-\mathbf{r}')\cdot\sigma(\mathbf{r})(\mathbf{r}-\mathbf{r}')\cdot\sigma(\mathbf{r}')}{|\mathbf{r}-\mathbf{r}'|^5} \right], \tag{2.10}$$

where γ is the gyromagnetic ratio.

However, in practice the above form is not entirely obvious for two quasiparticles. For this reason, Leggett introduced a renormalization factor R; expanding the spin operator $\sigma(\mathbf{r})$ in terms of spin density operators, introducing a renormalization factor, and using a certain approximation he expressed the dipole interaction energy between quasiparticles in the form

$$H_D = \tfrac{1}{2}\pi\gamma^2\hbar^2(g(0)\lambda)^2 R^2 \int \frac{d\Omega}{4\pi}\,[3|\hat{k}\cdot\mathbf{d}(k)|^2 - |\hat{d}(\hat{k})|^2], \tag{2.11}$$

where the integration is carried over the Fermi surface. $g(0)$ is the density of states at the Fermi surface, R is a renormalization factor, and

$$\lambda = \int dE(2E)^{-1} \tanh \beta E/2, \tag{2.12}$$

where the integration is carried out with a suitable cut-off. Further calculations yield the following excess orientational energies for the superfluid states:

$$\Delta H_D^{\text{ABM}} = -\tfrac{3}{5} g_D(T)(\hat{d} \cdot \hat{l})^2, \tag{2.13}$$

$$\Delta H_D^{\text{BW}} = \tfrac{4}{5} g_D(T)(\cos \theta + 2 \cos^2 \theta), \tag{2.14}$$

where $\hat{d}$ and $\hat{l}$ are spin and orbital unit vectors, θ is a rotational angle about the direction $\hat{n}$, and

$$g_D = \tfrac{1}{2} \pi \gamma^2 \hbar^2 (g(0)\lambda)^2 R^2 |\Delta|^2. \tag{2.15}$$

Δ is a gap. Near T_c, this parameter is estimated to be

$$g_D \sim 10^{-3}(1 - T/T_c) \text{ erg/cm}^3. \tag{2.16}$$

It is roughly equivalent to the energy due to susceptibility anisotropy. The dipole energy of the BW state is minimized at a magic angle:

$$\theta_0 = \cos^{-1}(-1/4) \sim 104°. \tag{2.17}$$

When $\mathbf{d}$ is disturbed out of equilibrium, the dipole torque $\mathbf{R}_D$ counteracts to restore equilibrium. This torque depends on the magnitude of an order parameter and is given by

$$\mathbf{R}_D = \frac{1}{4\pi} \int d\Omega \left(\mathbf{d} \times \frac{\partial H_D}{\partial \mathbf{d}} + \text{c.c.} \right). \tag{2.18}$$

For the A phase, this torque becomes

$$\mathbf{R}_D = \frac{24}{5} g_D(T)(\mathbf{d} \times l)(\mathbf{d} \cdot l). \tag{2.19}$$

Due to this torque, spin precession takes place. In a magnetic field $\mathbf{H}$, the equation of motion of the total spin $\mathbf{S}$ is

$$\frac{d\mathbf{S}}{dt} = \gamma \mathbf{S} \times \mathbf{H} + \mathbf{R}_D. \tag{2.20}$$

Instantaneously, S takes the equilibrium value $\chi H/\gamma$, where the susceptibility χ is assumed to be isotropic, but **d** precesses due to an effective field

$$\frac{d\mathbf{d}}{dt} = \mathbf{d} \times \gamma \left[\mathbf{H} - \frac{\gamma \mathbf{S}}{\chi} \right]. \tag{2.21}$$

Equations (2.10) and (2.21) are coupled, and the spin dynamics depends on their solutions. The resonance frequency in a transverse NMR experiment in the ABM state is given by

$$\omega^2 = \gamma^2 H^2 + \Omega_A^2, \tag{2.22}$$

where

$$\Omega_A^2 = \frac{6\gamma^2}{5\chi_A} g_D(T), \tag{2.23}$$

χ_A being the susceptibility. The longitudinal resonance occurs in this state at the same frequency. The last term in Eq. (2.22), due to the dipolar coupling, represents a frequency shift characteristic of the A phase. The characteristic frequency of the B phase is given by

$$\Omega_B^2 = \Omega_A^2 \frac{5}{2} \frac{\chi_A}{\chi_B} \tag{2.24}$$

under the assumption that the gap and therefore $g_D(T)$ is the same as in the A phase.

Figure 10.5 depicts the data of Gully et al [9] on the characteristic frequency Ω_A in comparison with the data of several others. Note that this

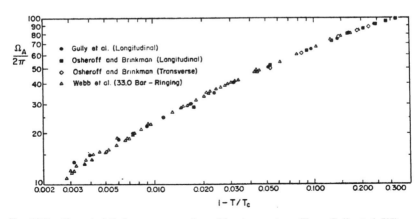

FIG. 10.5. Characteristic frequency near the melting temperature. (From Gully et al. [9]).

frequency varies approximately linearly with the abscissa. The maximum frequency is of order 120 kHz.

Let us now discuss nonlinear spin oscillations. If spins are disturbed, the orientation of **d** is caused to change relative to l. In an A liquid with **H** in the z direction, **S** can also be directed in the same direction even though its direction may change instantaneously. The orbital wavefunction and consequently the vector l are assumed not to change. According to Eq. (2.21), $d^2\mathbf{d}/dt^2$ depends on $d\mathbf{S}/dt$, which in turn is determined by Eq. (2.20) or $\mathbf{R}_D$. Therefore, the equation of motion of **d** about a fixed l is given on the average by

$$\frac{d^2\mathbf{d}}{dt^2} = -\Omega_A^2\, \mathbf{d} \sin\theta \cos\theta. \qquad (2.25)$$

That is, the motion follows the pendulum equation of motion. For very small angles, the oscillations are sinusoidal with angular frequency Ω_A, but in general they are nonlinear. Such nonlinear oscillations have been observed experimentally, in good agreement with theoretical considerations [10, 11].

10.3. Vortices

A rotational motion in superfluid ^{3}He is sustained by vortices as in superfluid ^{4}He. Superfluids cannot rotate as a whole without vortices. However, the ^{3}He superfluids are p-state paired-fermion condensates with a coherence length much longer than the interparticle distance, and the effect of hard-core repulsion on the Cooper pairing is large. The anisotropic pairing causes textures similar to those in liquid crystals. Moreover, the spin–magnetic field coupling results in complex vortex phenomena in the presence of an applied magnetic field. In fact, even in the isotropic B phase there are two different vortices with distinctively different vortex–core structures.

Vortex phenomena have been studied by NMR since 1981, and four stable and one metastable vortices have been observed in rotating superfluid ^{3}He. The metastable vortex depends on the cooling rate during rotation.

In the A phase continuous vortices with two circulation quanta and singular vortices with one circulation quantum have been observed. In the B phase there are two regions in the p–T plane with different stable vortex structures. Since the vortex phenomena in the A and B phases are different, we discuss them separately.

(1) *B phase.* In the B phase the vortex structure at intermediate magnetic fields are characterized by a distinct signature in the $\hat{n}$ texture which can be resolved in the NMR spectra. Two different vortex–core structures originate from the singular behavior of the order parameter.

In the B phase Cooper pairing occurs when the total angular momentum

$J = 0$. Hence, the order parameter is a solution of

$$\hat{J}A_{\alpha i} = 0, \tag{3.1}$$

where the amplitude $A_{\alpha i}$ of the order parameter, specified by the orbital and spin indices i and α, constitutes a 3×3 matrix. Note that the above property is preserved only under simultaneous orbital and spin rotations. That is, under a spin rotation $R_{\alpha\beta}$, A_{α_i} is transformed such that

$$A_{\alpha i} \rightarrow R_{\alpha\beta}A_{\beta i}, \tag{3.2}$$

The transformed order parameter belongs to a different state in the same B phase. Hence, both the original and transformed order parameters must be proportional to the energy gap of the B phase. More explicitly, the order parameter in the B phase can be expressed as

$$A_{\alpha i} = \Delta_B R_{\alpha i}(\hat{n}, \theta)e^{i\phi}. \tag{3.3}$$

Δ_B is the energy gap in the absence of a superflow and a magnetic field. $R_{\alpha i}(\hat{n}, \theta)$ is the αi element of the 3×3 orthogonal matrix R specified by angle θ about axis $\hat{n}$, its explicit form being

$$R_{\alpha i}(\hat{n}, \theta) = (1 - \cos\theta)n_\alpha n_i + \delta_{\alpha i}\cos\theta - \varepsilon_{\alpha i j}n_j\sin\theta. \tag{3.4}$$

The stable direction of θ is $104°$ due to dipolar forces, as discussed in the previous section. ϕ is a phase factor. The superfluid velocity is determined by this phase factor such that

$$\mathbf{v}_s = \left(\frac{\hbar}{2m_3}\right)\nabla\phi. \tag{3.5}$$

Thus, the flow around a vortex in the B phase is quantized in units of $h/2m_3$. Note the appearance of a factor 2 in the denominator.

In general, the vortices in ^{3}He superfluids depend on a magnetic field and rotation. When ^{3}He-B is subject to a magnetic field, a slight spin deviation from the direction perpendicular to H causes a quadratic orientational energy. It is given per unit volume by

$$F_H = -a(\mathbf{H}\cdot\hat{n})^2. \tag{3.6}$$

The parameter a is of order 10^{-12}.

Two additional energies arise in a rotating state when $\hat{n}$ deviates from $\hat{H}$; one is due to the superflow and the other to a vortex–core interaction. First, superflow causes a depopulation asymmetry with respect to the orbital pairing and tilts $\hat{n}$ so that the liquid is threaded by a lattice of vortices. If the rotation is specified by Ω the equilibrium density of vortices is determined by its magnitude:

$$n_v = \frac{2\Omega}{(h/2m_3)}. \tag{3.7}$$

Second, when the magnetic length exceeds the intervortex distance, a vortex–core interaction becomes important. The energies corresponding to these two mechanisms are essentially of the same functional dependency, and therefore can be parameterized together in the form

$$F_R = \tfrac{2}{5}a\lambda(\hat{\Omega}\cdot\vec{R}\cdot\mathbf{H})^2, \tag{3.8}$$

where λ is related to the flow (Ω) and vortex core and is given by

$$\lambda = \lambda_{\text{flow}} + \lambda_{\text{core}}. \tag{3.9}$$

On the other hand, the gyromagnetic energy is linear in magnetic field and can be expressed as

$$F_G = \tfrac{4}{5}a\kappa(\hat{\Omega}\cdot\vec{R}\cdot\mathbf{H}). \tag{3.10}$$

This energy originates from spin–orbit coupling. It corresponds to the small magnetization that compensates the circulating superflow around a vortex so that the original $J = 0$ orbital angular momentum state is retained. When the field or rotational direction is reversed, its sign change makes the energy distinctively different from the former two energies.

When the vector $\hat{n}$ deviates from its equilibrium direction, NMR can be used to probe the resonance frequency of the motion of $\hat{n}$ about $\mathbf{H}$. Figure 10.6 illustrates the lowest three spin-wave resonance frequencies measured at 29.3 bar in a magnetic field of 28.4 mT by Hakonen and Nummila [12]. The three lines on the right-hand side correspond to the stationary state, while the three lines on the left-hand side are observed in a rotating state at $\Omega = 1.4$ rad/s. These three lines abruptly change at $0.60T_c$, showing that splitting due to the direction of Ω has taken place. The open circles correspond to the case when $\mathbf{H}$ is parallel to Ω, while the solid circles represent the antiparallel case. The temperature T_c of the discontinuity is independent of Ω. Since there is hysteresis, the discontinuity is identified as a first-order phase transition.

This phase transition is mapped on the p–T plane as in Fig. 10.7. The open circles correspond to 28.4 mT while the filled circles correspond to 56.9 mT. The dashed line represents the transition points measured in persistent-current experiments in a powder-filled ac-gyroscope.

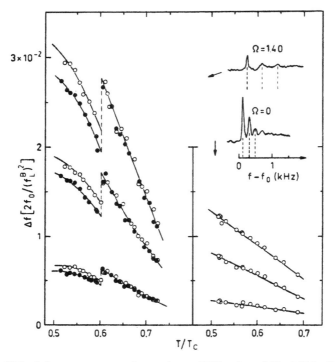

FIG. 10.6. Spin-wave resonance frequencies at 29.3 bar in a field of 28.4 mT. Right side: stationary. Left side: rotated at $\Omega = 1.40$ rad/s. Open and solid circles correspond respectively to **H** parallal and antiparallel to Ω. The inset is the measured NMR absorption spectrum at $T = 0.51 T_c$. (From Hakonen et al. [12])

The parameter λ in Eq. (3.8) is illustrated in Fig. 10.8 as a function of temperature near T_c for three pressures. At 15.5 bar (bottom curve) no jump is observed. At 17.1 bar two distinctive jumps occur. At 18.1 bar only one jump is found, but there is uncertainty near T_c due to scatter of the data points. The two jumps indicate that the vortex–core transition curve with a well-developed minimum is crossed. Note that different ordinates are used, as indicated by the arrows; the ordinate for 15.5 bar is on the left-hand side, and those for 17.1 and 18.1 bar on the right-hand side are shifted from each other.

(2) *A phase.* In the A phase continuous vortices with a soft core and a 4π circulation have been observed. However, it is difficult to generate vortices with a singular hard core. The normalized order parameter of the A phase is given by

$$A_{\alpha i} = d_\alpha(e_{1i} + ie_{2i})e^{i\phi}, \tag{3.11}$$

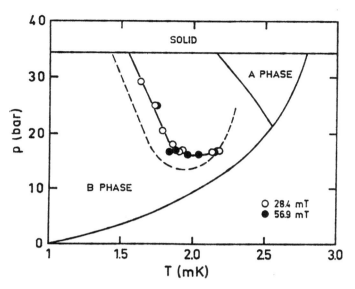

FIG. 10.7.   First-order vortex–core transition in rotating ³He-B. (From Krusius et al. [12])

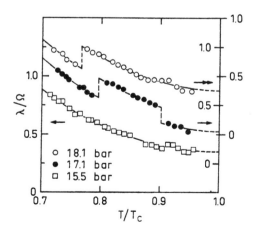

FIG. 10.8.   Behavior of λ/Ω close to the minimum pressure of the vortex–core transition curve. (From Krusius et al. [12])

where $d_\alpha \sim \Delta$ is in spin space, and e_1 and e_2 are the orbital orthonormal vectors that define the direction of the orbital angular momentum l of a Cooper pair:

$$l = e_1 \times e_2. \tag{3.12}$$

The superfluid velocity is given by

$$v_s = \frac{\hbar}{2m_3} (\nabla\phi + e_{1i}\nabla e_{2i}). \tag{3.13}$$

Hence, the texture in the A phase is orbital. The vorticity $\nabla \times v_s$ is determined by l such that [13]

$$(\nabla \times v_s)_z = \frac{\hbar}{2m_3} l \times \left[\frac{\partial l}{\partial \phi} \times \frac{\partial l}{\partial r}\right]. \tag{3.14}$$

Thus, bending of l will generate vortices continuously. It has been shown that each vortex has an even number of circulation quanta.

In a magnetic field, vortices with an odd number of circulation can be observed. In particular, for relatively weak fields and rotation a vortex lattice with circulation number $N = 1$ can be generated. At higher rotations, vortices with an even number circulation can be expected. The cases $N = 1$ and 2 have been identified experimentally.

According to the experiment of Krusius et al. [12] the NMR vortex line associated with a field of 28.4 mT is considerably reduced and broadened at $\Omega = 1.21$ rad/s. In addition, a small satelite peak appears immediately below the main broadened peak. This satellite peak is proportional to the rotational speed and therefore to the vortex density. It is nearly independent of temperature.

The frequency shift $v_v - v_0$ of the vortex line from the Larmor frequency v_0 given by Eq. (2.22) can be expressed as

$$v_v^2 - v_0^2 = R_T^2 [v_l^A(T)]^2. \tag{3.15}$$

Here

$$v_l^A = \frac{\Omega_A}{2\pi} \tag{3.16}$$

is the characteristic longitudinal resonance frequency of the A phase. In Eq. (3.15) R_T represents a fraction in reference to the total characteristic frequency.

Figure 10.9 represents the temperature dependence of R_T for a set of three tilted angles of the magnetic field [12]. The frequency shift can be calculated theoretically if the vortex type is specified. The arrows on the

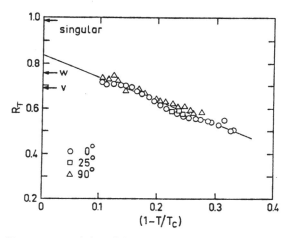

FIG. 10.9. Temperature variation of the vortex satellite frequency expressed by a fraction R_T from the total A-phase shift. (From Krusius et al [12])

graph represent theoretical values for the vortices specified as v, w, and singular [14]. When extrapolated, the data line comes close to these theoretical values, but it does not identify a particular vortex type. Theoretical values of spin-wave resonance frequencies depend in principle on the value of the dipolar coherence length ξ_D. Although for a given soft-core texture the lowest spin-wave eigenmode does not depend on ξ_D and thus Fig. 10.9 is not affected, ξ_D is 6.0 μm in the theoretical calculation.

10.4. Solid ³He

Due to the low mass and weak attractive forces ³He atoms solidify only under approximately 30 times atmospheric pressure. Solid ³He is a localized spin-1/2 Fermi system and is an ideal nuclear magnet in the sense that the characteristic ordering temperature is in the accessible millikelvin range. Nuclear spin ordering in this solid is caused by atomic exchange rather than magnetic dipole interactions. The strong zero-point motion allows the atoms to exchange positions. The thermodynamic properties at millikelvin temperatures are dominated by the nuclear spin contributions.

Magnetic resonance is a very powerful tool for probing nuclear spin ordering in the solid [15]. The resonance lines are very narrow, enabling precise measurements. It has been revealed that the solid is surprisingly complex. In fact, at around 0.4 mK, three pairs of resonance lines are observed, one below and the other above the Larmor frequency. These resonances have been associated with three separate magnetic domains.

Specific heat measurements have also been used very effectively to investigate the nature of the phase transitions in this solid [16]. The global

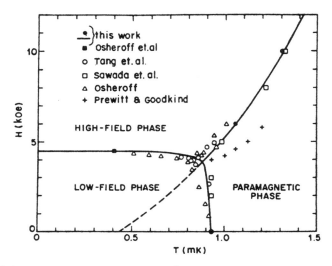

FIG. 10.10. Magnetic phase diagram of solid ³He. (After Greywall and Busch [16])

nature of the magnetic behavior can be seen in the phase diagram of Fig. 10.10 [16]. This graph includes several experimental data as indicated. On the high-temperature side there is a paramagnetic phase. This phase is bordered by the two distinct low- and high-field phases with a tricritical point at the corner near 0.4 T. Two other special fixed points are $T_N = 0.93$ mK at $H = 0$, and $H_{lh}(0.4 \text{ mK}) = 0.449$T. The transition between the two magnetic phases is second order. The bordering lines between these two phases in the limit $T \to 0$ are given by

$$H_{lh}(T) - H_{lh}(0) \propto T^4, \tag{4.1}$$

and in the limit $H \to 0$ by

$$T_N(H) - T_N(0) \propto H^2. \tag{4.2}$$

The transition between the paramagnetic and the high-field phase appears to be second-order.

Below the Néel temperature T_N of around 1 mK and for a field less than 0.4 T, the solid is ordered antiferromagnetically. This is the low-field phase. Although there is no direct confirmation, the spin arrangement shown in Fig. 10.11 has been proposed. In this spin arrangement called u2d2 (up-up and down-down), the ferromagnetic planes in the [100] direction of the sublattice shown in heavy lines alternate antiferromagnetically their spin directions. Another sublattice of similar spin arrangement is at the body-center position of the former, as shown by light lines. The spin axis $\hat{d}$ and the real-space axis

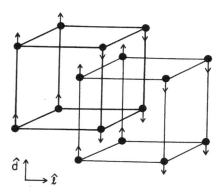

FIG. 10.11. Spin configuration in the lower-field phase of solid ^{3}He.

$\hat{l}$ are indicated in the figure. Note that there are longer and shorter distances for exchanges between atoms with parallel spins so that the Heisenberg Hamiltonian with a single parameter,

$$H = 2J \sum_{(ij)} \mathbf{s}_i \cdot \mathbf{s}_j,$$

is not suitable to describe the u2d2 spin configuration. Below T_N the specific heat shows a T^3 variation. This variation follows an empirical relation:

$$C/R = [0.43 + 0.003\,17H^3](T/T_N)^3. \tag{4.3}$$

The cubic variation is consistent with spin waves in a transverse antiferromagnet. The corresponding melting pressure follows

$$p(T) - p_A = 51.092 - AT^4, \tag{4.4}$$

where $A = 2.66$ is related to the spin-wave velocity. The spin-wave velocity is of order 8 cm/s. In the absence of a magnetic field, the transition into the paramagnetic phase is first-order. An experimental magnetic equation of state of the lower field phase is

$$M(T, H) - M(0, H) = 0.00065RT^4H^2/T_N(0)^3. \tag{4.5}$$

As the field exceeds 0.4 T a second-order phase transition into a high-field phase takes place. The magnetization is enhanced in this phase; at around 0.4 mK, the magnetization is already 0.6 of the saturation magnetization. Hence, the high-field phase is somewhat ferromagnetic. The transition into this phase is difficult to understand in terms of the above nearest-neighbor spin-$\frac{1}{2}$ Heisenberg model. Indications are that the high-field

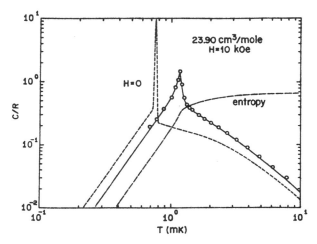

FIG. 10.12. Specific heat of solid ^{3}He at a molar volume of 23.90 cm^3. (After Greywall and Busch [16])

phase has a cubic symmetry. For molar volume V the specific heat appears to follow the empirical relation:

$$C(V, H) = 0.86(T/T_c(V, H))^3. \tag{4.6}$$

The specific heats in the two magnetic phases are compared in Fig. 10.12. The dashed curve corresponding to $H = 0$ has a sharp peak indicative of a first-order transition, while the solid curve for $H = 10\,\text{kOe}$ shows a λ-peak. Correspondingly, the entropy jumps in the former case while it does not in the latter case.

Although a single-parameter Heisenberg model is inadequate to describe solid ^{3}He, calculations with many exchanges are difficult. The spin configuration such as shown in Fig. 10.11 suggests a model with nearest-neighbor and next-nearest-neighbor exchanges. Note that the configuration includes several exchanges so that a dominant exchange loop must be specified. Although a classical mean-field theory has achieved some success, quantum fluctuations are expected to modify classical results significantly. A satisfactory quantum treatment is yet to be developed.

10.5. ^{3}He–^{4}He solutions

Dilute solutions of liquid ^{3}He in liquid ^{4}He are not ideal solutions and have complex properties. Their practical application to dilute refrigeration has made millikelvin temperatures accessible to experiment. The solutions are unique quantum mixtures with favorable experimental conditions such as the absence of impurities. In particular, the solutions provide ^{3}He Fermi

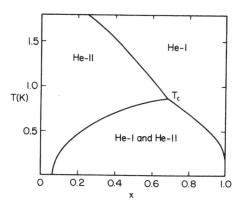

FIG. 10.13.   Phase diagram of ³He–⁴He solutions.

liquids with variable degeneracy and ⁴He Bose liquids with a λ-line instead of a λ-point.

As the most obvious effect, dissolving of ³He into liquid ⁴He causes a decrease in the λ-point. This decrease can be seen in the phase diagram of Fig. 10.13. The abscissa is mole fraction (x) of ³He at the saturated vapor pressure. The phases ⁴He-I and ⁴He-II are separated by the λ-line, and below the tricritical point T_c is the two-phase region. The λ transition in pure liquid ⁴He can now be studied with an advantageous extra dimension along the λ-line. Interestingly, this line is not straight but is curved.

For low concentrations ³He atoms can be treated as free quasiparticles moving in the background of liquid ⁴He. In fact, the specific heat of ³He varies linearly with temperature for low concentrations. As Landau and Pomeranchuk pointed out [17], the corresponding energy spectrum is given by

$$\varepsilon = -\varepsilon_0 + \frac{p^2}{2m^*},\qquad(5.1)$$

where ε_0 represents the depth of the potential well and m^* is an effective mass. At nominal concentrations ³He atoms constitute a normal Fermi liquid.

As their concentration increases the interection between ³He quasi-particles becomes important and causes deviations from the above description. Although only weakly, ³He quasiparticles do attact each other, as indicated by the phase separation. Two effects contribute to their effective interaction. First, a pair of ³He atoms may attract each other by exchanging a phonon in liquid ⁴He. At very low temperatures, a ³He atom does not excite a real phonon but induces density fluctuations as it travels through the solution. Equivalently, it excites virtual phonons. A second ³He atom will be scattered by the density fluctuations, leading to an effective interaction. Second, two ^{3}He atoms can also have a direct interaction because of the mass difference between ^{3}He and ^{4}He atoms. The force fields of the ^{3}He and ^{4}He atoms are similar, but the lighter mass of the former produces in the liquid a slightly larger volume than the latter. Hence, a ^{3}He atom will feel the presence of another ^{3}He that replaces a ^{4}He atom. This type of interaction is primarily repulsive, resulting in a decrease in the phonon-mediated attractive interaction.

In the so-called BBP theory Bardeen et al. (18) derived an effective interaction for the above two types of interaction in the long-wavelength limit. The effective interaction arises as the ^{3}He atom moves through the liquid, displacing ^{4}He atoms. The ^{4}He, driven by local changes in the chemical potential μ_4, moves but tries to maintain μ_4.

Thus, the effective interaction between two ^{3}He atoms at long wavelengths is simply related to the net change in energy of the system. The effective interaction V_0 in the long-wavelength limit is given by the change in the chemical potential $\mu_{3\uparrow}$ of up-spin ^{3}He when the density $n_{3\downarrow}$ of down-spin ^{3}He is changed. However, since $\mu_{3\uparrow}$ depends on $n_{3\downarrow}$ and n_4, this change can be evaluated first at constant n_4 and second via n_4:

$$V_0 = \left(\frac{\partial \mu_{3\uparrow}}{\partial n_{3\downarrow}}\right)_{n_4} + \left(\frac{\partial \mu_{3\uparrow}}{\partial n_4}\right)_{n_3}\left(\frac{\partial n_4}{\partial n_{3\downarrow}}\right)_{\mu_4}. \tag{5.2}$$

The first term on the right-hand side corresponds to the direction interaction when no density change in the background ^{4}He liquid takes place. The second term involves ^{4}He density changes and corresponds to the phonon-exchange type effective interaction. Both of these terms have a similar form except that the former is proportional to $(1 + 2\alpha)$ and the latter $-(1 + \alpha)^2$, where

$$\alpha \sim 0.28 \tag{5.3}$$

is the relative increase in volume resulting from the replacement of a ^{4}He by a ^{3}He atom. Hence, if c is the sound velocity in pure liquid ^{4}He, the effective potential is given by

$$V_0 = -\frac{\alpha^2 m_4 c^2}{n_4}. \tag{5.4}$$

For dilute solutions, a linear form of the ground-state energy may be assumed such that

$$\varepsilon(n) = \varepsilon_0(n) + n_3 \varepsilon_1(n), \tag{5.5}$$

where $\varepsilon_0(n)$ is the ground-state energy per volume of pure ^{4}He at density n, and

$$\varepsilon_1(n) = \langle p^2/6m_4 \rangle \tag{5.6}$$

represents one-third the average kinetic energy per particle in pure ^{4}He. Noting

$$\mu_3 - \mu_4 = \varepsilon_1,$$

one can confirm that V_0 is given by Eq. (5.4). At the same time, one finds

$$\alpha = \frac{n_4}{m_4 c^2} \frac{\partial}{\partial n_4} \left\langle \frac{p^2}{6m_4} \right\rangle. \tag{5.7}$$

If each ^{4}He is replaced by a hard sphere of diameter d and if a is an average nearest-neighbor distance, the hard sphere moves as a point particle in a shell of radius $(a-d)$. Hence, for a rough estimate of α, one can use

$$\langle p^2/2m_4 \rangle = \frac{\pi^2 \hbar^2}{[2m_4(a - d)^2]}. \tag{5.8}$$

Thus,

$$\alpha \sim \left(\frac{\pi \hbar}{3m_4 ca} \right)^2 \left(1 - \frac{d}{a} \right)^{-3}. \tag{5.9}$$

For $d = 2$ Å and $a = 3.8$ Å, this expression gives 0.31, which is not very far from the experimental value of 0.28.

For small but finite values of wavenumber q, the effective potential is expected to vary with q. This variation is important particularly for transport phenomena, which depend on scattering. Note that the above consideration does not specify the momentum dependence of the effective potential. BBP assumed a quadratic variation for small q:

$$V_q = -V_0[1 - bq^2]. \tag{5.10}$$

Since the BBP theory involves phenomenological assumptions, its microscopic justification has been attempted in several ways. In principle, the effective mass of interaction depends on just one ^{3}He atom or two placed in liquid ^{4}He, enabling quantum-mechanical approaches [19]. However, for the thermodynamic properties of the solutions, including the chemical potential, statistical mechanical treatments with a well-defined Hamiltonian for arbitrary concentrations of ^{3}He are desirable.

The model of hard-sphere mixtures is well-defined but does not reproduce the thermodynamic properties correctly. However, with a soft-core potential model [20] the chemical potential can be derived in the form of the BBP theory. This model produces α and m_3^* that depend on the range of interactions. The average theoretical value of the effective mass m_3^* is around 2.6 in comparison with an experimental value of 2.5. For similar interaction potentials among and between the two species, the osmotic pressure π is given by

$$\pi = \frac{2}{5}\varepsilon_F n_3 - \frac{1}{2}\alpha^2 n_3^3 \frac{m_4 c^2}{\bar{n}_4}, \tag{5.11}$$

where ε_F is the ideal Fermi energy of ^{3}He atoms, α is a renormalized volume expansion coefficient, and c is the sound velocity in pure liquid ^{4}He with density $\bar{n}_4$. The osmotic pressure is dominated by the first ideal term, which is proportional to $n_3^{n/3}$. The second term is a correction. There are other microscopic theories [21] with different expressions for the osmotic pressure. However, experimentally, deviations from the $x^{5/3}$ proportionality of π have been found to be small [22], so that an experimental comparison of various theoretical formulas is difficult.

A soft-core model yields the elementary excitations in the solutions such that

$$\varepsilon(q) = \frac{1 + y(1 - \delta m/2)}{1 + y}\frac{\hbar^2 q^2}{2m_4^*}[q^2 + \gamma(1 + y)]^{1/2}, \tag{5.12}$$

where

$$y = \frac{m_3^* n_3}{m_4^* n_4}; \qquad \delta m = \frac{2(m_3^* - m_4^*)}{m_4^*}; \qquad \gamma^2 = \frac{4m_4^* n_4 u(q)}{\hbar^2}. \tag{5.13}$$

$u(q)$ is the Fourier transform of the interaction potential of ^{4}He quasiparticles with effective mass m_4^*, and m_3^* is the effective mass of ^{3}He quasiparticles.

For $\alpha = 0.28$ and the effective masses given by

$$m_4^* = 1.69 m_3; \qquad m_3^* = 2.4 m_3,$$

Equation (5.12) yields the sound velocity as a function of ^{3}He fraction x:

$$c(x) = c(1 + 0.44x)^{-1/2}. \tag{5.14}$$

Here, $c = 238$ m/s and x is the ^{3}He fraction. For $x = 5\%$ and 13.4%, this formula yields 236.4 and 232.2 m/s in comparison with the experimental values of 236.2 and 231.5 m/s. The ratio of the sound velocities in the ^{3}He and ^{4}He liquids is

$$\frac{c_3}{c} = \left(\frac{m_4^* \bar{n}_3}{m_3 \bar{n}_4}\right)^{1/2}. \tag{5.15}$$

The ratio $\bar{n}_3/\bar{n}_4$ of the number of densities in the pure liquids is $1/1.362$ so that

$$\frac{c_3}{c} = (1.67)^{-1/2} = 0.773. \tag{5.16}$$

This compares favorably with the experimental ratio of 0.765.

The solutions show a variety of interesting nonequilibrium phenomena [23]. In general, on the normal side of T_λ, scaling properties are found, but on the superfluid side deviations from scaling behavior takes place. For instance, the attentuation coefficient α of sound is expected to be a function of $\omega\tau$, where ω is the frequency and τ is a characteristic time. If τ scales as $|T - T_\lambda|^{-s}$ with a parameter s, a plot of α as a function of ω/ε^s should show a scaling behavior. This is indeed so above T_λ, as shown in Fig. 10.14(a)

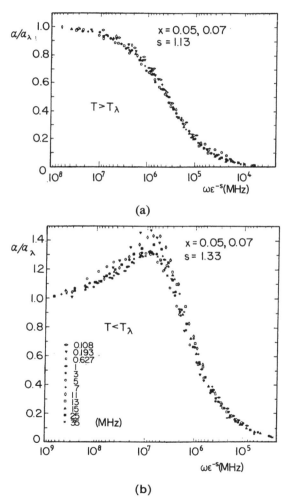

(a)

(b)

FIG. 10.14. Scaling behavior of the attenuation coefficient of sound in ^{3}He–^{4}He solutions at ^{3}He fractions 0.05 and 0.07. (a) $T > T_\lambda$; (b) $T < T_\lambda$. The graphs include several authors' data. $\varepsilon = |T - T_\lambda|/T_\lambda$. (From Ikushima [23], Roe et al. [24]).

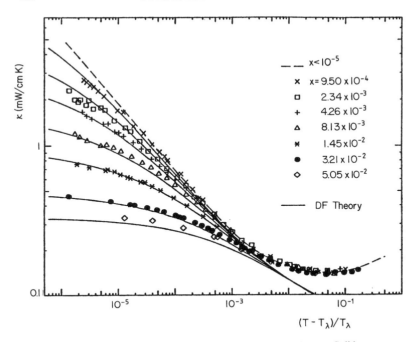

FIG. 10.15. Thermal conductivity for $T \to T_\lambda$ for seven dilute mixtures. Solid curves represent theory by Dohm and Folk [25]. (From Meyer [23])

which includes several sources of data [24]. However, the behavior below T_λ is different; the same plot shows a maximum around which scattering is large, as shown in Fig. 10.14(b).

The difference between ^{3}He solutions and pure liquid ^{4}He is most clearly seen in the behavior of the thermal conductivity κ; at finite concentration of solutions, κ does not diverge at T_λ, but it does in pure liquid ^{4}He. The variation of T_λ with the concentration of ^{3}He and the existence of the tricritical point make the solutions unique in testing critical dynamics [25].

On the normal fluid side of the transition the dynamic scaling theory reproduces well the data of Meyer [23] as shown in Fig. 10.15. However, there is no theory at present for high concentrations. On the superfluid side of the transition the experimental κ does not seem to show a singularity at finite concentrations. In the limit of vanishing ^{3}He concentration κ is expected to diverge. Theoretically, this dependence may follow

$$\kappa \sim x^{-1}. \tag{5.17}$$

Although such a dependence is observed in the range $10^{-2} \gtrsim x \gtrsim 5 \times 10^{-4}$, a much weaker divergence has been observed at lower concentrations as

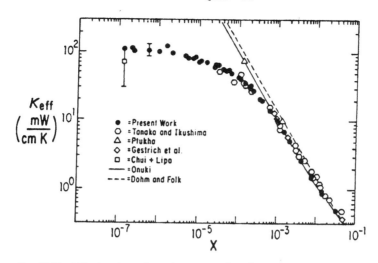

FIG. 10.16. Effective thermal conductivity of ^{3}He–^{4}He solutions as a function ^{3}He mole fraction x at $|T - T_\lambda|/T_\lambda = |\varepsilon| = 10^{-3}$. Key indicates various sources of data. The solid and dashed lines are theoretical. (From Dingus et al. [24])

shown in Fig. 10.16. The dashed and solid lines represent the theoretical results by Dohm and Folk and by Onuki respectively [25]. However, since as x is reduced the linear conduction region decreases, experimental measurements become increasingly difficult.

REFERENCES

1. D. D. Osheroff, R. C. Richardson, and D. M. Lee, *Phys. Rev. Lett.* **28**, 885 (1972). D. D. Osheroff et al. *Phys. Rev. Lett.* **29**, 920 (1972). H. Kojima, D. N. Paulson, and J. C. Wheatley, *Phys. Rev. Lett.* **32**, 141 (1974).

2. J. C. Wheatley, *Rev. Mod. Phys.* **47**, 415 (1975). T. J. Greytak, R. T. Johnson, D. N. Paulson, and J. C. Wheatley, *Phys. Rev. Lett.* **31**, 452 (1973). D. N. Paulson, H. Kojima, and J. C. Wheatley, *Phys. Rev. Lett.* **32**, 1098 (1974). A. I. Ahonen, M. T. Haikala, and M. Krusius, *Phys. Lett.* **47A**, 215 (1974).

3. P. W. Anderson and P. Morel, *Phys. Rev.* **123**. 1911 (1961). P. W. Anderson and W. F. Brinkman, *Phys. Rev. Lett.* **30**, 1108 (1973).

4. R. Balian and N. R. Werthamer, *Phys. Rev.* **131**, 1553 (1963).

5. H. Kojima, D. N. Paulson, and J. C. Wheatley, *Phys. Rev. Lett.* **32**, 141 (1964).

6. John C. Wheatley, *Proc. LT-14*, Vol. 5, ed. by Matti Krusius and Matti Vuorio (North-Holland, Amsterdam, 1975), p. 6.

7. O. Avenel and E. Varoquaux, *Jpn J. Appl. Phys.* **26**, Suppl. 26-3 [*Proc. Int. Conf. LT-18* 1798 (1987)].

8. A. J. Leggett, *Rev. Mod. Phys.* **47**, 331 (1975). W. F. Brinkman and M. C. Cross, in *Progress in Low Temperature Physics VIIA*, ed. by D. F. Brewer (North-Holland, Amsterdam, 1978), p. 105.

9. W. J. Gully, C. M. Gould, R. C. Richardson, and D. M. Lee, *J. Low Temp. Phys.* **24**, 563 (1976).
10. R. A. Webb, R. L. Kleinberg, and J. C. Wheatley, *Phys. Lett.* **48A**, 421 (1974). K. Maki and T. Tsuneto, *Prog. Theor. Phys.* **52**, 773 (1974).
11. D. D. Osheroff and L. R. Corruccini, *Phys. Lett.* **51A**, 447 (1975). W. F. Brinkman and H. Smith, *Phys. Lett.* **51A**, 449 (1975). H. Smith, W. F. Brinkman, and S. Engelberg, *Phys. Rev.* **B 15**, 199 (1977).
12. M. Krusius, P. J. Hakonen, and J. T. Simola, *Physica* **126B**, 22 (1984). P. J. Hakonen et al., *Phys. Rev. Lett.* **55**, 1362 (1983). P. J. Hakonen and K. K. Nummila, *Jpn J. Appl. Phys.* **26**, Suppl. **26-3**, 1814 (1987).
13. N. D. Mermin and T.-L. Ho, *Phys. Rev. Lett.* **36**, 594 (1976).
14. H. K. Seppälä and G. E. Hakonen, *J. Low Temp. Phys.* **51**, 279 (1983).
15. E. D. Adams, E. A. Schbert, G. E. Haas, and D. M. Bakalyar, *Phys. Rev. Lett.* **44**, 789 (1980). D. D. Osheroff, M. C. Cross, and D. S. Fisher, *Phys. Rev. Lett.* **44**, 792 (1980). M. C. Cross and D. S. Fischer, *Rev. Mod. Phys.* **57**, 881 (1985).
16. D. S. Greywall and P. A. Busch, *Phys. Rev.* **B 36**, 6853 (1987).
17. L. D. Landau and I. Pomeranchuk, *Dokl. Nauk, SSSR* **59**, 669 (1948).
18. J. Bardeen, G. Baym, and D. Pines, *Phys. Rev.* **156**. 27 (1967).
19. W. F. Saam, *Ann. Phys.* **53**, 219 (1969). G. Baym, *J. Low Temp. Phys.* **18**, 335 (1975).
20. A. Isihara and D. Y. Kojima, *Z. Phys.* **B 37**, 1 (1980); *Physica* **103B**, 247 (1981); *Physica* **94B & C**, 391 (1978).
21. C. Ebner and D. O. Edwards, *Phys. Rep.* **2C**, 77 (1971). Y. Disatonik and H. Brucker, *J. Low Temp. Phys.* **7**, 491 (1972).
22. J. Landau, J. T. Tough, N. R. Brubaker, and D. O. Edwards, *Phys. Rev. Lett.* **23**, 283 (1969).
23. See for instance: G. Ahlers, in *The Physics of Liquid and Solid Helium*, Part I, ed. by K. H. Bennemann and J. B. Ketterson (Wiley, New York, 1976). A. Ikushima, *Jpn J. Appl. Phys.* **19**, 2315 (1980). H. Meyer, *J. Low Temp. Phys.* **70**, 219 (1988). C. Ebner and D. O. Edwards, *Phys. Rep.* **2C**, 77 (1971).
24. M. Tanaka, A. Ikushima, and K. Kawasaki, *Phys. Lett.* **61A**, 119 (1977). D. B. Roe, H. Meyer, and A. Ikushima, *J. Low Temp. Phys.* **32**, 67 (1978). C. Buchal and F. Pobell, *Phys. Rev.* **B 14**, 1103 (1976). M. Dingus, F. Zhong, J. Tuttle, and H. Meyer, *J. Low Temp. Phys.* **65**, 213 (1986). F. Zhong, J. Tuttle, and H. Meyer, *J. Low Temp. Phys.* **79**, 9 (1990). D. Williams and I. Rudnik, *Phys. Rev. Lett.* **25**, 276 (1970).
25. K. Kawasaki, in *Phase Transitions and Critical Phenomena*, Vol. 5A, ed. by C. Domb and M. S. Green (Academic Press, New York, 1976), p. 165. E. D. Siggia and D. R. Nelson, *Phys. Rev.* **B 15**, 1427 (1977). V. Dohm and R. Folk, *Phys. Rev.* **B 28**, 1332 (1983). A. Onuki, *J. Low Temp. Phys.* **53**, 189 (1983).

11

LIQUID CRYSTALS

It has been known since approximately one hundred years ago that certain liquids exhibit peculiar crystal-like anisotropy while maintaining fluidity. They are called *liquid crystals* although they are not crystals but rather anisotropic liquids. Today some thousands of liquid crystals have been identified. They are liquids, either molten or dissolved, consisting of anisotropic molecules with various orientational or spatial ordering. Due to the presence of several new phases that are between solids and liquids and to relations with phase transitions in other systems, they are attracting strong attention. In addition, their temperature- or field-sensitive phase changes are utilized industrially for display devices.

11.1. Classification

In general, liquid crystals are isotropic liquids at high temperatures and as the temperature is reduced exhibit phase transitions into anisotropic phases. The liquid crystals consist of geometrically asymmetric, often long rodlike molecules, and the molecular distribution function becomes anisotropic in the new phases characteristics of liquid crystals. Typically, there are three anisotropic phases. They are called *nematic*, *smectic*, and *cholesteric* phases. Figure 11.1 illustrates the distribution of rodlike molecules in isotropic and anisotropic phases.

  In the nematic phase these molecules have an orientational order but there is no positional order. In the smectic phase an additional spatial order appears. In the cholesteric phase the molecular orientation in a plane rotates spirally in a direction perpendicular to the plane. If both ends of the molecule are the same, the pitch of the cholesteric molecular rotation is just one-half of the spiral period. The pitch, which is sensitive to environmental factors such as temperature, flow, or applied field, is comparable to the wavelength of visible light. Hence, Bragg reflection causes peculiar colors characteristic of the cholesteric phase. Note that the nematic phase represents a cholesteric phase of infinite pitch.

  The term "nematic" comes from the Greek *nematos*—threadlike. Such a texture is visible between crossed polaroids. The word "smectic" means soaplike, which originates from the Greek *smectos*. Among several smectic modifications, the molecular arrangements shown in Fig. 11.2 are typical.

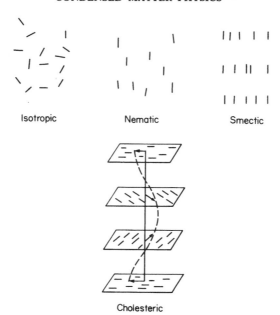

FIG. 11.1. Four basic phases of liquid crystals.

In smectic-A, the molecular orientation is perpendicular to the molecular layers. In smectic-C, it is inclined. In smectic-B, a hexagonal molecular arrangement is seen. However, in contrast to a crystal, there is no shear modulus in smectic-B, and the layers slip easily on each other. There are several other arrangements, including smectic-D with a cubic structure.

Cholesterol or other sterol molecules are the majority that form the cholesteric phase. In view of the existence of many molecules that form liquid crystals and their long names with often complicated chemical structures, we shall not present here a detailed discussion of their molecular structures, but remark only that there is a relatively simple family of liquid crystals with

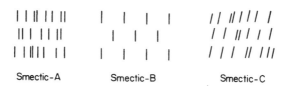

FIG. 11.2. Typical smectic subphases.

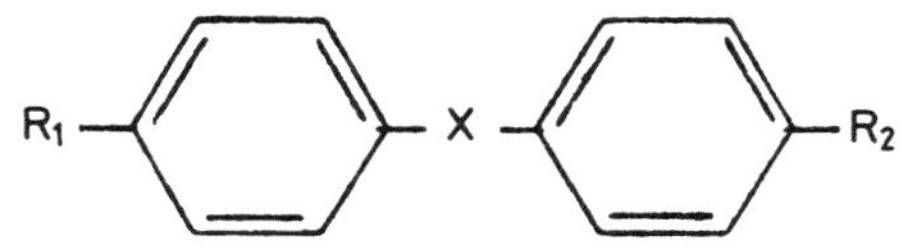

FIG. 11.3. A simple liquid crystal molecule.

formula shown in Fig. 11.3. Here, R_1 and R_2 may or may not be the same. For instance, $R_1 = R_2$ can be simple alkyl chains or alkoxy chains:

$$-C_nH_{2n+1}; \qquad -OC_nH_{2n+1}.$$

X can be the base:

$$-CH{=}N-$$

or compounds such as

azoxy: $-N{=}N;$ ester: $-CO-O-$
                $\downarrow$
                O

The molecular structures of two typical liquid crystals PAA (paraazoxy-anisole) and MBBA (2-p-methoxybenzylidene n-butylaniline) are shown in Fig. 11.4. The properties of liquid crystals depend on the chemical structure. The dipole moment, the polarizability, and the rigidity of the molecule are important factors in determining the properties. The molecular shape is mostly rodlike but can be flat disklike.

From the high-temperature side, sequences of phase transitions such as

Isotropic → nematic → solid
Isotropic → nematic → smectic → crystal
Isotropic → cholesteric → solid

are observed. Generally, not all the phases are exhibited by a single liquid crystal, but there can be phase transitions within the smectic phase.

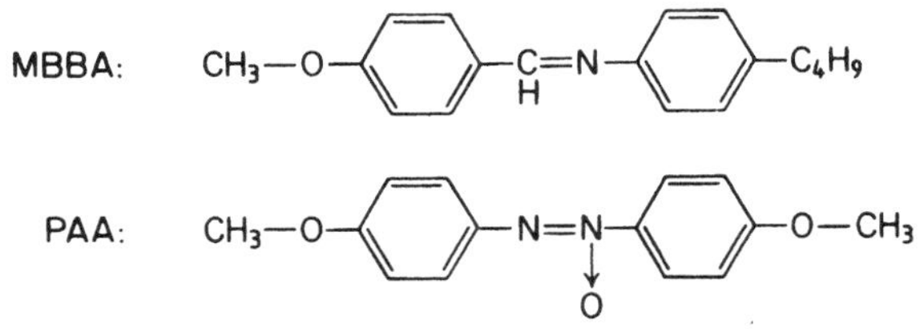

FIG. 11.4. Typical liquid crystal molecules, MBBA and PAA.

The transitions involving liquid crystals occur generally in the range from a few tens to hundreds of degrees Celsius. For instance, MBBA becomes nematic at 47°C and crystallizes at 20°C. In the case of cholesteryl benzoate, the crystal melts first into an opaque cholesteric liquid crystal at 145.5°C, which is then transformed into an isotropic liquid at 178.5°C. The phase transitions include both first and second orders.

11.2. Nematic phase

The transition from the isotropic phase to the nematic phase is due to a lowering of the free energy; at high temperatures, a random orientation of the molecules is entropically favored, while at low temperatures an oriented phase lowers the energy. The transition is known to occur in solutions of anisotropic molecules at low concentrations of order a few percent. Thus, we consider a dilute solution in order to describe the transition. For this purpose, we adopt a model in which the molecules are rigid and convex with the shape of a rod or an ellipsoid of revolution. Although real molecules are not completely rigid, this model allows a molecular approach.

The nematic phase characterized by a molecular orientation may be described by an orientational distribution function $g(\omega)$. Let this function be normalized such that $Ng(\omega) \, d\omega$ is the number of molecules with axis in the solid angle $d\omega$ in the direction of ω specified by (θ, ϕ), N being the total number of molecules. Hence, the normalization of $g(\omega)$ is

$$\int g(\omega) \, d\omega = 1. \tag{2.1}$$

where $d\omega = \sin \theta \, d\theta \, d\phi$.

The partition function can be constructed by treating the molecules in different orientations as if they were different species. The number of the "species" s is given by

$$N_s = Ng(\omega) \, d\omega.$$

In dilute solutions, two-particle interaction is most important. From the condition of minimal free energy, we obtain [1]

$$\ln 4\pi g(s) = \lambda - c \int \beta(s, t)g(t) \, d\omega(t), \tag{2.2}$$

where λ is a normalization constant, c is the number density of the molecules, and the kernel $\beta(s, t)$ is the effective interaction energy between species s and t.

Equation (2.2) is a self-consistent equation from which the distribution function can be determined. However, its nonlinearity requires a certain approximation. We resort to the diluteness of the solution and the highly elongated molecular shape.

An explicit form of the effective kernel $\beta(s, t)$ can be derived by specifying the molecular shape. If the shape is rotationally symmetric about its long axis, it may be expanded in the Legendre polynomials such that

$$\beta(\gamma) = 2v_0 + \beta_0(\varepsilon) - \sum_1^\infty B_l(\varepsilon)P_{2l}(\cos \gamma), \tag{2.3}$$

where γ is the solid angle between the orientations of a pair of molecules, v_0 is the molecular volume, and the $B_s(\varepsilon)$ depend on a particular molecular shape, which may be specified by a parameter ε. More specifically, in the case of a prolate ellipsoid of revolution, ε is the eccentricity given by

$$\varepsilon = \left(1 - \frac{a^2}{b^2}\right)^{1/2}. \tag{2.4}$$

Here, a and b are the short and long axes. For elongated molecules, b is considerably longer than a.

In correspondence to Eq. (2.3), the distribution function is expanded such that

$$g(\theta) = \sum_0^\infty C_{2n}P_{2n}(\cos \theta). \tag{2.5}$$

The integral equation (2.2) becomes an equation for determining the coefficient C_2, which is most important for dilute solutions. The finiteness of C_2 means that the molecules are oriented. One can show that this coefficient exists if

$$cB_1(\varepsilon) \geqq 5. \tag{2.6}$$

When the molecules are elongated such that $b \gg a$ and ε is nearly 1, it is found that

$$B_1(\varepsilon) = \frac{(5\pi/8)a^2b}{(1 - \varepsilon^2)^{1/2}}. \tag{2.7}$$

Therefore, the critical volume concentration is given by

$$c_c = \frac{32}{3\pi}\frac{a}{b}$$

$$= 3.4\frac{a}{b}. \tag{2.8}$$

This is indeed small.

In the case of tobacco mosaic virus, $2a = 150$ Å and $2b = 2700$ Å. Hence at a few percent, their solutions can be anisotropic. A similar result is obtained for a cylindrical shape. The critical concentration for long rodlike molecules is given by

$$c_c = \frac{4a}{b}. \tag{2.9}$$

Note that the numerical coefficients in Eqs. (2.8) and (2.9) are close to each other. That is, long rigid molecules tend to align to form a nematic phase when the concentration reaches a certain value. The same is true if the molecular shape is flat, like in bentonite.

On the other hand, a molecular-field approach can be made by assuming that a single order parameter determines the alignment of the molecules. A simple choice of an order parameter is [2]

$$s = \langle P_2(\cos \theta) \rangle$$
$$= \langle 3 \cos^2 \theta - 1 \rangle / 2. \tag{2.10}$$

The average $\langle \cdots \rangle$ is taken in terms of an as yet unknown orientational distribution function that will be introduced shortly. If this distribution is uniform, s vanishes. If it is anisotropic, s can be finite.

In order to determine s in a self-consistent way, let us assume that a pair of molecules with a relative orientation θ has an interaction energy given by

$$\Phi(\theta) = -\Phi_0 s \left(\frac{3 \cos^2 \theta - 1}{2} \right), \tag{2.11}$$

where Φ_0 is an unknown constant that may vary with an average distance between molecules. For a van der Waals-type interaction, Φ_0 can be proportional to $1/V^2$, where V is the molar volume. Note that $\phi(\theta)$ is referred to a polar coordinate fixed in space. In the case of a mutual interaction of a pair of molecules, Φ should depend on the angle between the two. Moreover, for a given intermolecular distance, $\Phi(\theta)$ changes its sign. For a positive Φ_0, $\Phi(\theta)$ is minimum at $\theta = 0$ and maximum at $\theta = \pi/2$.

In terms of the interaction energy, the distribution function for the direction $\theta(0)$ is given in the form

$$g(\theta) = \lambda \exp\left(-\frac{\Phi(\theta)}{kT} \right), \tag{2.12}$$

where λ is the normalization constant. Hence, for self-consistency, the order parameter must satisfy

$$s = 2\pi\lambda \int_0^\pi \left(\frac{3 \cos^2 \theta - 1}{2} \right) \exp\left(\frac{-\Phi(\theta)}{kT} \right) \sin \theta \, d\theta. \tag{2.13}$$

The average internal energy is then given by

$$U = \frac{N}{2} \frac{\int \Phi(\theta) \exp[-\Phi(\theta)/kT] \sin \theta \, d\theta}{\int \exp[-\Phi(\theta)/kT] \sin \theta \, d\theta}$$
$$= -\frac{NAs^2}{2}. \tag{2.14}$$

Here, the definition of the order parameter has been used. The entropy can be written down in terms of the distribution function:

$$S = -2\pi Nk \int g(\theta) \ln g(\theta) \sin \theta \, d\theta + \text{const.} \tag{2.15}$$

The constant can be chosen such that S vanishes for a uniform distribution. The free energy then becomes

$$F(s) = N[\tfrac{1}{2}\Phi_0 s^2 + kT \ln(\lambda/4\pi)]. \tag{2.16}$$

For equilibrium, F must be stationary with respect to variation in s. This condition is automatically satisfied, as can be confirmed easily. Furthermore, $s = 0$ is a solution of Eq. (2.13). Moreover, if $s = 0$, $\lambda = 4\pi$ so that

$$F(0) = 0.$$

By plotting $F(s)$ against s, one finds that $F(s)$ can have two minima at temperatures below a critical temperature. This temperature is found to be given by

$$kT_c = 0.22\Phi_0, \tag{2.17}$$

corresponding to the critical value of the order parameter:

$$s_c = 0.43. \tag{2.18}$$

for PAA, $T_c = 408$ K. The critical molar volume V_c of the nematic phase is 225 cm³. If Φ_0 is proportional to $(1/V_c^2)$ we find

$$\Phi_0 = \frac{13}{V_c^2} \quad \text{erg cm}^6.$$

The nematic order is established with little energy in an analogous way to the formation of spin waves in a ferromagnet. The critical point may be considered to correspond to the Curie point of a ferromagnet, and in fact there are many similarities between the two critical phenomena. However, in a nematic arrangement the molecules may not have complete symmetry with respect to inversion. The order parameter can be positive or negative, with the sign corresponding to that of birefringence. If a Landau-type expansion of the free energy is used, there is a cubic invariant that causes the transition to be first order. However, since the latent heat is less than kT per molecule, the first-order behavior is weak. In any case, molecular alignments promote scattering of light so that the nematics appear turbid.

11.3. Smectic phase

As many as nine subphases have been reported in the smectic phase, but smectic-A and smectic-C are most common. The characteristic molecular arrangement in a smectic corresponds to one-dimensional density waves. For instance, the A-phase is characterized by flat layers in which molecules are aligned parallel to each other. However, boundary or cooling conditions may result in bending of the layers into focal conic textures even though the interlayer distance is still kept fixed.

The transition from nematic to smectic-A, which is nearly second-order, may be described by the local density function $\rho(\mathbf{r})$. This function is nearly isotropic in the nematic phase. It varies sinusoidally in the molecular direction z in the smectic phase. This direction is perpendicular to the layers. We can express $\rho(\mathbf{r})$ as follows:

$$\rho(\mathbf{r}) = \rho_0 + \frac{\Psi(\mathbf{r}) \exp(i\mathbf{q}_0 \cdot \mathbf{r})}{V^{1/2}}. \tag{3.1}$$

Here, ρ_0 is the average density, and $\mathbf{q}_0$ is in the direction of the director $\hat{n}$ and is given by

$$\hat{n}(2\pi/d) = \mathbf{q}_0.$$

$\Psi(\mathbf{r})$ is a complex order parameter. It can be specified as

$$\Psi = |\Psi| \exp[i\mathbf{q}_0 \cdot \mathbf{u}(\mathbf{r})] = |\Psi| \exp(i\phi). \tag{3.2}$$

$\mathbf{u}(\mathbf{r})$ represents the layer displacement away from equilibrium and ϕ is a phase.

In general, displacements in a smectic-A can be classified into the longitudinal and transverse modes. If the displacement wavevector $\mathbf{q}$ is parallel to the director $\hat{n}_0$ of the molecular orientation, the layer distance is compressed or expanded and the mode is longitudinal. This type of propagation is similar to ordinary sound. If the corresponding wavevector is denoted by q_p, the energy is proportional to $q_p^2 u^2(q)$. In addition, there can be two viscously damped shear modes. If the displacement wavevector is normal to the director and hence transverse to the molecular orientation, the molecular layers bend. If the wavevector is denoted by q_v, the corresponding energy is proportional to $q_v^2 n^2$ or $q_v^4 u^2(q)$. In addition, there can be one overdamped shear mode and a very slowly damped undulation mode.

In general, a solid can accommodate three propagating modes, a liquid one propagating and two overdamped shear wave modes. Smectic-A can be considered as an intermediate between a solid and a liquid.

In correspondence to Eq. (3.1), a phenomenological free energy can be constructed in terms of powers and derivatives of Ψ in a Ginzburg–Landau fashion. Near the critical temperature, the free energy density is expressed to fourth order as follows [3]:

$$F = F_0 + a|\Psi|^2 + \tfrac{1}{2}b|\Psi|^4 + (\nabla + iq_0\,\delta\hat{n})\Psi^* \cdot \frac{\hbar^2}{2M}\,(\nabla - iq_0\,\delta\hat{n})\Psi + F_n, \quad (3.3)$$

where M is a tensor, and $\delta\hat{n}$ is the fluctuation in the director $\hat{n}$ about its equilibrium direction $\hat{n}_0$:

$$\hat{n} = \hat{n}_0 + \delta\hat{n}.$$

$\hat{n}_0$ is a unit vector perpendicular to the plane of the molecular orientation. The "mass-tensor" M may be characterized by two elements: M_T for a twisting deformation and M_v for a bending deformation. F_n represents the free energy of the nematic phase. In terms of the displacement, its elastic energy is a sum of the above two terms proportional to $q_p^2 u^2(q)$ and $q_v^4 u^2(q)$. The elastic energy density can be expressed in terms of changes in $\hat{n}$ such that

$$F_n = \tfrac{1}{2}K_1(\operatorname{div}\hat{n})^2 + \tfrac{1}{2}K_2(\hat{n}\cdot\operatorname{rot}\hat{n})^2 + \tfrac{1}{2}K_3(\hat{n}\times\operatorname{rot}\hat{n})^2. \quad (3.4)$$

The derivatives such as $\partial n_x/\partial x$, $\partial n_y/\partial y$ correspond to splaying deformation, $-\partial n_y/\partial x$, $\partial n_x/\partial y$ to twisting deformation, and $\partial n_x/\partial z$, $\partial n_y/\partial z$ to bending deformation, respectively. These three distinct strains of a liquid crystal are shown in Fig. 11.5.

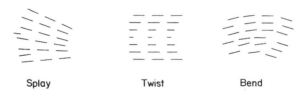

Splay   Twist   Bend

FIG. 11.5.   Three distinct types of deformation in the nematic phase.

The free energy expression (3.3) is similar to that for a superconductor. Note that in that case the superfluid velocity v_s is given by

$$v_s = \frac{1}{m}\left[\frac{\hbar}{i}\nabla - \frac{e}{c}A\right]\Psi,$$

where A is the vector potential. The corresponding kinetic energy is proportional to square of this velocity. The analogy with a superconductor suggests that smectic-A can be characterized by a "coherence length" $\xi(T)$ and a "penetration depth" $\lambda(T)$. In the neglect of any anisotropy, the former is given by

$$\xi(T) = \frac{\hbar}{(2M|a|)^{1/2}}. \tag{3.5}$$

This is the distance in which the amplitude of Ψ is perturbed. The latter is

$$\lambda(T) = \frac{d}{2\pi\hbar}\left[\frac{Mkb}{|a|}\right]^{1/2}. \tag{3.6}$$

The penetration depth represents the distance over which a weak bending deformation imposed on the free surface of a smectic-A decays. That is, over the distance $\lambda(T)$, $\hat{n}$ decays into $\hat{n}_0$. Depending whether $K = K_2$ or K_3, this length describes a twisting or bending deformation.

The "Ginzburg–Landau" parameter:

$$\kappa = \frac{\lambda}{\xi} \tag{3.7}$$

is temperature-independent and determines the basic category of smectic-A; similarly to the case of a superconductor there are Type-I or Type-II depending on $\kappa < 1/\sqrt{2}$ or $\kappa > 1/\sqrt{2}$. The phase diagram for a given category depends on the amplitude of an imposed twisting or bending deformation. This amplitude plays the role of a magnetic field for a superconductor. If the deformation is represented by

$$H = (2\pi/d)\,\mathrm{rot}(\delta\hat{n}), \tag{3.8}$$

the transition point T_{NA} is reduced when the amplitude $H = |H|$ increases.

A schematic phase diagram is given in Fig. 11.6. Note that the contour integral

$$\frac{1}{2}\oint \hat{n}\cdot dl$$

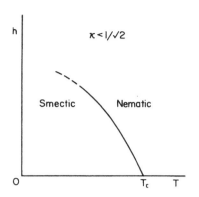
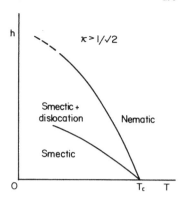

FIG. 11.6. Nematic–smectic phase transition.

along a closed circuit is zero if no dislocation exists within the loop. If there is, it is an integer. That is, with $\hat{n}$ playing the role of the vector potential, the above integral represents an analog of flux quantization in a superfluid. Fluctuations in the vector potential lead to first-order transition. Therefore, the transition from nematic to a nematic-A is also expected to be first-order, but experiments show that it is nearly second-order. This character can be seen in the specific heat data of octyloxycyanobiphenyl (8OCB) near the nematic–smectic-A transition in Fig. 11.7 [4]. The critical exponent is found to be

$$\alpha = \alpha' = 0.25 \pm 0.05. \tag{3.9}$$

As usual, the prime is used here below the transition point. Note that this value is too high to convert the specific heat variation into a logarithmic divergence. The data at 500, 1000, 1500, and 2000 bar are shifted down by 5, 10, 15, and 20 units, respectively.

The phase diagram of octyloxycyanobiphenyl due to Cladis et al. [5] is shown in Fig. 11.8. The chemical structure of the molecule is given at the top of the graph. The nematic–smectic transition includes data from three samples. Note that there is reentrant nematic phase. Such a reentrance phenomenon has been observed in other liquid crystals.

Near a second-order transition T_0, the parameter a in Eq. (3.3) is expected to vary as

$$a(T) = a_0|T - T_0|$$
$$= \frac{H\hbar^2}{2M}, \tag{3.10}$$

where $M = M_T$ for twist (H parallel to $\hat{n}_0$) and $M = (M_T M_v)^{1/2}$ for bending

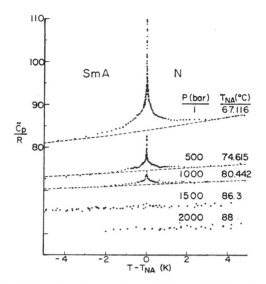

FIG. 11.7. Specific heat C_p at the nematic–smectic-A transition in octyloxycyanobiphenyl. The data are at 500, 1000, 1500 and 2000 bar. The dashed lines represent the background specific heat. (From Kasting et al. [4])

(H normal to $\hat{n}_0$). Being proportional to $a(T)^{-1/2}$, K_2 and K_3 are expected to diverge as $|T - T_0|^{-1/2}$, but as in the case of any other mean-field theory, the exponent 0.5 is expected to be incorrect. A more realistic variation in the form

$$a(T) = a_0|T - T_0|^{\gamma} \tag{3.11}$$

with $\gamma = 1.3$ yields an exponent of order 0.66 for the exponent of the elastic constants.

Since there is anisotropy at a nematic–smectic-A transition point T_{NA}, it is appropriate to introduce two exponents ν_p and ν_v in the parallel and vertical directions:

$$\xi_p \sim (T - T_{NA})^{-\nu_p}; \qquad \xi_v \sim (T - T_{NA})^{-\nu_v}.$$

Experimental values of the exponent ν are spread from around 0.57 to 0.83, even though the power law itself is well obeyed. Such a spread may at least partially be due to nonuniformity in molecular structure. The anisotropy $(\nu_p - \nu_v)$ is of order 0.13. The exponents have been determined not only by specific heat but also by scattering experiments [6]. With $\gamma = 1.33$, ξ and the elastic constants vary as

$$\xi \sim K \sim (T - T_{NA})^{-0.66}. \tag{3.12}$$

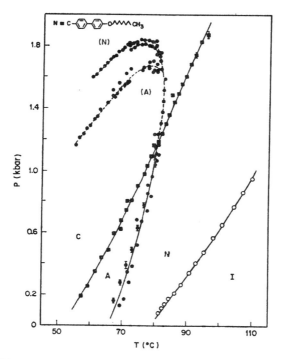

FIG. 11.8. Phase diagram of octyloxycyanobiphenyl. The data for the nematic–smectic-A transition correspond to three different samples. I, isotropic; N, nematic; (N), supercooled and reentrant nematic; A, smectic-A; (A), supercooled smectic-A; C, crystal.

The transition from the nematic to smectic phases can be described in terms of molecular field theory [7] with an additional positional order. A phenomenological interaction potential may be chosen such that

$$\Phi(r, \theta) = \Phi_0(r)P_2(\cos \theta)$$

or

$$\Phi(r, \theta) = \Phi_0(r) + \Phi_1(r)P_2(\cos \theta)$$

with some phenomenological parameters.

The transition from smectic-A to nematic is interesting because it represents probably the simplest form of smelting. It is also interesting that the elastic coefficients K_2 and K_3 show pretransitional anomalies.

11.4. Defects

Schlieren photographs of a nematic in crossed polarizers show brushes that are the regions where the director is either parallel or perpendicular to the plane of polarization of the incident light. Usually, two to four brushes converge upon a point like twists. The focal point of each twist represents a line of defects where the molecular inclination changes discontinuously. It is called a *disclination*. Moving around a disclination defect, the director $\hat{n}$ may rotate $2\pi s$, where s, called the *strength*, is a positive or negative integer. Because $\hat{n}$ and $-\hat{n}$ are equivalent, $s/2$ defines the strength of a disclination. Disclinations with s of $\pm\frac{1}{2}$ and ±1 are generally observed.

Four dark brushes are often visible when a nematic is placed in crossed polarizers. These brushes correspond to $s = \pm1$. The director can rotate for this type of disclination, lying in the central region so that the core size is rather large. Neighboring disclinations connected each other by brushes are of opposite signs. That is, they cancel each other's elastic fields. This cancellation results in an attractive force between the defects. The force per unit length between two lines is inversely proportional to their separation.

Let us assume that a disclination line is in the direction perpendicular to the xy plane in which the nematic director $\hat{n}$ lies. The molecules are aligned parallel to the xy plane and can have splaying and bending deformations. If the director is specified by angle ϕ,

$$n_x = \cos\phi, \qquad n_y = \sin\phi,$$

the elastic energy is given per unit volume by

$$F = \tfrac{1}{2}K\left[\left(\frac{\partial\phi}{\partial x}\right)^2 + \left(\frac{\partial\phi}{\partial y}\right)^2\right]. \tag{4.1}$$

The elastic energy is minimized if ϕ is a solution of the Laplace equation:

$$\Delta\phi = 0.$$

Liquid crystal molecules in capillary tubes can be aligned with the axis of the cylinder, but the director is radial at the walls. Therefore, point defects are created in order for the director $\hat{n}$ to escape into different directions. In general, two neighboring point defects cancel each other's elastic fields. Figure 11.9 illustrates positive (a) and negative (b) point defects. It is interesting that their interaction increases linearly with separation for short distances.

Smectic layers can form cylinders concentric to the glass tube in which a smectic is placed. Cylindrical samples of a liquid crystal with a larger latent heat at a nematic–smectic-A transition frequently produce a focal conic domain texture regardless the cooling rate of the sample. Such focal conic defects appear as confocal pairs of conic section curves. In addition to this type of defect there can be edge dislocations in the layer spacing. That is, a layer may start from a certain point between a pair of neighboring layers.

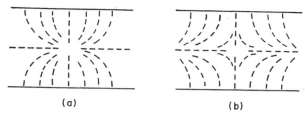

FIG. 11.9. Point defects in a nematic in a tube: (a) positive and (b) negative.

The molecular order in a smectic-A is described by the director $\hat{n}$ and a phase for the position in the layer. Those defects that involve $\hat{n}$ can be described in a way similar to those in a nematic. However, disclinations of a nematic type can cause a persistent change in the layer spacing far from the singular line and thus may not be realized energetically. In the case of the edge defects, the phase of the order parameter is decisive.

The cholesteric phase is recognizable for its selective optical reflection. The molecules are spirally aligned about an axis perpendicular to the director $\hat{n}$. The pitch p of the spiral is typically of order a few thousand angstroms. In the vicinity of the smectic phase, the pitch is temperature-dependent. A power law such as

$$p = p_0 + p_1(T - T_{cs})^{-\nu} \tag{4.2}$$

can be used to describe the variation. As in Eq. (3.12),

$$\nu \sim 0.66. \tag{4.3}$$

The cholesteric phase can be specificed by the director $\hat{n}$ and another unit vector $\hat{t}$ that represents the spiral axis. The cholesteric phase can be considered as a twisted nematic phase that is biaxial.

Three line defects known as λ, χ, and τ, can be defined in a cholesteric. These defects involve 180° rotations about their respective axes and line defects of equivalent 360° rotations. These arise because an orthogonal set of three coordinates is needed to specify $\hat{n}(r)$; for a constant pitch, this is represented by

$$\hat{n}(r) = \hat{n}_0(\mathbf{r}) \cos[(2\pi/p)\hat{q}\cdot\mathbf{r}] + \hat{n}_0(\mathbf{r}) \times \hat{q}(r) \sin[(2\pi/p)\hat{q}\cdot\mathbf{r}]. \tag{4.4}$$

Note that the vectors $\hat{n}_0(\mathbf{r})$, $\mathbf{q}$ and $\hat{n}_0 \times \mathbf{q}$ are involved for a given point $\mathbf{r}$. These three vectors are the axes of rotation of the above three line defects, λ, χ, and τ respectively. The χ defect corresponds to a nematic disclination with a strength $\frac{1}{2}$.

The constraint of a constant layer spacing in a cholesteric plays a similar role to that in a smectic. For this reason, focal conic defects are also seen in a cholesteric. In addition, edge dislocations can be observed rather well. However, the cholesteric phase is different from the smectic phase due to the anisotropy in the planes of the layers that causes the χ defects. Since χ defects involve a rotation about $\hat{q}$ and correspond to a nematic-type disclination, the cholesteric phase may be considered as a smectic with in-plane nematic behavior.

Orientational fluctuations in a cholesteric cause scattering of light. There is a typical mode corresponding to twisting of the spiral arrangement. It involves a twisting motion about the spiral axis of neighboring molecular planes. Its restoring force is provided by the twist elastic constant K_2. This mode is overdamped and causes broadening of scattered light. Its damping constant is proportional to K_2. Another typical mode is caused by viscous-splay motion about the director. This mode is also overdamped. Its damping constant is determined by the splay and bend elastic constants K_1 and K_2.

When a cholesteric liquid crystal is held between crossed polarizers near the cholesteric–isotropic transition point, three *blue phases* can be observed if the cholesteric pitch is less than 7000 Å. These blue phases are interesting because they represent stable ordered arrays of defects. They are intermediate between the regular helical phase and the isotropic phase and are thermo-dynamically stable in a narrow temperature range. The ordered networks of defects are of cubic symmetry, at least in several cases with unit-cell dimensions of the order of the cholesteric pitch.

Let us now discuss one possible mechanism for a blue phase under the assumption of a first-order cholesteric–isotropic transition [8]. We examine the behavior of the elastic free energy for this purpose and try to see that the ordinary planar helix structure of a cholesteric becomes unstable at temperatures near the transition point.

The elastic energy density of a cholesteric is essentially of the same form as that of a nematic given by Eq. (3.4) except that the twist energy associated with the elastic constant K_2 is slightly modified:

$$F_1 = \tfrac{1}{2}K_1(\mathrm{div}\ \hat{n})^2 + \tfrac{1}{2}K_2(\mathbf{q} + \hat{n}\cdot\mathrm{rot}\ \hat{n})^2 + \tfrac{1}{2}K_3(\hat{n} \times \mathrm{rot}\ \hat{n}). \tag{4.5}$$

This energy is estimated to be roughly equal to the elastic energy for a nematic surrounding an $s = -\tfrac{1}{2}$ disclination. This is given per unit length by

$$f_1 = \tfrac{1}{4}\pi K \ln\!\left(\frac{R_{\max}}{R}\right). \tag{4.6}$$

There are three additional energies to be introduced for the present case. *First*, it is appropriate to include the following energy density:

$$F_2 = \tfrac{1}{2}[(K_2 + K_{24})\nabla\cdot[(\hat{n}\cdot\nabla)\hat{n} - \hat{n}(\nabla\cdot\hat{n}))]. \tag{4.7}$$

This energy density is usually neglected because, due to the presence of a divergence, its integral can be transformed into a surface integral. In the present case, however, the surface integration must be carried over the core of the disclinations as well. Since the core area increases as the third power of the sample dimension, it does contribute. For strength $s = -\frac{1}{2}$, its integral over the surface of a cylinder surrounding the disclination is given by

$$f_2 = -\pi(K_2 + K_{24})$$
$$= -\pi K. \tag{4.8}$$

This is the energy per unit length along the disclination.

That this energy is negative turns out to be crucial for the appearance of the blue phases. This negative sign may be understood by visualizing stacking of ordinary screws. They can be packed together more tightly if their long axes are tilted slightly relative to one another. That is, the system gains energy if the molecules are stacked on the average at some small angle, rather than parallel as in the nematic case.

Second, the disclination core causes an excess free energy f_{core} per unit length. For a first-order cholesteric–isotropic transition, this energy can be represented by two lines intersecting at the transition point T_i. Hence, it can be expressed as

$$f_{\text{core}} = a(T_i - T)\pi R^2, \tag{4.9}$$

where R is the core radius and a is a constant.

Third, the interface between core and cholesteric has an extra energy. This energy is represented by a surface tension σ such that

$$f_s = 2\sigma\pi R. \tag{4.10}$$

Collecting all these energies, we arrive at

$$f = a(T_i - T)\pi R^2 + 2\sigma\pi R - \pi K + \tfrac{1}{4}\pi K \ln\left(\frac{R_{\max}}{R}\right). \tag{4.11}$$

For simplicity, let us choose $\sigma = 0$. The minimum of f occurs at

$$R = \left(\frac{K}{8a(T_i - T)}\right)^{1/2}. \tag{4.12}$$

With this R, $f = 0$ if

$$\ln\left(\frac{R_{\max}}{R}\right) = 3.5 \qquad \text{or} \qquad \frac{R_{\max}}{R} \sim 33. \tag{4.13}$$

Hence, if $R_{max} = 1000$ Å, $R = 333$ Å at the transition. An estimation from the latent heat of the cholesteric–isotropic transition yields $a = 8 \times 10^4$ erg/deg·cm³. Also, $K = 3 \times 10^{-7}$ dyne may be used. We then obtain $T_i - T \sim 5°C$, which is a reasonable theoretical estimate.

11.5. Two-dimensional liquid crystals

For miniaturized display devices and for the study of phase transitions, two-dimensional liquid crystals are very useful. The transition from a smectic-A to B represents melting in two dimensions. A smectic-C has similarities with an XY ferromagnet. Moreover, two-dimensional liquid crystals are interesting in relation to biological membranes. Therefore, even though many of their theoretical descriptions parallel those for bulk, it is worth examining the behavior of liquid crystals in two dimensions.

The transitions from the isotropic phase to the nematic phase can be treated as in the case of bulk [9]. The orientational distribution function $g(s)$ can be determined as in Eq. (2.2). Let us adopt a rigid-rod model with minor and major axes a and b respectively. When a second molecule moves around a first molecule keeping a relative orientational angle γ, one can use the following expression for the kernel of the integral equation (2.2):

$$\beta(\gamma) = B_0 + B_1 \cos 2\gamma, \tag{5.1}$$

where

$$B_0 = -2ab - 2(a + b)^2/\pi,$$
$$B_1 = 4(b - a)^2/3\pi. \tag{5.2}$$

We find that the critical areal concentration to be given by

$$c_c = \frac{3\pi}{2} \frac{ab}{(b - a)^2}. \tag{5.3}$$

For $b \gg a$, this critical concentration is proportional to a/b.

At this point, the pressure changes its slope. Figure 11.10 illustrates this change. Here, the ordinate is a reduced pressure with $p_0 = 2kT/B_1$, and the abscissa represents the area per molecule in units of the critical molecular concentration

$$c_{i \to n} = \frac{c_c}{ab}. \tag{5.4}$$

When the concentration increases beyond the critical value, a transition to a smectic phase may be expected. Since the molecules are already aligned, the distribution function may be assumed to have the form [9]:

$$g(\mathbf{r}, \hat{n}) = f(\mathbf{r}) \, \delta(\theta), \tag{5.5}$$

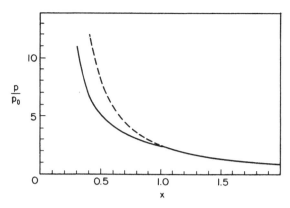

FIG. 11.10. Reduced pressure of a 2D liquid crystal as a function of reduced area. $p_0 = 2kT/B_1$ and $x = 2/cB_1$, where B_1 is given by Eq. (5.2) and c is the number concentration. (From Wadati and Isihara [9])

where $\delta(\theta)$ is a Dirac delta function. Let us use the y coordinate for the layered direction. In terms of a translational order parameter t, the function $f(r)$, as the solution of an integral equation with the same form as Eq. (2.2), is given by

$$f(\mathbf{r}) = A^{-1} \exp\left[c \int \beta(\mathbf{r}, \mathbf{r}') f(r')_0 \, dr' \right]. \tag{5.6}$$

A is the normalization constant. The kernel $\beta(\mathbf{r}, \mathbf{r}')$ can be expressed as

$$\beta(\mathbf{r}) = B_0 + B_1 \cos(2\pi y/y_0), \tag{5.7}$$

For a hard rod, it is appropriate to require $\beta(\mathbf{r}) = -1$ for $|x| \leqq a$ and $|y| \leqq b$ and use B_0 and B_1 given by

$$B_0 = -\frac{4ab}{L^2}, \qquad B_1 = -\frac{2ay_0}{\pi L^2} \sin(2\pi b/y_0), \tag{5.8}$$

where $V = 4L^2$ is the total area.

The nematic–smectric transition is determined by the consistency equation for the order parameter. The transition is found to take place at

$$c_{n \to s} = \frac{\pi}{5.6ab}. \tag{5.9}$$

The critical concentration is much larger than that for the isotropic–nematic transition. In fact, their ratio is

$$\frac{c_{i \to n}}{c_{n \to s}} \sim \frac{a}{b}.$$ (5.10)

This equation shows that the anisotropy in the shape of the molecules determines the concentration ratio of the two transitions.

Two-dimensional melting was discussed in Section 1.5.2. In a two-dimensional lattice dislocations destroy local positional order but with only a small effect on the local orientation of the lattice. Hence, the crystal can melt first into an intermediate phase with short-range positional order, maintaining quasi-long-range bond order, before becoming a liquid at a high temperature. This intermediate phase is called a *hexatic phase*. Liquid crystals provide opportunities to observe such a phase.

The smectic-B phase is a layered phase in which the molecules are oriented perpendicularly to the layer planes. Usually, there is hexagonal crystalline order in each layer and positional correlations are long-ranged. However, with free-standing liquid crystal film techniques, a hexatic phase has been observed with short-range in-plane positional correlations but, as opposed to the A phase, with long-range 6-fold bond-orientational order.

Figure 11.11 depicts the X-ray data of Pindak et al. [10] on the liquid crystal n-hexyl-4'-n-pentyloxybiphenyl-4-carboxylate (65OBC). This material exhibits a sequence of phase transitions; from isotropic to smectic-A at 85°C, then to smectic-B at 67.9°C, and finally to crystal at 60°C. Bond orientational order was probed by χ-scans that rotate the film about the layer normal. χ-Scans at three different temperatures are shown in this figure. Above the A $\to$ B transition point of 67.9°C, the scattering is constant as in (a). Just below the smectic A $\to$ B transition, sinusoidal modulations develop as in (b), indicating a 6-fold periodicity. This scattering provides direct evidence for hexagonal bond-orientational correlations. The correlations are three-dimensionally ordered. That is, the structure corresponds to a stacked hexatic phase. As the temperature is cooled below 66.4°C, a single sharp peak indicative of a single well-oriented domain is observed as in (c).

In the smectic-C phase the molecules are tilted and not perpendicular to the layers. The in-plane projection of the molecular orientation plays the role of the magnetic moment in the xy model. It is the order parameter for this phase and can be determined by measuring the amount of optical rotation. In contrast to the case of an xy ferromagnet, the in-plane projection of the tilt direction varies in magnitude with the tilt angle. However, this variation is not very important in the vicinity of a defect-mediated transition.

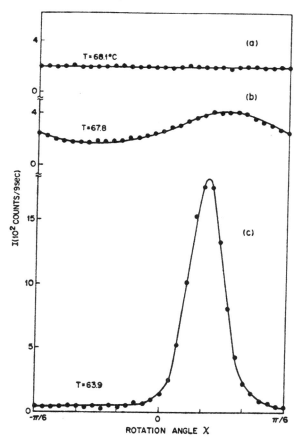

FIG. 11.11. χ-scans for three different temperatures of 65 OBC. Curve a, smectic-A phase; curve b, just below the smectic A → B transition curve; c, B phase.

REFERENCES

1. L. Onsager, *Ann. N.Y. Acad. Sci.* **51**, 627 (1949). A. Isihara, *J. Chem. Phys.* **19**, 1142 (1951).
2. W. Maier and A. Saupe, *Z. Naturforsch.* **15**, 287 (1960).
3. C. W. Oseen, *Trans. Farad. Soc.* **29**, 883 (1933). J. A. Geurst, *Phys. Lett. A* **34**, 283 (1971). P. G. de Gennes, *J. Phys. (Paris)* **30**, C4–65 (1969); *Solid State Commun.* **10**, 753 (1972).
4. G. B. Kasting, K. J. Lushington, and C. W. Garland, *Phys. Rev. B* **22**, 321 (1980).
5. P. E. Cladis, R. K. Bogardus, and D. Aadsen, *Phys. Rev. A* **18**, 2292 (1978).

6. J. D. Litster, R. J. Birgeneau, M. Kaplan, C. R. Safinya, and J. AlsNielsen, in *Ordering in Strongly Fluctuating Condensed Matter Systems*, ed. by T. Riste (Plenum Press, New York, 1980), p. 357.
7. W. L. McMillan, *Phys. Rev. A* **4**, 1238 (1971). K. Kobayashi, *Mol. Cryst. Liq. Cryst.* **13**, 137 (1971).
8. S. Meiboom, J. P. Sethna, P. W. Anderson, and W. F. Brinkman, *Phys. Rev. Lett.* **46**, 1216 (1981).
9. M. Wadati and A. Isihara, *Mol. Cryst. Liq. Cryst.* **17**, 95 (1972).
10. R. Pindak, D. E. Moncton, S. C. Davey, and J. W. Goodby, *Phys. Rev. Lett.* **46**, 1135 (1981).

12

POLYMERS

Despite their abundance in nature, the existence of molecules with very high molecular weights was recognized only around 1935. They are called *macromolecules*. These molecules include cellulose, proteins, DNA, starch, rubber, and many other molecules that are polymerized from monomers. Their molecular weights can be of the order of millions in comparison with tens for the molecular weights of ordinary molecules. Their wide variety of sizes and shapes make them the basic ingredients in living as well as inanimate objects.

12.1. Excluded volume effect

Many properties of chain polymers hinge upon their shape, which in turn depends on the internal structure of their constitutional unit, that is, the monomers. The chains can be stiff like cellulose molecules or flexible like rubber molecules. The chemical formulas of several common polymers are given in Table 12.1.

12.1.1. *End-to-end distance*

The shape of a chain polymer molecule in solutions depends on the molecular weight, its repeating units, the temperature, and the solvent. For a long flexible chain, it is convenient for theoretical consideration to introduce an effect repeating unit called a *segment*. The segments are more or less free from each other in orientation although they are connected. As such, each segment may consist of several monomers. The essential feature of their connectivity can be studied in terms of a simple *pearl necklace model*. The main point in using such a simple model comes from the expectation that as far as the dependence on the chain length is concerned, the internal structure of a segment will be irrelevant in the long-chain limit. This expectation is similar to the universality of the scaling laws near the critical points of second-order phase transitions, as we shall discuss later.

If the molecular rotation within each monomer is strong, the entire polymer can be randomly coiled. In this case, the polymer molecule is spherical. In the ideal case in which the segment does not have a volume, the shape is determined by a *random walk model*. Then, the end-to-end square

Table 12.1.

Polymer	Chemical Formula
Polyisoprene	$-CH_2-\overset{\displaystyle CH_3}{\underset{\displaystyle \vert}{C}}=CH-CH_3-$ and $-CH_3-\overset{\displaystyle CH_3}{\underset{\displaystyle \vert}{\underset{\displaystyle CH=CH_2}{C}}}-$
Polyisobutyrene	$-CH_2-\overset{\displaystyle CH_3}{\underset{\displaystyle CH_3}{C}}-$
Polystyrene	$-CH_2-CH-$ with phenyl ring
Polymethylmethacrylate	$-CH_2-\overset{\displaystyle CH_3}{\underset{\displaystyle O=C-O-CH_3}{C}}-$
Polyvinylchloride	$-CH_2-\underset{\displaystyle Cl}{CH}-$
Cellulose	ring structure with CH_2OH, $-O-CH$, $HC-$, OH, OH

average distance of a chain is given by

$$\langle r^2 \rangle_0 = b^2 N. \tag{1.1}$$

Here, N is the number of segments and b is their bond length.

In actuality, however, each segment has a finite volume so that the *self-avoiding walk* model is more realistic. In this case, the square average end-to-end distance has to be evaluated under the condition that the segments are not allowed to be on the same spot simultaneously. This self-avoiding walk problem can be studied in several ways [1]. One can start with the ideal case without volume exclusion and then consider the effect of volume exclusion by perturbation. In this case, Feynman graphs can be drawn to represent segmental interactions because the ideal case is determined by the diffusion equation, which corresponds to the Schrödinger equation of an ideal particle. However, it becomes increasingly difficult to analyze higher-order graphs, and hence it is virtually impossible to investigate the limit $N \to \infty$.

In an approach based on renormalization-group theory, a portion of a long chain, and a portion of this portion, and so on, are considered to form a semi-group. That is, they are all similar to each other in shape as $N \to \infty$. The term *fractal* is used to describe this kind of similarity. As a result, the square average end-to-end distance $\langle r^2 \rangle$ is expected to follow a power law:

$$\langle r^2 \rangle = b^2 N^{2\nu}, \tag{1.2}$$

where b is the bond length, N is the number of repeating units, and $\langle \cdots \rangle$ represents a statistical average.

For simplicity, we adopt the following notation for $\langle r^2 \rangle^{1/2}$:

$$R = \langle r^2 \rangle^{1/2}. \tag{1.3}$$

In order to determine the exponent ν, the square average distance of a portion that has N_1 segments is expressed as [2]

$$R_1 = b N_1^{1/2}[1 + h(N_1, \phi)]. \tag{1.4}$$

Here the first factor is the random-walk result. The function h represents the effect of intrachain interaction, and as such depends on the number N_1 of segments and the excluded volume parameter ϕ. The latter is a function of the diameter a of a segment considered as a hard sphere.

We remark that if the excluded volume effect is treated as a small perturbation, the first-order result is given by

$$R = b N^{1/2}\left[1 + \frac{8}{(6\pi)^{1/2}} N^{1/2}\left(\frac{a}{b}\right)^3\right]. \tag{1.5}$$

This is the exact first-order result with a representing the diameter of a segment considered to be a hard sphere. However, since N is as large as 10^4 it is clear that this result is invalid unless the ratio a/b is extremely small. The second-order term can be evaluated, but in the limit of large N such a perturbational result does not work.

The second term in Eq. (1.5) represents the ratio of the total excluded volume due to intersegmental interaction which is proportional to $N^2 a^3$ and the "free volume" $N^{3/2} b^3$. Making use of this observation let us now try to convert Eq. (1.4) into a recurrence form.

For this purpose, we introduce the process of dividing the original chain into N/p chains, each with p segments. Since we are interested in the limit $N \to \infty$, each divided chain is still long enough to allow a new division into chains of N/p^2 segments. We repeat the same process to form a renormalization group.

In order to express this process mathematically, first consider the interaction parameter ϕ in Eq. (1.4) as the ratio of the total excluded volume in the presence of volume exclusion to the "free volume" of that part of the self-avoiding chain that is under consideration. Due to the volume exclusion, the number of segmental pairs is reduced from p^2. Hence, the total interaction varies as

$$v_1 = vp^2[1 - g(p, \phi)], \tag{1.6}$$

where v is an effective interaction constant, and g is an unknown reduction function that can depends on p and ϕ. The "free volume" of the corresponding chain is proportional to R_1^3. Hence, we can introduce a new interaction parameter that has been reduced from ϕ as follows:

$$\phi_1 = \frac{vp^2(1 - g)}{b^3 p^{3/2}(1 + h)^3} = \phi p^{1/2}[1 - f(t, \phi)]. \tag{1.7}$$

The function $f = (1 - g)/(1 + h)^3$ is unknown. If $v = a^3$ was used in Eq. (1.6), $\phi \sim (a^3/b^3)N^{1/2}$ in Eq. (1.7). However, Eq. (1.7) is expressed with a general excluded volume parameter v. Before going further, it is worth noting that the exponent of p in Eq. (1.6) becomes $(2 - d/2)$ for d dimensions. This indicates that $d = 4$ is special.

Equation (1.4) can now be rewritten in a recurrence form:

$$R_s = R_{s-1} p^{1/2}[1 + h(\phi_{s-1})]. \tag{1.8}$$

The interaction ϕ_s satisfies a similar recurrence equation:

$$\phi_s = \phi_{s-1} p^{1/2}[1 - f(\phi_{s-1})]. \tag{1.9}$$

Note that Eqs. (1.8) and (1.9) generalize but somewhat simplify Eqs. (1.4) and (1.7).

We now require that in the limit $s \to \infty$, there is a "fixed point" such that

$$\lim_{s \to \infty} \phi_s = \lim_{s \to \infty} [v_s/R_s^3] = \phi^*. \tag{1.10}$$

Here, ϕ^* is a solution of

$$p^{1/2}[1 - f(\phi^*)] = 1. \tag{1.11}$$

At the fixed point ϕ^*, Eq. (1.8) yields

$$R_s = R_{s-1} p^{1/2}[1 + h(\phi^*)]. \tag{1.12}$$

This equation shows that R agrees in form with Eq. (1.2). Indeed, if we write for large s

$$R = R_s \Phi(N/p^s, \phi_s), \tag{1.13}$$

$$\frac{R_s}{R_{s-1}} = \mu$$

$$= \frac{\Phi(N/p^{s-1})}{\Phi(N/p^s)}. \tag{1.14}$$

Here, μ is determined by the fixed point in accordance with Eq. (1.12). Equation (1.14) leads us to a power law:

$$\Phi(N) \sim N^\nu, \tag{1.15}$$

where

$$\nu = \frac{\ln \mu}{\ln t}. \tag{1.16}$$

The exponent ν can be obtained explicitly from a numerical analysis of finite chains or by making use of higher dimensions. One can understand the utility of approaching from a higher dimension because in d dimensions the relation corresponding to Eq. (1.9) becomes

$$\frac{\phi_s}{\phi_{s-1}} \leqq p^{2-d/2}, \tag{1.17}$$

as remarked earlier. Hence, if $d > 4$, the fixed point is zero. That is, the chain is free. Hence, an approach with a parameter

$$\varepsilon = 4 - d \tag{1.18}$$

becomes possible.

Moreover, the self-avoiding walk problem is related to a magnetic problem. As a result, it has been shown that for small ε the exponent ν varies as [3]

$$v = \frac{v}{2 - \eta},$$

$$\gamma = 1 + \frac{\varepsilon}{8} + \frac{13\varepsilon^2}{2^8} + O(\varepsilon^3),$$

$$\eta = \frac{\varepsilon^2}{64}\left(1 + \frac{17\varepsilon}{16}\right) + O(\varepsilon^4). \tag{1.19}$$

For $\varepsilon = 1$ and $d = 3$, the exponent is given by [3, 4]

$$v = 0.598. \tag{1.20}$$

Let us now determine v in the simplest way by rewriting the first-order result in Eq. (1.5) in a self-consistent way in terms of a uniform expansion parameter α. As originally introduced by Flory it is defined by the ratio

$$\alpha^3 = \frac{R^3}{N^{3/2}b^3}. \tag{1.21}$$

We assume that the bond length b in Eq. (1.5) is also "expanded" in the same proportion as the entire chain and rewrite that equation in terms of α. The result is

$$\alpha^5 = \alpha^3 + \frac{AN^{1/2}}{b^3}. \tag{1.22}$$

The bond length b has been renormalized, and b in this equation is now independent of N. Because of the renormalization we consider that this equation is valid for all N, and take the limit $N \to \infty$ in which α is large. We arrive at

$$\alpha \sim N^{1/10}. \tag{1.23}$$

Equation (1.23) results in Flory's value [5]:

$$v = 3/5. \tag{1.24}$$

Despite the simplicity of derivation, this exponent is close to the 0.598 of Eq. (1.19).

12.1.2. *Analogy with a magnetic system*

It is well known in statistical mechanics that for $T > T_c$ the partition function of an Ising lattice can be obtained by considering closed polygons of all sizes. For an Ising lattice each spin takes on values that are either 1 or -1. More generally in the n-component vector spin model, the spin $\mathbf{S}_i$ at site i has n components, $S_{i\alpha}$, with $\alpha = (1, 2, \ldots, n)$. Its normalization can be conveniently defined by

$$\mathbf{S}_i^2 = \sum_{\alpha=1}^{n} (S_i^\alpha)^2 = n. \tag{1.25}$$

The partition function is defined formally by the configurational integral:

$$Z(K, h) = \left\langle \exp\left[K \sum_{(ij)} \mathbf{S}_i \cdot \mathbf{S}_j - \mathbf{h} \cdot \sum_i \mathbf{S}_i \right] \right\rangle, \tag{1.26}$$

where K and $\mathbf{h}$ are proportional to the exchange energy and an external magnetic field respectively and the average is taken over all spin directions. The number n of components must be a positive integer. However, starting from a finite n one can take the limit of $n = 0$, which turns out to be the interesting case for the self-avoiding walk problem [2]. Clearly $n = 0$ is peculiar in that there is no spin component. Therefore, the limit $n = 0$ must be taken as if $\mathbf{h}$ and $\mathbf{S}$ are n-dimensional vectors.

In order to show these points, let us investigate the behavior of the partition function of a single spin S in a magnetic field:

$$Z = \langle \exp(-\mathbf{h} \cdot \mathbf{S}) \rangle. \tag{1.27}$$

We treat $\mathbf{k}$ and $\mathbf{S}$ as if they are n-dimensional vectors, and then after calculation n is brought to zero.

For $\mathbf{h} \cdot \mathbf{h} = |h|^2 = \sum_\alpha h_\alpha^2$ we obtain:

$$\sum_\alpha \frac{\partial^2}{\partial h_\alpha^2} = \frac{\partial^2}{\partial h^2} + \sum_\alpha \frac{1}{h} \frac{\partial}{\partial h} - \frac{1}{h} \frac{\partial}{\partial h}.$$

In the limit $n \to 0$,

$$\nabla^2 Z = \frac{\partial^2 Z}{\partial h^2} - \frac{1}{h} \frac{\partial Z}{\partial h} = 0.$$

The right-hand side of this equation must vanish in accordance with Eq. (1.25).

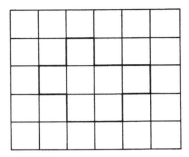

FIG. 12.1. A closed polygon on a lattice.

The function f that satisfies this differential equation can be determined uniquely under the boundary conditions $Z(0) = 1$ and $\nabla Z(0) = 0$ as follows:

$$Z = 1 + \frac{h^2}{2}.$$ (1.28)

We learn from this result that

(i) $\langle S^2 \rangle = \langle \nabla^2 Z \rangle = 1.$

(ii) The averages of all other powers of S vanish. (1.29)

The result $\langle S^2 \rangle = 1$ is unexpected since $\mathbf{S}^2 = 0$ in accordance with Eq. (1.25).

For high temperatures, the partition function of Eq. (1.26) can be evaluated by expanding the Boltzmann factor. Each term in the expansion can be illustrated by graphs as in the case of an Ising lattice. Due to the properties (i) and (ii), only closed single-loop polygons contribute to the partition function. These graphs represent self-avoiding walks on a lattice as illustrated in Fig. 12.1. As a result the spin–spin correlation function for the component α is given by

$$\langle S_{0\alpha}S_{j\alpha} \rangle = \sum_N N_N(0j)K^N,$$ (1.30)

where $N_N(ij)$ is the number of self-avoiding walks with N steps between the two sites 0 and j. This form suggests that the number is related to a magnetic susceptibility.

Numerical studies show that the total number N_T of steps obtained by summing $N_N(0j)$ over all j varies as [6]

$$N_T \sim \tilde{z}^N N^{\gamma-1},$$ (1.31)

where $\tilde{z}$ is a constant that is related to but smaller than the actual coordinate number. $\tilde{z}^N$ gives the probability for N steps. The exponent v depends only on the dimensionality. In three dimensions, $v = 7/6$. The susceptibility is given by

$$\chi \sim \sum_N K^N N^{\gamma-1}. \qquad (1.32)$$

For small $\varepsilon = (T - T_c)/T_c$ near a critical temperature T_c, it varies as

$$\chi \sim \sum_N \exp(-N\varepsilon)N^{\gamma-1} \sim \varepsilon^{-\gamma}, \qquad (1.33)$$

where $1/N$ plays the role of ε.

We have now shown that the self-avoiding walk problem on a lattice can be related to critical behavior of a magnetic system. Hence, the known critical scaling laws of a magnet can be transferred to polymer statistics. The most important point here is that these laws are universal in the limit $N \to \infty$ as in the case $\varepsilon = (T - T_c)/T_c \to 0$ in a magnet. Near T_c, the scaling form of the magnetic correlation function between the local magnetization $\mathbf{M(r)}$ at $\mathbf{r}$ and that at origin is given by

$$\langle \mathbf{M(r)} \cdot \mathbf{M(0)} \rangle = \frac{1}{r^{d-2+\eta}} f(r/\xi), \qquad (1.34)$$

where ξ is the correlation length and the function f depends only on r/ξ. This form ensures the correct variation of the susceptibility. The correlation length varies as

$$\xi \sim \varepsilon^{-v}. \qquad (1.35)$$

Correspondingly, the average distance R is expected to vary as $R \sim N^v$. Also, the probability $P(\mathbf{r})$ of arriving at $\mathbf{r}$ with N self-avoiding walks is of the form

$$P_N(r) = \frac{N_N(0j)}{N_T}$$
$$= R^{-d}\phi(r/R), \qquad (1.36)$$

where $r = a|j|$ and $R \sim N^v$. The function ϕ that corresponds to f in Eq. (1.34) can be shown to vary as

$$\phi(x) \sim x^{1/(1-v)}. \qquad (1.37)$$

Table 12.2. Scaling correspondences

	Magnet	Self-avoiding walk
Criticality	ε	$1/N$
Spin dimension	n	0
Correlation length	$\xi \sim |\varepsilon|^{-\nu}$	$R \sim N^\nu$
Susceptibility	$\chi \sim |\varepsilon|^{-\gamma}$	$N_T \sim \bar{z}^N N^{\gamma-1}$
Energy	$E \sim -c\varepsilon^{1-\alpha}$	$N_N(a) \sim N^{-2+\alpha}$

For small r the number $N_N(01)$ is given by the total number of polygons that will close with a final step to the origin. This number is expected to be determined by the volume ratio a^d/R^d and the probability for N steps:

$$N_N(01) = \bar{z}^N(a/R)^d. \tag{1.38}$$

In terms of the specific heat exponent α and Kadanoff's relation,

$$\alpha = 2 - \nu d,$$

the number of walks varies as

$$N_N(01) = N^{-2+\alpha}. \tag{1.39}$$

Using Eq. (1.31) and the first equality of Eq. (1.36), one obtains

$$P_N(a) = \frac{N^{1-\gamma}}{R^d}. \tag{1.40}$$

Omitting further details, some of the corresponding scaling relations are listed in Table 12.2. In this table, the coordination number $\bar{z}$ is less than the number of the nearest neighbors. For instance, in a simple cubic lattice it is 4.68 instead of 6. $N_N(a)$ is the total number of steps to return to the site nearest to the starting point.

In actual polymer solutions, the number N of segments is not uniform but is distributed about an average $\langle N \rangle$. It has been shown that such an average can be related to a grand-ensemble average in a magnetic system.

12.2. Polymer solutions

The term "polymer" comes from the Greek words "polys meros", meaning many parts. Therefore, in principle, the term "polymer molecules" has a more restrictive meaning than "macromolecules." In what follows, we shall be concerned with two limiting models, a compact molecular model for hemoglobin, tobacco mosaic virus, or liquid crystal molecules, and a flexible-chain model for rubber, polystyrene, etc. In general, these molecules are identified in dilute solutions because attempted heating into a gas phase causes decomposition. These solutions have unique properties that differ in many respects from those at ordinary solutions.

In an ideal solution of two similar species, the properties depend on the molecular fraction

$$x = \frac{N_1}{N_1 + N_2}.$$

For instance, the mixing entropy is given by

$$\Delta S = Nk[x \ln x + (1 - x) \ln(1 - x)], \tag{2.1}$$

where N is the total number of molecules and k is the Boltzmann constant. However, when huge macromolecules are dissolved in a solvent of low-molecular-weight molecules, the size difference between the solute and solvent molecules necessitates the use of a volume fraction ϕ defined by

$$\phi = \frac{rN_r}{N_1 + rN_r}, \tag{2.2}$$

where N_1 and N_r are the numbers of the solvent molecules and polymers respectively. The solvent molecules are considered to be monomers, r being the relative size of the polymers. This volume fraction represents the probability of finding the solute macromolecules in the solution. For $r = 1$, $\phi = x$, but since r can be as large as 10^3–10^4, ϕ differs from x in general. In fact, for $r \gg 1$, the mixing entropy becomes

$$\Delta S = Nk[x \ln \phi + (1 - x) \ln (1 - \phi)]. \tag{2.3}$$

12.2.1. Dilute solutions

The huge size of polymer molecules causes large deviations from ideal solution behavior. For instance, the osmotic pressure π deviates from van't Hoff's law in the manner

$$\frac{\pi}{N_0 kTc} = \frac{1}{M} + A_2 c + A_3 c^2 + \cdots, \tag{2.4}$$

where N_0 is Avogadro's number, M is the molecular weight of a polymer molecule, c is the weight concentration of the solute molecules, and the "second virial" coefficient A_2 is given by [7]

$$A_2 = -\frac{1}{2Mm} \int [g_{22}(r) - 1] \, d\mathbf{r}, \tag{2.5}$$

where $g_{22}(r)$ is the radial distribution function of the solute molecules and m is the mass of a polymer.

The second virial coefficient is generally large for macromolecules, even for dilute solutions, making its determination important. It can be determined from the slope of π/RTc plotted against c. It can also be determined by light scattering based on the formula [8]:

$$\frac{Hc}{\tau} = \frac{1}{M} + 2A_2c + \cdots, \tag{2.6}$$

where τ is the turbidity, and

$$H = \frac{32\pi^3}{3N_0} \frac{n^2}{\lambda_0^4} \left(\frac{\partial n}{\partial c}\right)^2_{p,T}. \tag{2.7}$$

Here, n is the refractive index, and λ_0 is the wavelength of the incident light.

The tobacco mosaic virus, many liquid crystal molecules, hemoglobin, and others, are compact and elongated. For example, tobacco mosaic virus molecules are cylindrical with diameter 150 Å and length 2700 Å. For these molecules, considered to be rigid bodies, it is convenient to express A_2 such that [9]

$$A_2 = \frac{4N_0v_0}{M^2} f. \tag{2.8}$$

Here, v_0 is the volume of the solute molecule, N_0 is the Avogadro number, and f is a dimensionless function of the molecular shape. The function f is 1 for a spherical molecule, but is always larger than 1 if the shape deviates from a sphere. The exact formulas for rigid convex molecules and dumbbell molecules are known.

For rigid-rod molecules with length $2b$ and radius a, it is given exactly by

$$f = \frac{1}{4}\left[1 + \frac{b}{a}\left(1 + \frac{a}{2b}\right)\left(1 + \frac{\pi a}{2b}\right)\right]. \tag{2.9}$$

Hence, for $b \gg a$, f becomes very large. The corresponding value of A_2 is

$$A_2 = \frac{2\pi N_0 a b^2}{M^2}. \tag{2.10}$$

The second virial coefficient A_2 of long rigid molecules is generally proportional to the square of their length, although its numerical factor depends on the particular shape. Hence, A_2 can be very large for long molecules. One can show that it is also large for flat molecules.

Chain polymer molecules tend to take on a randomly coiled shape, particularly in poor solvents. In good solvents, they are stretched. The molecular size and shape depend also on the properties of the repeating units, called *segments*. In rubber molecules such as polyisoprene, the molecular rotation about the double bond in each isoprene C_5H_8 group is easy. As a consequence the entire chain can be randomly coiled, allowing the use of the simplest random walk model. In this case, A_2 is determined in the following form [10]:

$$A_2 = \frac{N_0}{m^2} b_2 F(x), \qquad (2.11)$$

where m is the molecular weight of the repeating unit (segment),

$$x = 2\left(\frac{9}{2\pi}\right)^{3/2} N^{1/2} b^{-3} \frac{m^2}{N_0} b_2 \qquad (2.12)$$

is a dimensionless variable, N is the number of segments that are separated from each other by b, and

$$b_2 = 2\pi \int_0^\infty [1 - e^{-u(s)/kT}) s^2 \, ds. \qquad (2.13)$$

$u(s)$ is the potential of average force between the segments of the flexible chain molecules. $F(x)$ is a somewhat complicated function that depends on the distribution of the segments within the molecule. Near $x \sim 1$ its approximate form is

$$F(x) = \frac{1}{1 + 1.01x}. \qquad (2.14)$$

In the literature, a parameter

$$z = 2^{3/2} x$$
$$= (3/2\pi)^{3/2} (N^{1/2} b_2) b^{-3}$$

is sometimes used, with b_2 denoted as β.

Note that $b_2(T)$ is the second virial coefficient of the segments, and when $b_2(\Theta) = 0$ at the *theta-point* $T = \Theta$, $A_2(\Theta)$ also vanishes independently

Table 12.3. Exponent of the second virial coefficient

t	Polymer–solvent	Range of $M \times 10^{-4}$
0.23	Polystyrene–dichlorethane	0.25–178
0.23	Polystyrene–butanone	0.25–178
0.22, 0.25	Polystyrene–toluene	3–61
0.14, 0.17	Polyisobutyrene–cyclohexane	3.7–72.2
0.35	Polymethylmethacrylate–acetone	7.2–158

of the molecular weight M:

$$b_2(\Theta) = 0 \rightarrow A_2(\Theta) = 0. \tag{2.15}$$

This Θ-point property is theoretically approximate but holds rather well experimentally.

The second virial coefficient A_2 of a chain polymer molecule may be expressed in powers of the number N of segments such that

$$A_2 \sim N^{-t}. \tag{2.16}$$

The exponent μ is then given by

$$t = -\frac{1}{2} \frac{d \ln F}{d \ln x}. \tag{2.17}$$

α vanishes for $x = 0$ and increases rather slowly as x increases.

The experimental values of t range from around 0.15 to 0.35, the average value being around 0.23. Some examples of the exponents are listed in Table 12.3. Note that high exponents such as 0.25 and 0.35 in Table 12.3 have been questioned.

If a randomly coiled molecule is approximated by a sphere of radius R, Eq. (1.2) can be used with R proportional to N^ν. In the random walk model, $\nu = 1/2$ and $t = 0.5$. However, in the presence of volume exclusion due to the finite size of the segments we use

$$R \sim \alpha N^{1/2}. \tag{2.18}$$

with a volume expansion coefficient α. The second virial coefficient is then given by

$$A_2 = K \frac{\alpha^3}{N^{1/2}}. \tag{2.19}$$

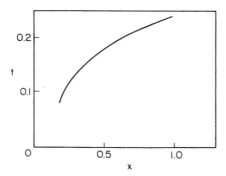

FIG. 12.2. The molecular weight dependence of the exponent t of the second virial coefficient, $A_2 \sim N^{-t}$. The parameter x is proportional to $N^{1/2}$.

Here K is a proportionality constant. In the limit $N \to \infty$, A_2 is expected to vary with exponent

$$t = 0.2. \tag{2.20}$$

Equation (2.19) shows that the molecular weight dependence of A_2 increases due to the presence of volume exclusion. This increase is of the same nature as that indicated by Eq. (2.18). Indeed, the exponent determined from Eq. (2.17) depends on N because if each segment has a volume represented by a hard sphere of diameter a we expect

$$x \sim N^{1/2}(a/b)^3. \tag{2.21}$$

For x of order 1 we expect that the exponent of A_2 varies as

$$t = 0.5\left(1 - \frac{1}{1 + 1.01x}\right). \tag{2.22}$$

This theoretical result is illustrated in Fig. 12.2 as a function of x. The exponent t is 0.25 at $x = 1$ and 0.21 at $x = 0.7$.

When $A_2(T = \theta) = 0$, the third virial coefficient A_3 becomes important. While A_3 can be evaluated as in the case of A_2, it is clear that the series of Eq. (2.4) becomes meaningless for high concentrations. Therefore, let us try to examine the behavior of the osmotic pressure without using expansion.

We remark first that for small concentrations polymer molecules are separated from each other. The volume occupied by each molecule is well defined, and the molecule takes on a shape that is determined by its composition, the temperature, and the solvent. This is the excluded volume limit, because the shape is essentially determined by the excluded volume effect.

12.2.2. *Concentrated solutions*

In the high-concentration limit, the segments of all the polymers will uniformly fill out the entire space. Therefore, all the segments, whether they belong to the same polymer molecule or not, become more or less equivalent. Hence, the polymer chain will behave like an ideal chain with a shape determined by a random walk model. This observation is very simple but important, so that it is worth examining the approach to this limit in a somewhat elaborate way.

We view each polymer a sphere of radius R. The critical volume concentration ϕ_c, corresponding to close-packing of the entire system by the polymer spheres, is given by [2]

$$\phi_c = \frac{Na^3}{R^3} = \left(\frac{a^3}{b^3}\right) N^{1-3\nu}.$$
$$\sim N^{-4/5}. \tag{2.23}$$

for $\nu = 3/5$. Note that for high-polymer molecules ϕ_c is generally small. In fact, the larger N the smaller ϕ_c.

When the critical concentration is reached the physical properties of the system may become independent of N. Although polymer molecules become immiscible before reaching such a limiting concentration, we continue defining an osmotic pressure as in dilute cases, and drop the first ideal gas term in Eq. (2.4). Furthermore, instead of the weight concentration c of polymer molecules, we use the number density n of solvent molecules because this is the quantity that is expected to determine the physical properties. Obviously, it is N times the number density of polymer molecules.

For high concentrations, and dropping the first term, we rewrite Eq. (2.4) such that

$$\frac{\pi}{kT} = \frac{n}{N} f\left(\frac{nR^3}{N}\right), \tag{2.24}$$

where n is the number density of the solvent molecules and $f(x)$ is an unknown function. We assume that a power law holds:

$$f(x) \sim x^m. \tag{2.25}$$

For $R \sim 3/5$, $m = 5/4$ makes the right-hand side of Eq. (2.24) independent of N. That is,

$$\frac{\pi}{kT} \sim n^{9/4}. \tag{2.26}$$

This represents an asymptotic variation of the osmotic pressure with the concentration of solvent molecules.

12.3. Viscosity of chain polymer solutions

Solutions of chain polymer molecules are highly viscous even at low concentrations. This is due to the largeness of the molecular weight. In fact, the discovery of the very existence of high-molecular-weight molecules stemmed from the high viscosity. Moreover, the viscosity of not only solutions but also that of melts depends on the molecular weight, as discussed in what follows.

12.3.1. Dilute solutions

In dilute solutions a viscosity increase over that of the solvent is expected to be proportional to the concentration of the solute polymer molecules. In the limit of zero concentration the increase is essentially due to a single molecule. It is customary to express this limiting viscosity increase in terms of the *intrinsic viscosity* defined by

$$\lim_{c \to 0} \frac{\eta - \eta_0}{\eta_0 c} = [\eta]. \tag{3.1}$$

Here η is the viscosity of the solution, η_0 is that of the solvent, and c is the weight concentration. The intrinsic viscosity has dimensions of cm^3/g. If the weight concentration is expressed in $g/100 \, cm^3$, as often done experimentally, $[\eta]$ expressed in cm^3/g must be given a factor $1/100$.

For high-polymer solutions, $[\eta]$ depends on the molecular weight of the polymer molecules as

$$[\eta] = KM^s. \tag{3.2}$$

The exponent varies with systems. While many polymer solutions have exponents below around 0.8, there are cases with higher values. Some examples of the exponent α are listed in Table 12.4 [11].

For spherical molecules Einstein derived a formula:

$$[\eta] = 2.5 N_0 \frac{v_0}{M}, \tag{3.3}$$

where v_0 is the volume of a solute molecule. For randomly coiled chain molecules, Flory used $v_0 \sim R^3$, where R is the radius. In the random walk model, $R \sim N^{1/2}$, and

$$s = 1/2. \tag{3.4}$$

Table 12.4. Viscosity exponent

s	Solute	Solvent	$M \times 10^{-4}$
0.74	Rubber	Benzene	8–28
0.85	Polyvinylchloride	Cychlohexanone	3–10
0.77	Polystyrene (isotactic)	Benzene	4–75
0.92	Polyvinylchloride	Tetrahydrofuran	3–17
0.83	Polymethylmethacrylate	Chloroform	41–326
0.905	Cellulose	Cupriethylenediamine	2–54
0.87	Poly-γ-benzyl-L-glutamate	Dichloroacetic acid	2–34

With the excluded volume effect, $R \sim \alpha N^{1/2}$, where α is a volume expansion coefficient. Hence,

$$[\eta] = k\alpha^3 M^{1/2}. \tag{3.5}$$

If we adopt Eq. (2.18), $\alpha \sim M^{1/10}$ is the maximum variation. Hence,

$$\text{Max } s = 0.8. \tag{3.6}$$

Experimentally, the viscosity exponent s is between 0.5 and 1. Table 12.4 includes stiff molecules such as cellulose, but there are deviations from this Flory's maximum exponent. In particular, the derivatives of polysaccharides have exponents close to 1. The exponent s varies, between polymer molecules and within the same molecule, with the solvent because in good solvents the molecular chain is stretched.

In Einstein's formula, the flow of the solvent acts directly on the molecule. However, in the case of a chain polymer molecule, the solvent flow that reaches interior segments may be "shielded" by the outer segments. As a result, the hydrodynamical force $F(r)$ on the segment at $\mathbf{r}$ referred to the center of gravity will be reduced. In a self-consistent approach, $\mathbf{F(r)}$ satisfies [12]

$$\mathbf{F(r)} = \mathbf{F}^0(\mathbf{r}) - \zeta \int \mathsf{T}(\mathbf{r} - \mathbf{r}') \cdot \mathbf{F(r')} \rho(\mathbf{r} - \mathbf{r}') \, d\mathbf{r}', \tag{3.7}$$

where $\zeta = 6\pi\eta_0 a$ is Stokes friction, when the segment is a sphere of radius a, and η_0 is the viscosity of the solvent. $\mathbf{F(r)}$ is the direct hydrodynamical force. $\mathsf{T(r)}$ is Oseen's tensor given by

$$\mathsf{T}(r) = \frac{1}{8\pi\eta_0 r}\left(1 + \frac{\mathbf{rr}}{r^2}\right). \tag{3.8}$$

$\rho(\mathbf{r} - \mathbf{r}')$ is a segment distribution function at $\mathbf{r}$ from another segment at $\mathbf{r}'$.

The Oseen tensor represents a hydrodynamical interaction. It decreases with r very slowly as in the case of an electron gas. This is the reason why one can talk about hydrodynamical shielding. The exact expression for the relative distribution function $\rho(\mathbf{r} - r')$ is unknown. However, on the average the hydrodynamical interaction is expected to vary with an average segment separation. In the presence of an excluded volume effect, this separation varies as $N^{0.5}/\alpha$.

The viscosity can be evaluated from the energy loss due to the force. A molecular-field result for the intrinsic viscosity is

$$[\eta] = [\eta]_0 \frac{1}{1 + AM^{0.5}/\alpha}, \tag{3.9}$$

where $[\eta]_0$ is the intrinsic viscosity in the absence of the hydrodynamical interaction and is proportional to $R^2 \sim M\alpha^2$, where R is the average radius of the entire chain. The constant A depends on a/b, which is the ratio of the segmental radius to the bond length. Hence, we arrive at

$$[\eta] = K \frac{M\alpha^2}{1 + AM^{0.5}/\alpha}. \tag{3.10}$$

This equation approaches the Flory form in the limit when the second term in the denominator is large. However, before reaching this limit, the exponent v can exceed 0.8. Especially, for small excluded volume effects, $\alpha \sim 1$ so that

$$\frac{M}{[\eta]} = A + BM^{1/2}. \tag{3.11}$$

That is, $M/[\eta]$ is a straight line when plotted against $M^{1/2}$: Formula (3.10) can then be considered to interpolate the two limiting expressions, Eq. (3.11) and Flory's (3.5).

12.3.2. Concentrated solutions and melts

The intrinsic viscosity represents the zero-concentration result. The viscosity theory becomes significantly difficult for a low but finite concentration. In fact, even for spherical molecules, the extension of Einstein's formula to second order involves serious mathematical difficulties. In a local field-type approximation, it may be shown that

$$\eta = \eta_0 \left(1 + \frac{5}{2} \frac{\phi}{1 - \phi}\right), \tag{3.12}$$

where ϕ is the volume fraction. However, the presence of a divergence indicates that this result is not quite self-consistent.

In the extreme limit, one is concerned with melt viscosity. This viscosity, denoted by η, depends on the molecular weight of the polymer molecules. Melt viscosity depends strongly on shear force. However, in the limit of zero shear, it can be expressed as

$$\eta = KM^{\kappa}, \tag{3.13}$$

where $\kappa = \kappa(M)$. According to Fox and Flory [13], the melt viscosity above and below a certain molecular weight M_c follows

$$\kappa = \begin{cases} 3.4 & M > M_c; \\ 1 & M < M_c. \end{cases} \tag{3.14}$$

The "critical" molecular weight M_c depends on the polymer. It is 17,000 for polyisobutylene and 38,000 for polystyrene.

The melt viscosity of several polymers is illustrated in Fig. 12.3. The curves have been shifted arbitrarily along the ordinate to avoid overlap. The abscissa is actually $\log(10^{17}X_w)$ where X_w is the weight average

$$X = \frac{Z(\langle s^2 \rangle_0 / M)\phi}{v_2}, \tag{3.15}$$

where Z is the number of atoms in the backbone of a polymer chain, $\langle s^2 \rangle_0$ is the unperturbed radius of gyration, ϕ is the volume fraction of the polymer, M is the molecular weight of the polymer, and v_2 is the polymer specific volume. Hence, the abscissa represents $\ln Nc$ with c for the concentration of polymer. In most cases, the critical value of X is given by

$$10^{17}X_c = 400.$$

Although Fig. 12.3 indicates that Eq. (3.14) is correct, there are cases in which deviations occur. Also, the transition between the exponents may not be sharp.

The molecular weight dependence of melt viscosity appears to contradict the view that at high concentrations the physical properties do not depend on molecular weight. However, nonequilibrium cases can be different from the case of equilibrium. If a segment is forced to move under a shear force, the other segments of the same polymer must also move to accommodate the motion. In order to avoid bond stretching, which costs energy, the entire chain moves in a coherent way following the external shear force. Therefore, the chain connectivity plays an essential role in viscosity even though the density of the segments may stay the same during the motion. Another

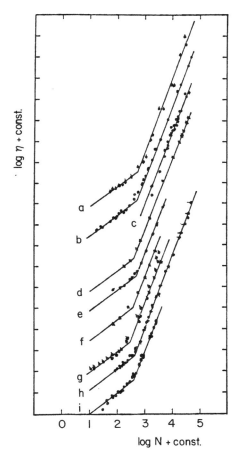

Fig. 12.3. Melt viscosity of polystyrenes. The curves have been shifted vertically for visibility. The abscissa is proportional to ln N. Curves: a, polydimethyl siloxane; b, polyisobutylene; c, polyethylene; d, polybutadiene; e, polytetramethyl p-siliphenyl siloxane; f, polymethylmethacrylate; g, polyethylene glycol; h, polyvinylacetate; i, polystyrene. (From Berry and Fox [13])

important factor is molecular entanglement. The possibility of entanglement in short chains is expected to be small, but it increases strongly with the chain length. This difference is schematically illustrated in Fig. 12.4.

The melt viscosity η can be obtained by evaluating the energy loss per second per unit volume [14]. For a fluid with a velocity gradient q, this is given by ηq^2. In order to determine η, we evaluate the same energy loss microscopically.

When the molecular weight is small, the effect of molecular entanglements is not strong. In this case, individual molecules can move independently of other molecules. Thus, all the polymers become equivalent.

FIG. 12.4. Polymer melts: (a) short chains; (b) entangled long chains.

In a polymer melt, the segments are distributed more or less uniformly. If each segment has radius a and bond length b, the segmental volume is approximately a^2b so that the segmental density is $1/(a^2b)$. In a given polymer, a segment at an average position R from the center of gravity moves with a speed Rq. Its resistance is ζRq, where ζ is the friction coefficient of the segment. Hence, its energy loss per unit volume per unit time is $(\zeta qR)(qR)/(a^2b)$. Equating this with ηq^2, we find

$$\eta = \frac{\zeta R^2}{a^2 b}. \tag{3.16}$$

The exponent 3/5 may not be used for this case because it is the result for $N \to \infty$. For a spherical polymer with a small excluded volume effect, one might expect $R^2 \sim N$.

As the chain length increases, the molecules start entangling with each other. The energy dissipation of a cluster of entangled molecules must now be investigated. Fortunately, one can assume that similar clusters fill the entire system more or less uniformly as the segments do for the case of short chains. The friction coefficient will now be of order $N\zeta$, while the average "distance" of a representative unit will be scaled up by a factor R/b to $R(R/b)$. As a result,

$$\eta = \frac{\zeta N R^4}{a^2 b^3}. \tag{3.17}$$

Hence, for $R = bN^\nu$ with $\nu = 3/5$,

$$\eta \sim N^{3.4}. \tag{3.18a}$$

However, $\nu = 1/2$ in concentrated solutions. Hence, we expect

$$\eta \sim N^3. \tag{3.18b}$$

Equation (3.17) can be rederived as follows. The energy loss for moving a mass of Nm at speed qR is $(qR)^2 Nm/\tau$ per unit time, where τ is a relaxation time. The number density of the segments of a polymer within its volume is N/R^3. Hence, there are $(N/R^3)\, d\mathbf{R}$ segments in a volume element $d\mathbf{R}$. We can use $d\mathbf{R} = 4\pi R^2 b$. On the average, the segments of other polymers in the system are distributed uniformly at a density $1/(a^2 b)$. Hence, there are $R^3/(a^2 b)$ segments within a polymer. The number density of polymers is $1/[N(a^2 b)]$. Hence, the number density of molecular entanglements at a distance R is

$$\frac{1}{Na^2 b}\frac{N}{R^3}\frac{R^3}{a^2 b} R^2 b = \left(\frac{1}{a^2 b}\right)^2 R^2 b = \frac{R^2}{a^4 b}.$$

The second equality shows that we are counting pairs of segments within the volume element $R^2 b$. The energy required to sustain the fluid flow is then proportional to $N(qR^2)^2$, which corresponds to ηq^2. Hence, η is proportional to NR^4:

$$\eta = \frac{mNR^4}{\tau(a^4 b)}, \tag{3.19}$$

as in Eq. (3.17).

At the same time, we see that the friction coefficient

$$\zeta = \frac{m(b/a)^2}{\tau}. \tag{3.20}$$

It is conceivable that ζ for a segment is proportional to a. It is now scaled up by a factor (b/a) due to molecular entanglements.

If molecular entanglements are independent of temperature, the main source of the temperature dependence of η is in τ. This quantity may be assumed to vary as $\exp[-E/(T - T_0)])$, where E is an activation energy and T_0 is a minimum temperature that is necessary for the dissipation process. The viscosity is expected to vary as

$$\ln \eta = A + \frac{B}{T - T_0}. \tag{3.21}$$

This equation is known as the Vogel equation.

12.4. Viscosity of rodlike molecules

The intrinsic viscosity of rodlike molecules can be evaluated hydrodynamically on the basis of respectively a compact model [15] or a bead model

[15, 16]. We shall adopt the latter model since it yields the result of the compact model in the continuous limit.

Let us consider a rodlike polymer consisting of $2N + 1$ segments, each being a sphere of radius a. It is placed in a laminar flow:

$$\mathbf{v} = (0, qx, 0), \tag{4.1}$$

where q is the velocity gradient and the origin of the coordinates is at the center segment. This velocity field can be decomposed into the rotational flow given by $((-q/2)y, (q/2)x, 0)$ and the shear flow $((q/2)y, (q/2)x, 0)$.

On the average, the ith segment of this molecule will rotate with velocity

$$\mathbf{u}_i = (-(q/2)y_i, (q/2)x_i, 0). \tag{4.2}$$

The frictional force $\mathbf{F}_i$ acting on the segment may be determined from the self-consistent equation due to Riseman and Kirkwood [15]:

$$\mathbf{F}_i = \mathbf{F}_i^0 - \zeta \sum_s{}' T_{is} \cdot \mathbf{F}_s \tag{4.3}$$

where $\zeta = 6\pi\eta_0 a$ is the Stokes friction coefficient, and

$$\mathbf{F}_i^0 = -\zeta(\mathbf{v}_i^0 - \mathbf{u}_i). \tag{4.4}$$

η_0 is the viscosity of the solvent. The second term in Eq. (4.3) represents the effect of hydrodynamical interaction between the beads with the Oseen tensor T_{is} defined by

$$T_{is} = \frac{1}{8\pi\eta_0 r_{is}}\left[1 + \frac{\mathbf{r}_{is}\mathbf{r}_{is}}{r_{is}^2}\right]. \tag{4.5}$$

The intrinsic viscosity can be obtained by evaluating the energy loss per second per unit volume. In terms of the solution $\mathbf{F}_i$ of Eq. (4.3), it is given by

$$[\eta] = -\frac{N_0}{\eta_0 q^2 M}\sum_i (\mathbf{F}_i \cdot \mathbf{v}_i^0), \tag{4.6}$$

where N_0 is the Avogadro number and M is the molecular weight of the polymer molecule.

In order to solve Eq. (4.3), it is convenient to define a vector and tensor in a $(2N + 1)$-dimensional molecular space as follows [16]:

$$\begin{aligned}
\mathbf{F} &= (\mathbf{F}_i) \qquad (i = -N, -N+1, \ldots, N) \\
\mathsf{T} &= \begin{cases} T_{ij} & (i \neq j) \\ 0 & (i = j). \end{cases}
\end{aligned} \tag{4.7}$$

For rodlike molecules, the Oseen tensor can be expressed as

$$T_{ij} = \frac{1}{\lambda|i-j|}(1 + \mathbf{ee}), \tag{4.8}$$

where $\mathbf{e}$ is a unit vector in the direction of the rod, and $\lambda = 8\pi\eta_0 b$.

We note that the tensor T satisfies

$$\mathsf{T}^s = \frac{t^s}{\lambda^s}(2^s\mathbf{ee} + 1 - \mathbf{ee}),$$

where t is a tensor in the molecular space with the elements

$$t_{ij} = |i-j|^{-1}.$$

By making use of this relation, the exact solution of Eq. (4.3) can be expressed as follows:

$$\mathbf{F} = [(1 - \mathbf{ee})/(1 + \lambda_0 t)]\cdot\mathbf{F}^0 + [\mathbf{ee}/(1 + 2\lambda_0 t)]\cdot\mathbf{F}^0, \tag{4.9}$$

Here, with b for the bond length the parameter λ_0 is defined by

$$\lambda_0 = \frac{\zeta}{\lambda}$$

$$= \frac{3a}{4b}$$

The force $\mathbf{F}$ given by Eq. (4.9) can be rewritten in a more convenient form in terms of a vector $\mathbf{G}$ defined by

$$\mathbf{G}(\lambda_0) = (1 + \lambda_0 t)^{-1}\cdot\mathbf{G}_0. \tag{4.10}$$

The ith component of the vector satisfies:

$$G_i = G_i^0 - \lambda_0 \sum_j{}' |i-j|^{-1}G_j. \tag{4.11}$$

The force is then

$$\mathbf{F} = -\tfrac{1}{2}\zeta qb[(1 - \mathbf{ee})\cdot\mathbf{s}G(\lambda_0) + \mathbf{ee}\cdot sG(2\lambda_0)], \tag{4.12}$$

where

$$\mathbf{s} = (\sin\theta\sin\phi, \sin\theta\cos\phi, 0). \tag{4.13}$$

is an orientational unit vector.

If the chain is uniformly oriented in space, the intrinsic viscosity is given by an average over the angles (θ, ϕ) as follows:

$$[\eta] = \frac{N_0 \zeta b^2}{10 \eta_0 M} \sum_i i[G_i(\lambda_0) + \tfrac{2}{3} G_i(2\lambda_0)]. \tag{4.14}$$

N_0 is Avogadro's number.

In the limit $N \to \infty$, Eq. (4.11) can be converted into an integral equation

$$G(x, \lambda_0) = Nx - \lambda_0 \int_{-1}^{1} |x - y|^{-1} G(y, \lambda_0)\, dy. \tag{4.15}$$

Similarly, the function $F(\lambda_0)$ defined by

$$F(\lambda_0) = \frac{3}{2N} \int_{-1}^{1} x G(x, \lambda_0)\, dx \tag{4.16}$$

can be used to give

$$[\eta] = \frac{N_0 \zeta l^2}{180 \eta_0 M_0} [\tfrac{3}{2} F(\lambda_0) + F(2\lambda_0)], \tag{4.17}$$

where $l = 2N$ represents the total length of the molecule and M_0 is the molecular weight of a segment.

The intrinsic viscosity thus obtained is the exact consequence of the original integral equation for the force $\mathbf{F}_i$ for the limiting case of uniform molecular orientation. In the derivation a diagonalization approximation of the Oseen tensor, which has frequently been adopted in the literature, is not used. The function $F(\lambda_0)$ can be obtained numerically. Its asymptotic form for large λ_0 is

$$F(\lambda_0) \to (2\lambda_0 \ln l/b)^{-1}. \tag{4.18}$$

Note that the parameter λ_0 determines the hydrodynamical interaction.

The above result corresponds the case in which the orientational diffusion of molecules is so strong that the molecules are uniformly oriented in space. In general, however, the orientation is determined by the balance between the diffusion and the fluid flow. If $f(\theta, \phi, t)$ is the orientational distribution function, it may be determined from

$$\frac{\partial f}{\partial t} = D\left[(\sin \theta)^{-1} \frac{\partial}{\partial \theta}\left(\sin \theta \frac{\partial f}{\partial \theta} \right) + (\sin^2 \theta)^{-1} \frac{\partial^2 f}{\partial \phi^2} \right]$$
$$- q[\tfrac{1}{4}(\sin 2\theta \sin 2\phi) \frac{\partial f}{\partial \theta} + \cos^2 \phi \frac{\partial f}{\partial \phi} - \frac{3}{2} f \sin^2 \theta \sin 2\phi], \tag{4.19}$$

where D is the rotational diffusion coefficient of the molecule.

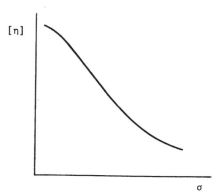

FIG. 12.5. Schematic stress dependence of $[\eta]$.

The solution to this differential equation may be obtained for small q/D. The second order, and for the stationary case,

$$f(\theta, \phi) = \frac{1}{4\pi} [1 + \tfrac{1}{2}\sigma \sin^2 \theta \sin 2\phi + \tfrac{1}{16}\sigma^2(\sin^4 \theta - \tfrac{8}{3} \sin^2 \theta^2 \cos 2\phi - \tfrac{8}{15}$$

$$- \sin^4 \theta \cos 4\phi)]. \tag{4.20}$$

Here,

$$\sigma = \frac{q}{2D}$$

is a dimensionless parameter that compares the effect of fluid flow to molecular diffusion.

The intrinsic viscosity can be obtained by taking the average of Eq. (4.5) in terms of $f(\theta, \phi)$. To second order in σ, it is given by

$$[\eta] = [\eta]_1 - \sigma^2[\eta]_2, \tag{4.21}$$

where

$$[\eta]_1 = \frac{N_0 \zeta l^2}{180 \eta_0 M_0} [F(2\lambda_0) + \tfrac{3}{2}F(\lambda_0)], \tag{4.22}$$

$$[\eta]_2 = \frac{N_0 \zeta l^2}{180 \eta_0 M_0} (\tfrac{1}{105}[19F(\lambda_0) - 4F(2\lambda_0)]). \tag{4.23}$$

The intrinsic viscosity $[\eta]$ decreases in proportion to σ^2 due to fluid flow. In such cases, the term *structural viscosity* is used and $[\eta]$ varies as illustrated schematically in Fig. 12.5. There is a decrease because the molecules can flow more easily if oriented in the direction of fluid flow.

We can take a continuous limit in the above results. For long rodlike molecules, we find

$$[\eta] = \frac{N_0 \zeta l^2}{180 \eta_0 M_0} \left[\lambda_0 \ln\left(\frac{l}{b}\right) \right]^{-1} \left[1 - \left(\frac{17\sigma^2}{210}\right) \right]. \tag{4.24}$$

A similar result has been obtained from a compact model also. The $l^2/\ln l$ variation of $[\eta]$ is characteristic of rodlike molecules. For large l, the viscosity can become very large in comparison with the globular case.

12.5. Condensation of a chain polymer

The excluded volume effect originates from the repulsive forces between the segments. These forces are important at short distances, but at large distances particularly in poor solvents, the van der Waals attractive forces between the segments can cause an opposite effect. In fact, the chain configuration determined by these forces can be examined in a way similar to the case of an imperfect gas.

In order to establish this similarity in a more formal way [17], we introduce the relative probability $I(r)$ that the Nth segment is at a position r from the first segment, which is fixed at the origin. It is given by

$$I(r) = \int \exp[-\beta \Phi(\mathbf{r}^N)] \, d\tau, \tag{5.1}$$

where $\beta = 1/kT$, and $\Phi(\mathbf{r}^N)$ is the total potential between the segments:

$$\Phi(\mathbf{r}^N) = \sum_{i<j}^{N} [\phi_R(r_{ij}) + \phi_A(r_{ij})].$$

Here, the subscripts R and A stand respectively for repulsive and attractive. The volume element $d\tau$ is defined by

$$d\tau = \prod_{i=1}^{N} dr_i \left\{ \prod_{i=1}^{N-1} \delta(|\mathbf{r}_{i+1} - \mathbf{r}_i| - b)(4\pi b^2)^{-N} \, \delta(\mathbf{r} - \mathbf{r}_N). \tag{5.2} \right.$$

We introduce a "partition function" of the chain by

$$Z(\lambda) = \int \exp(-\lambda r^2) I(r) \, d\mathbf{r}, \tag{5.3}$$

where λ is a parameter. The square average end-to-end distance of the chain is then given by

$$\langle r^2 \rangle = -\left(\frac{\partial \ln Z}{\partial \lambda}\right)_{\lambda=0}. \tag{5.4}$$

The differential operation on the right-hand side suggests that the square average distance plays the role of the pressure of an "imperfect" gas.

In analogy with a gas, the chain is in a "gaseous" phase at high temperatures, where $\langle r^2 \rangle$ is proportional to N if there is no volume exclusion and to $N^{2\nu}$ if there is. As the temperature is lowered, the chain might condense into a liquidlike state in which $\langle r^2 \rangle$ is expected to be much smaller. In fact, if N hard-sphere segments are closely packed in a sphere of radius r, we expect

$$r^3 \sim (a/2)^3 N. \tag{5.5}$$

Hence,

$$\langle r^2 \rangle \sim N^{2/3}. \tag{5.6}$$

In order to discuss the possible condensation of a chain, we note that the partition function $Z(\lambda)$ can be rewritten such that

$$Z(\lambda) = Z_R(\lambda)\langle \exp[-\beta\Phi_A]\rangle_R, \tag{5.7}$$

where $Z_R(\lambda)$ is the partition function of the chain with only the repulsive potential. Let us call this case as the reference chain. The average $\langle \cdots \rangle_R$ is taken in this reference chain as follows:

$$\langle \cdots \rangle = \frac{\iint e^{-\lambda r^2} e^{-\beta\Phi} R(\cdots)\, d\tau\, d\mathbf{r}}{\iint e^{-\lambda r^2} e^{-\beta\Phi_R}\, d\tau\, d\mathbf{r}}. \tag{5.8}$$

The average $\langle r^2 \rangle$ is given by

$$\langle r^2 \rangle = \langle r^2 \rangle_R + \frac{\langle r^2 e^{-\beta\Phi}A\rangle - \langle r^2\rangle\langle e^{-\beta\Phi}A\rangle}{\langle e^{-\beta\Phi}A\rangle}, \tag{5.9}$$

where the first term corresponds to the reference chain and the second term represents the effect of attractive forces. Note that Eq. (5.9) is exact.

To proceed further, let us adopt a linear attractive potential given by

$$\phi_A(r) = -\gamma[1 - \gamma\mu r], \tag{5.10}$$

where γ and μ are positive parameters. This form can be adopted if a chosen point r does not coincide with the minimum point of the potential. The linear form simplifies further calculations. It is based on the observation that the end-to-end distance is expected to be much longer than the distance of the potential minimum whether or not the chain is in a condensed state. Hence, the effect of an attractive potential can be investigated in a region that is beyond the potential minimum. The divergence at large distances of the linear potential does not cause any problem because the segment distribution function is expected to fall off rapidly. In fact, a numerical analysis shows that a possible form for the segment distribution function may be expressed by [18]

$$P_N(r) = \frac{c}{N^{9/5}b^3} \left(\frac{r}{N^{3/5}b}\right)^{1/3} \exp\left(-\frac{r^{5/2}}{N^{3/2}b^{5/2}}\right). \qquad (5.11)$$

Here c is a normalization constant.

At high temperatures where the chain is in a gaseous state, the effect of the attractive potential can be evaluated to first order in $\beta\phi_A$ in accordance with

$$_N C_2 \left\{ \int r^2 \phi_A(r) P_N(r) \, d\mathbf{r} - \int \phi_A(r) P_N(r) \, d\mathbf{r} \int r^2 P_N(r) \, d\mathbf{r} \right\}. \qquad (5.12)$$

We arrive at

$$\langle r^2 \rangle = r_0^2 (1 - 0.7\beta c\gamma^2 \mu r_0) \qquad (5.13)$$

with $r_0 = N^{3/5}b$. Hence, $\langle r^2 \rangle$ decreases from the limiting value of r_0^2 as the temperature decreases. Note that $\mu\gamma^2 r_0 < 1$ because of the form of Eq. (5.10). A similar result can be obtained even if a random walk model is used although r_0 in that case is proportional to $N^{1/2}$.

In the opposite low-temperature region the square average may be evaluated approximately from

$$\langle r^2 \rangle = \frac{\langle r^2 \exp(-\beta\phi_A) \rangle}{\langle \exp(-\beta\phi_A) \rangle}, \qquad (5.14)$$

where Φ_A has been replaced by ϕ_A. The contributions of the two averages on the right-hand side can be evaluated by a steepest-descent method. A numerical analysis shows that

$$\langle r^2 \rangle \sim (T/T_0)N^{0.6}b^2, \qquad (5.15)$$

where

$$T_0 = \frac{\gamma^2 \mu b}{k} \qquad (5.16)$$

is the condensation point.

Equation (5.15) shows that the square average end-to-end distance in the condensed state varies as $N^{0.6}$. This variation is significantly different from the variation $N^{1.2}$ in the gaseous phase and is nearly the close-packed result in Eq. (5.6). Hence, we may conclude that condensation has taken place.

The collapse problem has been discussed in several ways [19]. For instance, an attempt has been made based on Flory's free-energy expression:

$$\frac{F}{kT} = 3\left(\frac{\alpha^2}{2} - \ln \alpha\right) + \frac{N}{2}\left(\rho W_1 + \rho^2 W_2 + \cdots\right). \tag{5.17}$$

Here, ρ is the average monomer concentration, and W_1 and W_2 are interaction parameters that depend on temperature. The former parameter is the excluded volume coefficient, which is proportional to $(T - \theta)$. The free energy has an interesting form even though a serious question exists concerning the logarithmic term. In particular, it can be expressed as a function of the expansion coefficient α by using R,

$$\rho = \frac{kN}{R^3},$$

where k is a constant and the radius R is proportional to $aN^{1/2}b$. For equilibrium $F(\alpha)$ must be stationary with respect to a variation in α. Accordingly, α is determined by

$$\alpha^5 - \alpha^3 - \frac{y}{\alpha^3} = \frac{kN^{1/2}W_1}{2b^3} = x, \tag{5.18}$$

where b is a bond length, and $y = k^2 W_2 b^{-6}$.

The behavior of α as a function of x depends sensitively on y. The chain contracts smoothly upon cooling if $y > 0.038$, but a van der Waals-type loop appears if $y < 0.038$. Hence, a first-order phase transition may be expected. However, x must be negatively large for very small α. The expansion parameter α changes smoothly with x in this region, which is away from the unstable region. Such a variation of α occurs because the original free energy is based on a virial expansion. In fact, the next-order term, which is proportional to $\rho^3 \sim \alpha^{-9}$, is more strongly divergent than the term $\rho^2 \sim \alpha^{-6}$. Although the high-order virial coefficients may be small in actual solutions, the free energy must be renormalized when the segment density ρ is large. Accordingly, the Θ-point itself may be renormalized. The situation is similar to the case of an imperfect gas, where Mayer's virial expansion fails to describe condensation.

In addition to these theoretical works, several numerical approaches have been made to the condensation and other configurational problems of

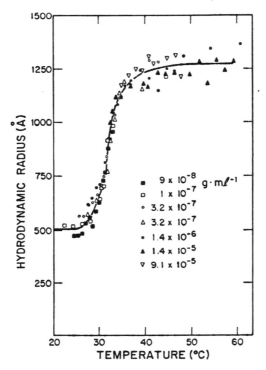

a polymer chain [20]. The existence of a collapsed state has long been expected theoretically, and it has actually been observed [21]. Figure 12.6 illustrates the light-scattering data obtained by Swislow et al. [22]. Here, the temperature dependence of the radius of polystyrene of a weight-average molecular weight 2.7×10^7 in cyclohexane is plotted for a series of concentrations. The transition occurs at 32°C, where the radius shows an abrupt change. The radius is around 1250 Å on the high-temperature side and 500 Å on the low-temperature side. This change is not very large in magnitude, but it is clear that a transition has taken place.

Thus, one can conclude that a single-chain polymer can be studied in analogy to a gas–liquid system or to a magnetic system. Such a relationship between apparently different systems makes condensed matter physics very interesting and attractive.

REFERENCES

1. A. Isihara, *Adv. Polymar Sci.* **7**, Band 4, 449 (1971).
2. P. G. De Gennes, *Riv. del Nuovo Cimento* **7**(3), 363 (1977); *Scaling Concepts in Polymer Physics* (Cornell University Press, 1979).
3. P. G. De Gennes, *Phys. Lett.* **38A**, 339 (1972).
4. M. Fisher and B. Hilley, *J. Chem. Phys.* **34**, 1253 (1961).
5. P. J. Flory, *Principles of Polymer Chemistry* (Cronell University Press, 1953).
6. D. S. McKenzie, *Phys. Rep.* **27C**(2) (1976).
7. W. G. McMillan and J. E. Mayer, *J. Chem. Phys.* **13**, 276 (1945). J. G. Kirkwood and F. P. Buff, *J. Chem. Phys.* **19**, 774 (1951).
8. P. Doty, B. H. Zimm, and H. Mark, *J. Chem. Phys.* **12**, 144 (1944), **13**, 159 (1945). B. H. Zimm, *J. Chem. Phys.* **16**, 1093 (1948).
9. A. Isihara, *J. Chem. Phys.* **18**, 1446 (1950). A. Isihara and T. Hayashida, *J. Phys. Soc. Jpn* **6**, 40, 46 (1951). B. H. Zimm, *J. Chem. Phys.* **14**, 164 (1946).
10. A. Isihara and R. Koyama, *J. Chem. Phys.* **25**, 712 (1956).
11. M. Kurata and W. H. Stockmayer, *Adv. Polymer Sci.* **3**, 196 (1963).
12. A. Isihara, *Adv. Polymer Sci.* **5**, 531 (1968).
13. T. G. Fox and P. J. Flory, *J. Appl. Phys.* **21**, 581 (1950). G. C. Berry and T. G. Fox, *Adv. Polymer Sci.* **5**, 261 (1968). J. D. Ferry, *Viscoeleastic Properties of Polymers* (Wiley, New York, 1980).
14. A. Isihara, *Polymer* (1991) to be published.
15. R. Simha, *J. Chem. Phys.* **13**, 188 (1945).
16. J. Riseman and J. G. Kirkwood, *J. Chem. Phys.* **18**, 512 (1950). J. G. Kirkwood and P. L. Auer, *J. Chem. Phys.* **19**, 281 (1951). Some of the errors in these articles have been pointed out by R. Ullman, *J. Chem. Phys.* **40**, 2422 (1964).
17. A. Isihara, *J. Chem. Phys.* **49**, 257 (1968).
18. A. Isihara and C. H. Isihara, *Physica* **81A**, 623 (1975).
19. D. S. McKenzie and M. A. Moore, *J. Phys. A* **4**, 282 (1971). C. Domb, A. J. Barrett, and M. Lax, *J. Phys.* **6A**, L82 (1973).
20. I. M. Lifshitz, *Zh. Eksp. Theor. Fiz.* **55**, 2408 (1968) [*Sov. Phys.—JETP* **28**, 1280 (1969)]. C. Domb, *Polymer* **15**, 259 (1974). P. G. de Gennes, *J. de Phys. Lett.* **36**, L55 (1975). A. R. Massih and M. A. Moore, *J. Phys. A* **8**, 237 (1975).
21. I. Webman, Joel L. Lebowitz, and M. H. Kalos, *Macromolecules* **14**, 1495 (1981). M. Bishop and J. P. J. Michels, *J. Chem. Phys.* **84**, 447 (1986). H. Meirovitch and H. A. Lim, *J. Chem. Phys.* **91**, 2544 (1989). D. C. Rapaport, *J. Phys. A* **10**, 637 (1977). A. Baumgartner, *J. Chem. Phys.* **72**, 871 (1980). F. L. McCrackin, J. Mazur, and C. M. Guttman, *Macromolecules* **6**, 859 (1973).
22. E. Slagowski, B. Tsai, and D. McIntyre, *Macromolecules* **9**, 687 (1976). M. Nierlich, J. P. Cotton, and B. Farnoux, *J. Chem. Phys.* **69**, 1379 (1978). G. Swislow, S. T. Sun, I. Nishio, and T. Tanaka, *Phys. Rev. Lett.* **44**, 796 (1980).

APPENDIX

MOLECULAR DISTRIBUTION FUNCTIONS

The statistical properties of N molecules are determined by the Boltzmann factor $\exp[-\beta H]$, where H is the Hamiltonian. However, a given physical quantity may be represented by a sum of terms that depend only on small numbers of molecules. For example, the density can be expressed by the sum:

$$n(\mathbf{r}) = \sum_{i-1}^{N} \delta(\mathbf{r} - \mathbf{r}_i),$$

and its statistical average is given in terms of the singlet distribution function. Likewise, those functions that depend on molecular pairs can be averaged by the pair distribution function. The case of the potential energy is a typical example. Therefore, it is important to determine these reduced distribution functions. In what follows we shall focus our attention on the pair distribution function since it plays a major role in describing equilibrium properties of liquids.

A.1. Graphical method

The pair distribution function can be determined by a graphical method. Since the number of particles in a graph is variable, it is natural to define the pair distribution function in a grand canonical ensemble rather than in a canonical ensemble.

The graphs for the pair distribution function for fluids in equilibrium can be generated from the Bloch equation for the density matrix:

$$\frac{\partial \rho}{\partial \beta} = -H\rho, \qquad (\text{A.1})$$

where $\beta = 1/kT$ and ρ is the Hamiltonian. Since this equation is of the form of the Schrödinger equation with β playing the role of time, it is convenient to draw graphs in β–r space. The density matrix and its graphs are generated by the propagator defined by

$$K(\mathbf{r}\beta'', \mathbf{r}'\beta') = \sum_{n} \exp[-(\beta'' - \beta')H]\psi_n^*(\mathbf{r})\psi_n(\mathbf{r}'). \qquad (\text{A.2})$$

where H is the Hamiltonian of the total system, ψ_n is its eigenfunction, and r is the coordinates of all the particles. Equation (A.2) represents the propagation of the particles from (r', β') to (r, β''). Such a propagation can be illustrated by a graph. The propagator yields the density matrix $\rho(r, r')$ if $\beta' = 0$ and $\beta'' = \beta$:

$$\rho(r, r') = K(r\beta, r'0). \tag{A.3}$$

When the Hamiltonian is split into two parts:

$$H = H_0 + H_1, \tag{A.4}$$

one can show that the propagator satisfies the integral equation [1]:

$$K(2, 1) = K_0(2, 1) - \int K_0(2, 3)H_1(3)K(3, 1)\,d3, \tag{A.5}$$

where K_0 belongs to H_0 and a simplified number-notation has been used for the variables in the (r, β) space. Equation (A.5) can be solved by iteration if K_0 is known, and the iterative series can be represented by Feynman graphs in the β–r space.

One can show rigorously that $\rho_2(r)$ is expressed in a linked cluster series [2]

$$\rho_2(r) = n^2 + \sum_{s=2}^{\infty} b_s(r)z^s, \tag{A.6}$$

where z is the absolute activity and $b_s(r)$ is the cluster integral represented by all the s-particle connected graphs that are labeled by two representative particles at r_1 and r_2 such that $r = |r_2 - r_1|$. The above formula is valid for any classical as well as quantum liquid. Graphical connections are made for classical systems through interaction potentials, and for quantum systems by either interaction potentials or quantum-mechanical exchanges.

The cluster integral $b_s(r)$ is normalized such that

$$V \int b_s(r)\,dr = s(s - 1)b_s, \tag{A.7}$$

where b_s is the constant cluster integral associated with all the graphs of s unlabeled particles. The factor $s(s - 1)$ on the right-hand side reflects the number of ways of labeling these graphs by the two representative particles for the graphs of $b_s(r)$.

In terms of b_s, the grand partition function can be expanded as follows:

$$\ln \Xi = \sum_{s=1}^{\infty} b_s z^s. \tag{A.8}$$

Note that z is determined from

$$\langle N \rangle = \left(\frac{\partial \ln \Xi}{\partial \ln z}\right)_{T,V}$$

$$= \sum_{s=1}^{\infty} s b_s z^s. \tag{A.9}$$

Note also that

$$\left(\frac{\partial^2 \ln \Xi}{\partial (\ln z)^2}\right)_{T,V} = \langle (N - \langle N \rangle)^2 \rangle$$

$$= \sum_{s=1}^{\infty} s^2 b_s z^s. \tag{A.10}$$

The above formulas are general and exact. Although any given graph can be treated theoretically, the result may not be simple. Besides, it may not be very meaningful unless all other graphs of the same order are also treated equally. As the order of graphs determined by the number of interaction lines increases, theoretical treatment becomes increasingly difficult. Thus, only by restricting the types of graphs can one obtain relatively simple and consistent results.

A.2. Simple and exchange-chain diagrams

Graphically, the two representative particles at $\mathbf{r}_1$ and $\mathbf{r}_2$ appearing in the pair distribution function $\rho_2(|\mathbf{r}_2 - \mathbf{r}_1|)$ may be connected linearly in the form of *chain diagrams*. When these chains are constructed by the same repeating unit, it may be possible to sum over such diagrams to infinite order. When this is achieved, the result is expected to show collective coupling effects of the particles.

For classical systems, the *simple chain diagrams* (Fig. A.1) and some of the more complicated graphs can be treated theoretically. The former is represented by a single chain of interaction potentials connecting the two representative particles at $\mathbf{r}_1$ and $\mathbf{r}_2$. In Fig. A.2, the box represents the total sum of the right-side diagrams. The signs attached to the graphs are

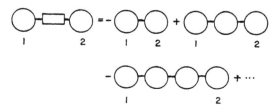

FIG. A.1. Simple chain diagrams.

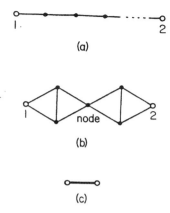

(a)

(b)

(c)

FIG. A.2.   Essentially chain diagrams: (a) simple chain; (b) simple chain with one node; (c) generalized chain. The heavy line represents an effective interaction line. Solid circles, unlabeled particles; open circles, labeled particles.

determined by the number of interactions in accordance with the iterative solution of Eq. (A.5).

Figure A.2(a) represents the same simple chain diagram as in Fig. A.1 except that unlabeled particles are simply shown by dots. In the *essentially chain diagrams* such as shown by Fig. A.2(b), several simple chains meet at unlabeled particles between the two representative particles, 1 and 2. These meeting points are called *nodes*. Any node where a chain branches out must be passed in order to go from r_1 to r_2. Without such a branching the essentially chains may be represented by heavy lines as in (c). We can repeat the same step to construct graphs with heavy lines. One example with three heavy lines is shown in Fig. A.3, where the box represents a structure such as node.

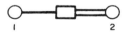

FIG. A.3.   Essentially chain diagram constructed by effective interaction lines.

On the other hand, the two representative particles may be connected by many chains without any node, resulting in the so-called *watermelon diagrams* such as shown in Fig. A.4.

Generally, it is most convenient to assign a momentum to an interaction line in correspondence to a Fourier transform:

$$u(q) = \int \phi(r)\, e^{i\mathbf{q}\cdot\mathbf{r}}\, d\mathbf{r}, \tag{A.11}$$

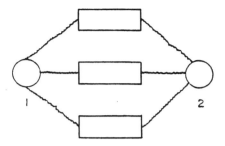

FIG. A.4. Watermelon diagram.

where $\phi(r)$ is the interaction potential. The vector $\mathbf{q}$ here is of the dimension of a reciprocal length and is a wavevector. However, $\mathbf{q}$ may conventionally be called momentum with the understanding that units have been taken such that $\hbar = 1$. As one can see easily by an integration over the coordinate of an unlabeled particle or a node, the total momentum entering and leaving such a particle must be the same. That is, a momentum conservation law holds. The use of this conservation law makes the use of a momentum space representation of graphs advantageous over a coordinate space representation.

Similar types of graphs can be introduced into quantum-mechanical cases except that graphical connections are made not only by interactions but also by quantum exchanges and that a particle with a momentum may not be represented by a point in a coordinate space. Instead, the propagation of a quantum particle from $(\mathbf{r}, 0)$ to $(\mathbf{r}, \beta)$ may be represented by a loop or a circle.

The ideal gas contribution to the pair distribution function due to quantum exchanges is given by [2]

$$I_2(r) = \pm \left[\frac{1}{(2\pi)^3} \int f(p)\, e^{i\mathbf{p}\cdot\mathbf{r}}\, d\mathbf{p} \right]^2, \tag{A.12}$$

where the $+$ $(-)$ sign is chosen for the Bose (Fermi) distribution function $f(p)$.

The chain diagrams are important because their contributions can be summed to infinite order. Moreover, their total contribution is expressed in terms of a single momentum variable because of momentum conservation. This results in a relatively simple formula for the contribution.

The classical limit of a quantum contribution can easily be taken. In a grand ensemble, all the unlabeled units of a simple chain can be represented by the same function because they consist of either a single particle, or two particles with one exchange, or three particles with two exchanges, and so on, all with one incoming interaction line and one outgoing interaction line. These interaction lines can be represented by the same momentum $\mathbf{q}$ due to the momentum conservation.

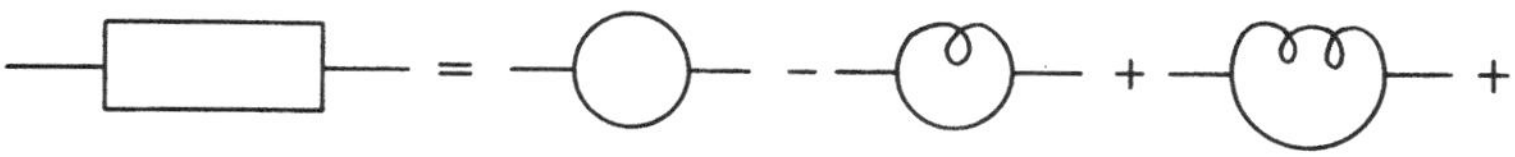

FIG. A.5. Quantum chain diagrams.

A quantum chain is illustrated in Fig. A.5. The graphs show exchanges, with each single loop corresponding to a propagation of one particle for interval β. When two loops are connected, it means that the two have one exchange. The box represents the entire sum of the right-side graphs. The signs are determined not only by the number of interaction lines but also by the number of exchanges. For instance, with one exchange and two interactions, the second graph is given a minus sign.

In momentum space, the probability of finding a particle with a momentum $\mathbf{p}$ is given by the Fermi or Bose distribution function $f(p)$. Upon absorbing momentum $\mathbf{q}$ that comes in through interaction, the particle makes a transition to the momentum $\mathbf{p} + \mathbf{q}$ state. The probability that this transition takes place depends not only on the initial probability $f(p)$ but also on the final probability $[1 \pm f(\mathbf{p} + \mathbf{q})]$. Here, a plus (minus) sign is used for Bose (Fermi) statistics.

If the two interaction lines entering and leaving a unit of a chain take place at intermediate "times" β' and β'' and if $\alpha = |\beta'' - \beta'|$, all the unlabeled units are represented by a single function given by

$$G(q, \alpha) = \frac{1}{(2\pi)^3} \int f(p)[1 \pm f(\mathbf{p} + \mathbf{q})] \exp\left(\frac{\alpha\hbar^2}{2m}[p^2 - (\mathbf{p} + \mathbf{q})^2]\right) d\mathbf{p}. \quad \text{(A.13)}$$

For convenience, we introduce the Fourier transform of $G(q, \alpha)$ by

$$\lambda_j(q) = \int_0^\beta G(q, \alpha) \exp(2\pi i j\alpha/\beta) \, d\alpha, \quad \text{(A.14)}$$

where j is an integer or zero and $\beta = 1/kT$. An arbitrary unlabeled unit of a simple chain is now represented by $\lambda_j(q)$, which we call an *eigenvalue function* because Eq. (A.14) can be expressed in the form of a Fredholm integral equation with λ_j as the eigenvalues.

Similar considerations apply for the graphs in which the two representative particles have an exchange. These graphs are called *exchange-chain diagrams*. These graphs are illustrated in Fig. A.6. They are represented by a new type of eigenvalue given by [3]

$$\lambda_j(q, r) = \frac{1}{(2\pi)^3} \int_0^\beta d\alpha \, e^{2\pi i j\alpha/\beta} \int f(p)[1 \pm f(\mathbf{p} + \mathbf{q})]$$

$$\times \exp\left(\frac{\alpha\hbar^2}{2m}[p^2 - (\mathbf{p} + \mathbf{q})^2] + i\mathbf{p}\cdot\mathbf{r}\right) d\mathbf{p}. \quad \text{(A.15)}$$

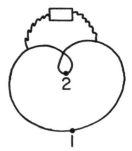

FIG. A.6. Exchange-chain diagram.

Note that

$$\lambda_j(q. 0) = \lambda_j(q). \tag{A.16}$$

Adding the contributions from the simple chain and exchange-chain diagrams, and also the ideal gas contribution given by Eq. (A.12), we arrive at a formula:

$$\rho_2(r) = n^2 \mp I_2(r) - \frac{1}{(2\pi)^3 \beta} \sum_j \frac{u(q)}{1 + \lambda_j(q)u(q)} \exp\left(\frac{\alpha\hbar^2}{2m}\left[p^2 - (\mathbf{p} + \mathbf{q})^2\right]\right)$$
$$\times \left[\lambda_j^2(q) \pm \lambda_j^2(q, r)\right] \exp(i\mathbf{q}\cdot\mathbf{r})\, d\mathbf{q}. \tag{A.17}$$

The first and second terms on the right-hand side represent respectively the classical and quantum ideal gas contribution. The first and second terms in the square brackets of the last term are respectively the simple chain and exchange-chain contributions. Note that, at $r = 0$, the two terms cancel with each other in the Fermi case $(-)$, while they enhance each other in the Bose case $(+)$. Moreover, since

$$I_2(0) = n^2$$

$\rho_2(0)$ for fermions vanishes in Eq. (A.17).

The above formula for the pair distribution function neglects spins. With spins, we must consider distribution functions $\rho_2^{\uparrow\uparrow}$ and $\rho_2^{\uparrow\downarrow}$ for parallel and antiparallel spin pairs. Since a quantum exchange takes place only between parallel spins, the above formula must be modified. For fermions, we arrive at

$$\rho_2^{\uparrow\uparrow}(r) = \frac{n^2}{2} - 2I_2(r) - \frac{1}{2}\sum_j \frac{1}{(2\pi)^3} \int \frac{u(q)\exp(i\mathbf{q}\cdot\mathbf{r})}{1 + \lambda_j(q)u(q)}\left[\lambda_j^2(q) - \lambda_j^2(q, r)\right] d\mathbf{q},$$

$$\rho_2^{\uparrow\downarrow}(r) = \frac{n^2}{2} - \frac{1}{2}\sum_j \frac{1}{(2\pi)^3} \int \frac{u(q)\exp(i\mathbf{q}\cdot\mathbf{r})}{1 + \lambda_j(q)u(q)}\lambda_j^2(q)\, d\mathbf{q}. \tag{A.18}$$

Formulas (A.17) and (A.18) include $\lambda_j(q, r)$ or $\lambda_j(q)$. These functions have been evaluated for all statistics. Therefore, once the Fourier transform $u(q)$ of the interaction potential of a system is given, they are ready for use.

In general, the pair distribution function can be used to generate the grand partition function. In particular, the pair distribution function in the (simple) chain diagram approximation yields the grand partition function in what is called the *ring diagram approximation* because an integration process of $\rho_2(r)\phi(r)$ over $\mathbf{r}$ is involved, which amounts to closing all the chain diagrams into rings. As one can find easily, this grand partition function is given by [3]

$$\ln \Xi_r = \frac{V}{2(2\pi)^3} \sum_j \int \{u(q)\lambda_j(q) - \ln[1 + u(q)\lambda_j(q)]\}\, d\mathbf{q}. \qquad (A.19)$$

The pair distribution function in the chain diagram and/or high-order approximations yields information on screening constants of Coulombic systems [2, 4]. For instance, the radial distribution function of a classical electron gas that improves the Debye approximation is

$$g(r) = 1 - \frac{\varepsilon}{\kappa_0 r}\, e^{-\kappa r}[1 + \varepsilon(\tfrac{3}{8}\ln 3 + \tfrac{1}{6}) + O(\varepsilon^2)], \qquad (A.20)$$

where $\kappa_0 = (4\pi n e^2/kT)^{1/2}$ is the *Debye screening constant*, $\varepsilon = e^2\kappa_0/(kT)$ is the plasma parameter, and

$$\kappa = \kappa_0\left(1 + \frac{\varepsilon}{8}\ln 3\right) \qquad (A.21)$$

is a new screening constant for low but finite densities. One can show also that the screening constant of ions with a certain size is modified from the Debye screening constant for point charges.

A similar result can be obtained for low temperatures. In this case the *Thomas–Fermi screening constant* $\mu_0 = 2(k_F/\pi a_0)^{1/2}$ is modified such that

$$\mu = \mu_0\left(1 + \frac{1}{6k_F a_0}\right). \qquad (A.22)$$

The corresponding radial distribution function is given by

$$g(r) = 1 - \frac{k_F^4}{\pi^3 a_0 n^2}\left(1 + \frac{1}{3\pi k_F a_0}\right)\frac{e^{-\mu r}}{r}. \qquad (A.23)$$

k_F is the Fermi wavenumber and a_0 is the Bohr radius. Some other interesting long-distance behaviors of $g(r)$ have been obtained [5].

By contrast, at short distances interaction effects are generally so strong that higher-order correlations have to be taken into consideration even though the exchange-chain diagram contribution gives improvements over the simple chain results. A consideration of short-distance correlations is given in the next section.

A.3. Ladder diagram contribution

At short distances, two-particle correlations are most important. These correlations can be represented by two types of graphs, direct and exchange ladder diagrams as illustrated in Fig. A.7(a) and (b) respectively. The corresponding pair distribution function can be obtained by summing over these ladder diagrams. However, the same pair distribution function can be obtained by making use of the solutions of the Schrödinger equation for two particles. Let us discuss the latter approach to the case of an electron system [6].

The contribution from the direct ladder diagrams of two electrons to the pair distribution function is given by taking the matrix element

$$\rho^L_{2D}(r) = n^2 \langle \mathbf{r}_1 \mathbf{r}_2 | [1 - f(p_1)][1 - f(p_2)] e^{-\beta H} | \mathbf{r}_1 \mathbf{r}_2 \rangle, \tag{A.24}$$

and that of the exchange ladder diagrams by

$$\rho^L_{2E}(r) = n^2 \langle \mathbf{r}_1 \mathbf{r}_2 | [1 - f(p_1)][1 - f(p_2)] e^{-\beta H} | \mathbf{r}_2 \mathbf{r}_1 \rangle. \tag{A.25}$$

In these equations, H is the Hamiltonian of two electrons.

The trade in Eq. (A.24) can be taken in momentum space by using the

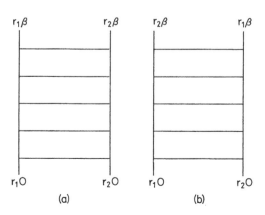

FIG. A.7. Ladder diagrams: (a) direct, (b) exchange.

center of mass and relative momenta:

$$\mathbf{P} = \mathbf{p}_1 + \mathbf{p}_2; \qquad \mathbf{p} = \tfrac{1}{2}(\mathbf{p}_2 - \mathbf{p}_1).$$

For low temperatures, the momenta $\mathbf{p}_1$ and $\mathbf{p}_2$ stay on the Fermi sphere. In terms of the solutions $R_{kl}(r)Y_{lm}(\theta, \phi)$ of the two-body Schrödinger equation the contribution from the direct ladder diagrams is expressed by

$$\rho_{2D}^{L}(r) = 4\left(\frac{32\pi}{3}\right)\frac{k_F^3}{(2\pi)^3}\int_0^{k_F} dp \sum_{lm} |R_{kl}(r)|^2 \, Y_{lm}^{*}(\theta, \phi)Y_{lm}(\theta, \phi)$$

$$\times \left(1 - \frac{3p}{2k_F} + \frac{1}{2}\frac{p^3}{k_F^3}\right), \tag{A.26}$$

where k_F is the Fermi wavenumber given by

$$k_F^3 = 3\pi^2 n. \tag{A.27}$$

To order r^3, the s- and p-waves' contributions must be taken into consideration. The d-wave contribution is of the order of r^4. To order r^3, $\rho_{2D}^{L}(r)$ is given by

$$\rho_{2D}^{L}(r) = \frac{8k_F^6}{3\pi^3}\alpha r_s \int_0^1 dy \, \frac{y}{\exp(\pi\alpha r_s/y) - 1}\left(1 - \frac{3y}{2} + \frac{y^3}{2}\right)$$

$$\times \left[1 + \frac{r}{a_0} + \frac{1}{2}\left(\frac{r}{a_0}\right)^2 + \frac{5}{36}\left(\frac{r}{a_0}\right)^3 - \tfrac{1}{9}(k_F r)^2 \frac{r}{a_0} y^2\right]. \tag{A.28}$$

A similar expression can be derived for the exchange ladder contribution.

An interesting consequence of these calculations is that the radial distribution function at the origin, that is, $g(0)$, does not vanish but is a constant. This constant depends on the interaction parameter r_s defined by

$$\frac{4\pi}{3}(r_s a_0)^3 = n^{-1}. \tag{A.29}$$

For small and large r_s, $g(0)$ is given by

$$g(0) = \begin{cases} \tfrac{1}{2} - \tfrac{3}{5}(\pi\alpha r_s) & (r_s \ll 1), \\[2mm] 12(\pi\alpha r_s)\int_0^1 dy \, y\left(1 - \frac{3y}{2} + \frac{y^2}{2}\right)\exp\left(-\frac{\pi\alpha r_s}{y}\right); & (r_s \gg 1), \end{cases} \tag{A.30}$$

where $\alpha = (4/9\pi)^{1/3}$. The above results show that for small r_s the radial distribution function decreases linearly with r_s starting from $1/2$, and for large

r_s approximately as

$$g(0) = \frac{36 \exp(-\pi\alpha r_s)}{(\pi\alpha r_s)^2} \qquad (r_s \gg 1). \tag{A.31}$$

Hence, in this strong coupling limit, $g(0)$ approaches 0.

A.4. Integral equation approach

For classical fluids, the Liouville equation for the total distribution function of N particles in the phase space of a system can be used as a starting equation. It is given by

$$\frac{\partial f^{(N)}}{\partial t} + [f^{(N)}, H_N] = 0, \tag{A.32}$$

where $[f^{(N)}, H_N]$ is the Poisson bracket of the distribution function with the total Hamiltonian H_N. By integrating over the phase space of $(N - 2)$ particles one can reduce this equation to that of two particles [7]:

$$\frac{\partial f^{(2)}}{\partial t} + [f^{(2)}, H_2] = \iint \sum_{i=1}^{2} \frac{\partial \phi_{i3}}{\partial \mathbf{r}_i} \cdot \frac{\partial f^{(3)}}{\partial \mathbf{p}_i} \, d\mathbf{r}_3 \, d\mathbf{p}_3, \tag{A.33}$$

where H_2 is the Hamiltonian of two particles. The molecular potential is denoted by $\phi(r)$. The integrodifferential equation (A.33) is not closed because the right-hand side includes the triplet distribution function $f^{(3)}$. Therefore, $f^{(2)}$ can be determined from the above equation but some simplifying approximations are introduced because the same problem repeats itself for $f^{(3)}$. The above types of differential equations for the reduced distribution functions form a hierarchy of equations known as *BBGKY* named after Bogoliubov, Born, Green, Kirkwood, and Yvon. The above equation can be integrated in momentum space to obtain a differential equation for the pair distribution function ρ_2, but the equation will depend on the triplet distribution function ρ_3.

In the *Kirkwood superposition approximation*, the triplet distribution function ρ_3 is assumed to be given by a product of pair distribution functions in a symmetric way. Symbolically, this assumption takes the form

$$\rho_3(1, 2, 3) = \frac{\rho_2(1, 2)\rho_2(2, 3)\rho_2(3, 1)}{n^3}, \tag{A.34}$$

where n is the number density that represents the singlet distribution function in equilibrium fluids.

Even if this Kirkwood superposition approximation is adopted, the integrodifferential equation for ρ_2 is still highly nonlinear. Hence, further approximations are necessary. While a number of approximations have been developed, the so-called *hypernetted chain* (HNC) approximation is frequently used. Under this approximation, the radial distribution $g(r)$ is determined from [7]

$$\ln g(r) + \beta\phi(r) = n \int [g(r_{13}) - 1 - \ln g(r_{13}) - \beta\phi(r_{13})][g(r_{32}) - 1] \, dr_3.$$

(A.35)

The HNC approximation is correct to first order in density, but misses one graph in second order. In general, it is not a particularly good approximation, but it yields reasonable results for systems with long-range forces.

Another approximation due to Percus and Yevick [8] is given by

$$g(r)e^{\beta\phi(r)} = 1 + n \int [1 - e^{\beta\phi(13)}]g(r_{13})[g(r_{32}) - 1] \, d\mathbf{r}_3.$$ (A.36)

In terms of graphs, this PY approximation is also correct to first order in density but misses two diagrams in second order. Nevertheless, the solution of the above integral equation produces very good results particularly for short-range potentials.

The above two equations stem from different assumptions about the contributions from graphs with nodes [1]. They are still nonlinear and require numerical solutions. However, along with the improvements in computer techniques and facilities, explicit determinations of the radial distribution functions have been made based on these and other approximations, and the results have been compared with results of direct Monte Carlo calculations or molecular dynamics.

A.5. Charged bosons and hard-sphere bosons

Molecular distribution functions have been studied extensively for classical or quantal liquids and electrons systems. Although each of these cases deserves detailed discussions, let us remark on a particular system consisting of charged bosons. This system is theoretically interesting because its Bose condensation competes with long-range Coulomb forces. In addition, it provides a lower bound for the ground-state energy of an electron gas. We shall investigate basic differences between Bose systems with long-range Coulombic forces and short-range hard-sphere type interactions. For simplicity, we use in this section the natural units in which $\hbar = 1$ and $2m = 1$.

The Fourier transform of the Coulomb potential is given by

$$u(q) = \frac{4\pi e^2}{q^2} \qquad \text{(Coulomb)}.$$

The long-rangedness of the potential is reflected in the divergence of this potential at $q = 0$. On the other hand, the Fourier transform of a simplified pseudopotential for a hard-sphere of diameter a is given by

$$u(q) = 8\pi a \qquad \text{(hard-sphere)}.$$

The short-rangedness of a hard-sphere potential is exhibited in the constancy of $u(0)$

The pair distribution function $\rho_2(r)$ can be evaluated for long distances based on Eq. (A.17). In that formula, the eigenvalues λ_j defined by Eq. (A.14) are given for bosons at low temperatures as follows:

$$\lambda_j = \frac{2nq^2}{q^4 + (2\pi j/\beta)^2}. \qquad (A.37)$$

Here $\beta = 1/kT$. The contribution from $\lambda_j(q, r)$ may be neglected for long distances, and $I_2(r)$ can be canceled out if we use

$$-\frac{1}{\beta} \sum_j \int \frac{d\mathbf{q}}{(2\pi)^3} \lambda_j(q) e^{i\mathbf{q}\cdot\mathbf{r}} = -n\delta(r) - I_2(r). \qquad (A.38)$$

Introducing Eq. (A.37) into Eq. (A.17) we find that the pair distribution function of charged bosons is given by a sum of screened functions [9]:

$$\rho_2(r) - n^2 = \frac{n}{\kappa_0^3} \sum_j \left(\frac{\exp\{-[2\pi\alpha^2 + (4\pi^2 j)^2]^{1/2}x\}}{x} \right) \cos([2\pi\alpha^2 + (4\pi^2 j)^2]^{1/4}x). \qquad (A.39)$$

Here $\lambda = (4\pi\beta)^{1/2}$ is the thermal de Broglie wavelength, $\kappa_0 = (4\pi\beta ne^2)^{1/2}$ is the Debye constant, $\alpha = \kappa_0\lambda$, and x is a reduced distance given by r/κ_0. Hence, $\rho_2(r)$ depends on r/κ_0.

The contribution from the $j = 0$ term is dominant for $r > \kappa_0$ but the screening due to the $j > 0$ terms becomes important if $r < \kappa_0$. In fact, at absolute zero the $j = 0$ term vanishes. We arrive at

$$\rho_2(r) - n^2 = -\frac{2n}{(2\pi)^2 r} \int_0^\infty dy \, y^3 \sin yr \, [y^{-2} - (y^4 + A^4)^{-1/2}], \quad (A.40)$$

where

$$A^4 = 8\pi ne^2.$$

An explicit asymptotic result is given by

$$\rho_2(r) - n^2 = -2^{-5/2} nA^2 e^{-Dr} x[B \cos Br + D \sin Br], \qquad (A.41)$$

where

$$2D^2 = (2^{1/2} + 1)A^2; \qquad 2B^2 = (2^{1/2} - 1)A^2.$$

The asymptotic behavior of the pair distribution function depends on a dimensionless parameter η defined by

$$\eta = \frac{4(\pi\delta)^{1/2}}{\kappa_0}. \qquad (A.42)$$

This is the ratio of distance r to the correlation length given by $\kappa_0/4(\pi\delta)^{1/2}$. Here $\delta = 1.105(T - T_c)/T_c$, where T_c is the condensation temperature. One can show that for r greater than the correlation length, the pair distribution function is characterized by an exponential decay. In the opposite limit, the pair distribution function decays asymptotically as $1/r^8$:

$$\rho_2(r) - n^2 \sim \frac{45a_0^2}{\pi^4}\left(\frac{1}{r^8}\right) \qquad \text{(charged),} \qquad (A.43)$$

where a_0 is the Bohr radius. This decay contrasts to the $1/r^4$ variation of a hard-sphere Bose gas:

$$\rho_2(r) - n^2 \sim -\frac{2n^2a}{\pi\gamma^{3/2}}\left(\frac{1}{r^4}\right) \qquad \text{(hard-sphere),} \qquad (A.44)$$

where $\gamma = 4\pi an$. Figure A.8 [10] shows $\psi(x) = [\rho_2/n^2 - 1]/(-4a/r)$ of this case with $x = 2\gamma^{1/2}r$.

The above difference in the behavior of $\rho_2(r)$ is related to the difference in the excitation energies; a plasmon excitation in the former case and a

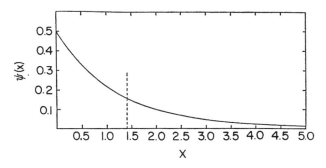

FIG. A.8.   Reduced pair correlation function $\psi(x) = [\rho_2(r)/n^2 - 1]/(-4a/r)$ of a hard-sphere Bose gas as a function of reduced distance $x = 2\gamma^{1/2}r$; $\gamma = 4\pi an$, n is the number density, and a is the hard-sphere diameter. (From Isihara and Jepsen [11])

phonon excitation in the latter case. Let us look into this interesting difference.

The pair distribution function generates the internal energy in accordance with the general formula

$$U(T) = U_0(T) + \frac{V}{2} \int_0^1 d\mathbf{r} \, \phi(r) \frac{\partial}{\partial \beta} [\beta \rho_2(r, \lambda)]. \tag{A.45}$$

$U_0(T)$ represents the energy of the ideal case and λ is a coupling parameter. An analogous formula has been given in Eq. (1.36) of Chapter 1. We arrive at an interesting result:

$$U(T) = \frac{V}{2} \int \frac{d\mathbf{q}}{(2\pi)^3} \left((q^4 + A^4)^{1/2} - q^2 - \frac{A^4}{2q^2} \right)$$
$$+ V \int \frac{d\mathbf{q}}{(2\pi)^3} (q^4 + A^4)^{1/2} f[\beta (q^4 + A^4)^{1/2}], \tag{A.46}$$

where $f(\varepsilon)$ is a Bose distribution function of quasiparticles:

$$f(\varepsilon) = [\exp(\beta\varepsilon) - 1]^{-1}.$$

The first term in Eq. (A.46) represents the ground-state energy while the second term represents the excitation energy. We find from the second term that the excitation spectrum is related to quasiparticles.

More explicitly, the ground-state energy per particle is given by

$$\varepsilon_0 = -\left(\frac{2^{5/2} 3^{1/4}}{\pi} I \right) r_s^{-3/4} \qquad \text{(charged)}, \tag{A.47}$$

where $I = (1/5)(\pi/2)^{1/2} \Gamma(3/4)/\Gamma(5/4)$. The excitation energy is

$$\varepsilon(q) = [q^4 + \omega_p^2]^{1/2} \qquad \text{(charged)}, \tag{A.48}$$

with $\omega_p^2 = A^4 = 8\pi n e^2$ as in Eq. (2.19). (Note $2m = 1$ in this section.) Hence plasmon excitations characterize the system for small q.

A similar calculation leads us to the ground-state energy of a hard-sphere Bose gas given by

$$\varepsilon_0 = 4\pi n a \left(1 + \frac{128}{15\pi^{1/2}} (a^3 n)^{1/2} \right) \qquad \text{(hard sphere)}, \tag{A.49}$$

and to the excitation energy:

$$\varepsilon(q) = q[q^2 + 2nu(0)]^{1/2} \qquad \text{(hard sphere).} \qquad \text{(A.50)}$$

For small q, $\varepsilon(q) = cq$, where $c = (4\pi an)^{1/2}$ represents the sound velocity. It is important to note that Eq. (A.50) has been obtained in a natural way from $U(T)$, which is expressed in terms of the Bose distribution of massless quasiparticles. Equation (A.50) has been obtained in the first approximation. It has been shown that a consideration of $\lambda_j(q, r)$ yields an effective mass and that a more elaborate treatment gives rise to a temperature-dependent excitation energy [11].

REFERENCES

1. E. W. Montroll and J. C. Ward, *Phys. Fluids* **1**, 55 (1958).
2. S. Fujita, A. Isihara, and E. W. Montroll, *Bull. Cl. Sci. Acad. Roy. Belg.* **44**, 1018 (1958). A. Isihara, *Prog. Theor. Phys.* Suppl. **44**, 1 (1969).
3. A. Isihara, *Phys. Rev.* **172**, 166 (1968).
4. A. Isihara, *Phys. Rev.* **178**, 412 (1969). A. Isihara and M. Wadati, *Phys. Rev.* **183**, 312 (1969).
5. A. Isihara and D. D. H. Yee, *Phys. Rev. A* **154**, 618 (1964). A. Isihara and A. K. Gupta, *Phys. Rev.* **160**, 225 (1967).
6. A. Isihara and M. Wadati, *Physica* **57**, 237 (1972). M. Wadati and A. Isihara, *Phys. Rev. A* **4**, 344 (1971).
7. See references 1, Chapter 1.
8. J. K. Percus and G. J. Yevick, *Phys. Rev.* **110**, 1 (1958).
9. T. Samulski and A. Isihara, *Physica* **82A**, 294 (1976). See also: L. L. Foldy, *Phys. Rev.* **124**, 649 (1961). A. L. Fetter, *Ann. Phys.* **64**, 1 (1971). C. I. Um and A. Isihara, *J. Kor. Phys. Soc.* **11**, 60 (1978).
10. A. Isihara and Donald W. Jepsen, *Phys. Rev.* **158**, 112 (1967).
11. A. Isihara and T. Samulski, *Phys. Rev. B* **16**, 1969 (1977). T. Samulski and A. Isihara, *Physica* **86A**, 257 (1977).

SUBJECT INDEX

Physics

OPTICAL RESONANCE AND TWO-LEVEL ATOMS, L. Allen and J. H. Eberly. Clear, comprehensive introduction to basic principles behind all quantum optical resonance phenomena. 53 illustrations. Preface. Index. 256pp. 5⅜ x 8½. 0-486-65533-4

QUANTUM THEORY, David Bohm. This advanced undergraduate-level text presents the quantum theory in terms of qualitative and imaginative concepts, followed by specific applications worked out in mathematical detail. Preface. Index. 655pp. 5⅜ x 8½. 0-486-65969-0

ATOMIC PHYSICS (8th EDITION), Max Born. Nobel laureate's lucid treatment of kinetic theory of gases, elementary particles, nuclear atom, wave-corpuscles, atomic structure and spectral lines, much more. Over 40 appendices, bibliography. 495pp. 5⅜ x 8½. 0-486-65984-4

A SOPHISTICATE'S PRIMER OF RELATIVITY, P. W. Bridgman. Geared toward readers already acquainted with special relativity, this book transcends the view of theory as a working tool to answer natural questions: What is a frame of reference? What is a "law of nature"? What is the role of the "observer"? Extensive treatment, written in terms accessible to those without a scientific background. 1983 ed. xlviii+172pp. 5⅜ x 8½. 0-486-42549-5

AN INTRODUCTION TO HAMILTONIAN OPTICS, H. A. Buchdahl. Detailed account of the Hamiltonian treatment of aberration theory in geometrical optics. Many classes of optical systems defined in terms of the symmetries they possess. Problems with detailed solutions. 1970 edition. xv + 360pp. 5⅜ x 8½. 0-486-67597-1

PRIMER OF QUANTUM MECHANICS, Marvin Chester. Introductory text examines the classical quantum bead on a track: its state and representations; operator eigenvalues; harmonic oscillator and bound bead in a symmetric force field; and bead in a spherical shell. Other topics include spin, matrices, and the structure of quantum mechanics; the simplest atom; indistinguishable particles; and stationary-state perturbation theory. 1992 ed. xiv+314pp. 6⅛ x 9¼. 0-486-42878-8

LECTURES ON QUANTUM MECHANICS, Paul A. M. Dirac. Four concise, brilliant lectures on mathematical methods in quantum mechanics from Nobel Prize-winning quantum pioneer build on idea of visualizing quantum theory through the use of classical mechanics. 96pp. 5⅜ x 8½. 0-486-41713-1

THIRTY YEARS THAT SHOOK PHYSICS: THE STORY OF QUANTUM THEORY, George Gamow. Lucid, accessible introduction to influential theory of energy and matter. Careful explanations of Dirac's anti-particles, Bohr's model of the atom, much more. 12 plates. Numerous drawings. 240pp. 5⅜ x 8½. 0-486-24895-X

ELECTRONIC STRUCTURE AND THE PROPERTIES OF SOLIDS: THE PHYSICS OF THE CHEMICAL BOND, Walter A. Harrison. Innovative text offers basic understanding of the electronic structure of covalent and ionic solids, simple metals, transition metals and their compounds. Problems. 1980 edition. 582pp. 6⅛ x 9¼. 0-486-66021-4

HYDRODYNAMIC AND HYDROMAGNETIC STABILITY, S. Chandrasekhar. Lucid examination of the Rayleigh-Benard problem; clear coverage of the theory of instabilities causing convection. 704pp. 5⅜ x 8¼. 0-486-64071-X

INVESTIGATIONS ON THE THEORY OF THE BROWNIAN MOVEMENT, Albert Einstein. Five papers (1905–8) investigating dynamics of Brownian motion and evolving elementary theory. Notes by R. Fürth. 122pp. 5⅜ x 8½. 0-486-60304-0

THE PHYSICS OF WAVES, William C. Elmore and Mark A. Heald. Unique overview of classical wave theory. Acoustics, optics, electromagnetic radiation, more. Ideal as classroom text or for self-study. Problems. 477pp. 5⅜ x 8½. 0-486-64926-1

GRAVITY, George Gamow. Distinguished physicist and teacher takes reader-friendly look at three scientists whose work unlocked many of the mysteries behind the laws of physics: Galileo, Newton, and Einstein. Most of the book focuses on Newton's ideas, with a concluding chapter on post-Einsteinian speculations concerning the relationship between gravity and other physical phenomena. 160pp. 5⅜ x 8½.
0-486-42563-0

PHYSICAL PRINCIPLES OF THE QUANTUM THEORY, Werner Heisenberg. Nobel Laureate discusses quantum theory, uncertainty, wave mechanics, work of Dirac, Schroedinger, Compton, Wilson, Einstein, etc. 184pp. 5⅜ x 8½. 0-486-60113-7

ATOMIC SPECTRA AND ATOMIC STRUCTURE, Gerhard Herzberg. One of best introductions; especially for specialist in other fields. Treatment is physical rather than mathematical. 80 illustrations. 257pp. 5⅜ x 8½. 0-486-60115-3

AN INTRODUCTION TO STATISTICAL THERMODYNAMICS, Terrell L. Hill. Excellent basic text offers wide-ranging coverage of quantum statistical mechanics, systems of interacting molecules, quantum statistics, more. 523pp. 5⅜ x 8½.
0-486-65242-4

THEORETICAL PHYSICS, Georg Joos, with Ira M. Freeman. Classic overview covers essential math, mechanics, electromagnetic theory, thermodynamics, quantum mechanics, nuclear physics, other topics. First paperback edition. xxiii + 885pp. 5⅜ x 8½. 0-486-65227-0

PROBLEMS AND SOLUTIONS IN QUANTUM CHEMISTRY AND PHYSICS, Charles S. Johnson, Jr. and Lee G. Pedersen. Unusually varied problems, detailed solutions in coverage of quantum mechanics, wave mechanics, angular momentum, molecular spectroscopy, more. 280 problems plus 139 supplementary exercises. 430pp. 6½ x 9¼. 0-486-65236-X

THEORETICAL SOLID STATE PHYSICS, Vol. 1: Perfect Lattices in Equilibrium; Vol. II: Non-Equilibrium and Disorder, William Jones and Norman H. March. Monumental reference work covers fundamental theory of equilibrium properties of perfect crystalline solids, non-equilibrium properties, defects and disordered systems. Appendices. Problems. Preface. Diagrams. Index. Bibliography. Total of 1,301pp. 5⅜ x 8½. Two volumes. Vol. I: 0-486-65015-4 Vol. II: 0-486-65016-2

WHAT IS RELATIVITY? L. D. Landau and G. B. Rumer. Written by a Nobel Prize physicist and his distinguished colleague, this compelling book explains the special theory of relativity to readers with no scientific background, using such familiar objects as trains, rulers, and clocks. 1960 ed. vi+72pp. 5⅜ x 8½. 0-486-42806-0

Math–Decision Theory, Statistics, Probability

ELEMENTARY DECISION THEORY, Herman Chernoff and Lincoln E. Moses. Clear introduction to statistics and statistical theory covers data processing, probability and random variables, testing hypotheses, much more. Exercises. 364pp. 5⅜ x 8½. 0-486-65218-1

STATISTICS MANUAL, Edwin L. Crow et al. Comprehensive, practical collection of classical and modern methods prepared by U.S. Naval Ordnance Test Station. Stress on use. Basics of statistics assumed. 288pp. 5⅜ x 8½. 0-486-60599-X

SOME THEORY OF SAMPLING, William Edwards Deming. Analysis of the problems, theory and design of sampling techniques for social scientists, industrial managers and others who find statistics important at work. 61 tables. 90 figures. xvii +602pp. 5⅜ x 8½. 0-486-64684-X

LINEAR PROGRAMMING AND ECONOMIC ANALYSIS, Robert Dorfman, Paul A. Samuelson and Robert M. Solow. First comprehensive treatment of linear programming in standard economic analysis. Game theory, modern welfare economics, Leontief input-output, more. 525pp. 5⅜ x 8½. 0-486-65491-5

PROBABILITY: AN INTRODUCTION, Samuel Goldberg. Excellent basic text covers set theory, probability theory for finite sample spaces, binomial theorem, much more. 360 problems. Bibliographies. 322pp. 5⅜ x 8½. 0-486-65252-1

GAMES AND DECISIONS: INTRODUCTION AND CRITICAL SURVEY, R. Duncan Luce and Howard Raiffa. Superb nontechnical introduction to game theory, primarily applied to social sciences. Utility theory, zero-sum games, n-person games, decision-making, much more. Bibliography. 509pp. 5⅜ x 8½. 0-486-65943-7

INTRODUCTION TO THE THEORY OF GAMES, J. C. C. McKinsey. This comprehensive overview of the mathematical theory of games illustrates applications to situations involving conflicts of interest, including economic, social, political, and military contexts. Appropriate for advanced undergraduate and graduate courses; advanced calculus a prerequisite. 1952 ed. x+372pp. 5⅜ x 8½. 0-486-42811-7

FIFTY CHALLENGING PROBLEMS IN PROBABILITY WITH SOLUTIONS, Frederick Mosteller. Remarkable puzzlers, graded in difficulty, illustrate elementary and advanced aspects of probability. Detailed solutions. 88pp. 5⅜ x 8½. 65355-2

PROBABILITY THEORY: A CONCISE COURSE, Y. A. Rozanov. Highly readable, self-contained introduction covers combination of events, dependent events, Bernoulli trials, etc. 148pp. 5⅜ x 8¼. 0-486-63544-9

STATISTICAL METHOD FROM THE VIEWPOINT OF QUALITY CONTROL, Walter A. Shewhart. Important text explains regulation of variables, uses of statistical control to achieve quality control in industry, agriculture, other areas. 192pp. 5⅜ x 8½. 0-486-65232-7

Math–Geometry and Topology

ELEMENTARY CONCEPTS OF TOPOLOGY, Paul Alexandroff. Elegant, intuitive approach to topology from set-theoretic topology to Betti groups; how concepts of topology are useful in math and physics. 25 figures. 57pp. 5⅜ x 8½. 0-486-60747-X

COMBINATORIAL TOPOLOGY, P. S. Alexandrov. Clearly written, well-organized, three-part text begins by dealing with certain classic problems without using the formal techniques of homology theory and advances to the central concept, the Betti groups. Numerous detailed examples. 654pp. 5⅜ x 8½. 0-486-40179-0

EXPERIMENTS IN TOPOLOGY, Stephen Barr. Classic, lively explanation of one of the byways of mathematics. Klein bottles, Moebius strips, projective planes, map coloring, problem of the Koenigsberg bridges, much more, described with clarity and wit. 43 figures. 210pp. 5⅜ x 8½. 0-486-25933-1

THE GEOMETRY OF RENÉ DESCARTES, René Descartes. The great work founded analytical geometry. Original French text, Descartes's own diagrams, together with definitive Smith-Latham translation. 244pp. 5⅜ x 8½. 0-486-60068-8

EUCLIDEAN GEOMETRY AND TRANSFORMATIONS, Clayton W. Dodge. This introduction to Euclidean geometry emphasizes transformations, particularly isometries and similarities. Suitable for undergraduate courses, it includes numerous examples, many with detailed answers. 1972 ed. viii+296pp. 6⅛ x 9¼. 0-486-43476-1

PRACTICAL CONIC SECTIONS: THE GEOMETRIC PROPERTIES OF ELLIPSES, PARABOLAS AND HYPERBOLAS, J. W. Downs. This text shows how to create ellipses, parabolas, and hyperbolas. It also presents historical background on their ancient origins and describes the reflective properties and roles of curves in design applications. 1993 ed. 98 figures. xii+100pp. 6½ x 9¼. 0-486-42876-1

THE THIRTEEN BOOKS OF EUCLID'S ELEMENTS, translated with introduction and commentary by Sir Thomas L. Heath. Definitive edition. Textual and linguistic notes, mathematical analysis. 2,500 years of critical commentary. Unabridged. 1,414pp. 5⅜ x 8½. Three-vol. set.
Vol. I: 0-486-60088-2 Vol. II: 0-486-60089-0 Vol. III: 0-486-60090-4

SPACE AND GEOMETRY: IN THE LIGHT OF PHYSIOLOGICAL, PSYCHOLOGICAL AND PHYSICAL INQUIRY, Ernst Mach. Three essays by an eminent philosopher and scientist explore the nature, origin, and development of our concepts of space, with a distinctness and precision suitable for undergraduate students and other readers. 1906 ed. vi+148pp. 5⅜ x 8½. 0-486-43909-7

GEOMETRY OF COMPLEX NUMBERS, Hans Schwerdtfeger. Illuminating, widely praised book on analytic geometry of circles, the Moebius transformation, and two-dimensional non-Euclidean geometries. 200pp. 5⅝ x 8¼. 0-486-63830-8

DIFFERENTIAL GEOMETRY, Heinrich W. Guggenheimer. Local differential geometry as an application of advanced calculus and linear algebra. Curvature, transformation groups, surfaces, more. Exercises. 62 figures. 378pp. 5⅜ x 8½. 0-486-63433-7

History of Math

THE WORKS OF ARCHIMEDES, Archimedes (T. L. Heath, ed.). Topics include the famous problems of the ratio of the areas of a cylinder and an inscribed sphere; the measurement of a circle; the properties of conoids, spheroids, and spirals; and the quadrature of the parabola. Informative introduction. clxxxvi+326pp. 5⅜ x 8½.
0-486-42084-1

A SHORT ACCOUNT OF THE HISTORY OF MATHEMATICS, W. W. Rouse Ball. One of clearest, most authoritative surveys from the Egyptians and Phoenicians through 19th-century figures such as Grassman, Galois, Riemann. Fourth edition. 522pp. 5⅜ x 8½.
0-486-20630-0

THE HISTORY OF THE CALCULUS AND ITS CONCEPTUAL DEVELOP-MENT, Carl B. Boyer. Origins in antiquity, medieval contributions, work of Newton, Leibniz, rigorous formulation. Treatment is verbal. 346pp. 5⅜ x 8½. 0-486-60509-4

THE HISTORICAL ROOTS OF ELEMENTARY MATHEMATICS, Lucas N. H. Bunt, Phillip S. Jones, and Jack D. Bedient. Fundamental underpinnings of modern arithmetic, algebra, geometry and number systems derived from ancient civilizations. 320pp. 5⅜ x 8½.
0-486-25563-8

A HISTORY OF MATHEMATICAL NOTATIONS, Florian Cajori. This classic study notes the first appearance of a mathematical symbol and its origin, the competition it encountered, its spread among writers in different countries, its rise to popularity, its eventual decline or ultimate survival. Original 1929 two-volume edition presented here in one volume. xxviii+820pp. 5⅜ x 8½.
0-486-67766-4

GAMES, GODS & GAMBLING: A HISTORY OF PROBABILITY AND STATISTICAL IDEAS, F. N. David. Episodes from the lives of Galileo, Fermat, Pascal, and others illustrate this fascinating account of the roots of mathematics. Features thought-provoking references to classics, archaeology, biography, poetry. 1962 edition. 304pp. 5⅜ x 8½. (Available in U.S. only.)
0-486-40023-9

OF MEN AND NUMBERS: THE STORY OF THE GREAT MATHEMATICIANS, Jane Muir. Fascinating accounts of the lives and accomplishments of history's greatest mathematical minds–Pythagoras, Descartes, Euler, Pascal, Cantor, many more. Anecdotal, illuminating. 30 diagrams. Bibliography. 256pp. 5⅜ x 8½.
0-486-28973-7

HISTORY OF MATHEMATICS, David E. Smith. Nontechnical survey from ancient Greece and Orient to late 19th century; evolution of arithmetic, geometry, trigonometry, calculating devices, algebra, the calculus. 362 illustrations. 1,355pp. 5⅜ x 8½. Two-vol. set.   Vol. I: 0-486-20429-4   Vol. II: 0-486-20430-8

A CONCISE HISTORY OF MATHEMATICS, Dirk J. Struik. The best brief history of mathematics. Stresses origins and covers every major figure from ancient Near East to 19th century. 41 illustrations. 195pp. 5⅜ x 8½.   0-486-60255-9

Mathematics

FUNCTIONAL ANALYSIS (Second Corrected Edition), George Bachman and Lawrence Narici. Excellent treatment of subject geared toward students with background in linear algebra, advanced calculus, physics and engineering. Text covers introduction to inner-product spaces, normed, metric spaces, and topological spaces; complete orthonormal sets, the Hahn-Banach Theorem and its consequences, and many other related subjects. 1966 ed. 544pp. 6⅛ x 9¼. 0-486-40251-7

ASYMPTOTIC EXPANSIONS OF INTEGRALS, Norman Bleistein & Richard A. Handelsman. Best introduction to important field with applications in a variety of scientific disciplines. New preface. Problems. Diagrams. Tables. Bibliography. Index. 448pp. 5⅜ x 8½. 0-486-65082-0

VECTOR AND TENSOR ANALYSIS WITH APPLICATIONS, A. I. Borisenko and I. E. Tarapov. Concise introduction. Worked-out problems, solutions, exercises. 257pp. 5⅝ x 8¼. 0-486-63833-2

AN INTRODUCTION TO ORDINARY DIFFERENTIAL EQUATIONS, Earl A. Coddington. A thorough and systematic first course in elementary differential equations for undergraduates in mathematics and science, with many exercises and problems (with answers). Index. 304pp. 5⅜ x 8½. 0-486-65942-9

FOURIER SERIES AND ORTHOGONAL FUNCTIONS, Harry F. Davis. An incisive text combining theory and practical example to introduce Fourier series, orthogonal functions and applications of the Fourier method to boundary-value problems. 570 exercises. Answers and notes. 416pp. 5⅜ x 8½. 0-486-65973-9

COMPUTABILITY AND UNSOLVABILITY, Martin Davis. Classic graduate-level introduction to theory of computability, usually referred to as theory of recurrent functions. New preface and appendix. 288pp. 5⅜ x 8½. 0-486-61471-9

ASYMPTOTIC METHODS IN ANALYSIS, N. G. de Bruijn. An inexpensive, comprehensive guide to asymptotic methods–the pioneering work that teaches by explaining worked examples in detail. Index. 224pp. 5⅜ x 8½ 0-486-64221-6

APPLIED COMPLEX VARIABLES, John W. Dettman. Step-by-step coverage of fundamentals of analytic function theory–plus lucid exposition of five important applications: Potential Theory; Ordinary Differential Equations; Fourier Transforms; Laplace Transforms; Asymptotic Expansions. 66 figures. Exercises at chapter ends. 512pp. 5⅜ x 8½. 0-486-64670-X

INTRODUCTION TO LINEAR ALGEBRA AND DIFFERENTIAL EQUATIONS, John W. Dettman. Excellent text covers complex numbers, determinants, orthonormal bases, Laplace transforms, much more. Exercises with solutions. Undergraduate level. 416pp. 5⅜ x 8½. 0-486-65191-6

RIEMANN'S ZETA FUNCTION, H. M. Edwards. Superb, high-level study of landmark 1859 publication entitled "On the Number of Primes Less Than a Given Magnitude" traces developments in mathematical theory that it inspired. xiv+315pp. 5⅜ x 8½. 0-486-41740-9

CALCULUS OF VARIATIONS WITH APPLICATIONS, George M. Ewing. Applications-oriented introduction to variational theory develops insight and promotes understanding of specialized books, research papers. Suitable for advanced undergraduate/graduate students as primary, supplementary text. 352pp. 5⅜ x 8½.
0-486-64856-7

COMPLEX VARIABLES, Francis J. Flanigan. Unusual approach, delaying complex algebra till harmonic functions have been analyzed from real variable viewpoint. Includes problems with answers. 364pp. 5⅜ x 8½. 0-486-61388-7

AN INTRODUCTION TO THE CALCULUS OF VARIATIONS, Charles Fox. Graduate-level text covers variations of an integral, isoperimetrical problems, least action, special relativity, approximations, more. References. 279pp. 5⅜ x 8½.
0-486-65499-0

COUNTEREXAMPLES IN ANALYSIS, Bernard R. Gelbaum and John M. H. Olmsted. These counterexamples deal mostly with the part of analysis known as "real variables." The first half covers the real number system, and the second half encompasses higher dimensions. 1962 edition. xxiv+198pp. 5⅜ x 8½. 0-486-42875-3

CATASTROPHE THEORY FOR SCIENTISTS AND ENGINEERS, Robert Gilmore. Advanced-level treatment describes mathematics of theory grounded in the work of Poincaré, R. Thom, other mathematicians. Also important applications to problems in mathematics, physics, chemistry and engineering. 1981 edition. References. 28 tables. 397 black-and-white illustrations. xvii + 666pp. 6⅛ x 9¼.
0-486-67539-4

INTRODUCTION TO DIFFERENCE EQUATIONS, Samuel Goldberg. Exceptionally clear exposition of important discipline with applications to sociology, psychology, economics. Many illustrative examples; over 250 problems. 260pp. 5⅜ x 8½.
0-486-65084-7

NUMERICAL METHODS FOR SCIENTISTS AND ENGINEERS, Richard Hamming. Classic text stresses frequency approach in coverage of algorithms, polynomial approximation, Fourier approximation, exponential approximation, other topics. Revised and enlarged 2nd edition. 721pp. 5⅜ x 8½. 0-486-65241-6

INTRODUCTION TO NUMERICAL ANALYSIS (2nd Edition), F. B. Hildebrand. Classic, fundamental treatment covers computation, approximation, interpolation, numerical differentiation and integration, other topics. 150 new problems. 669pp. 5⅜ x 8½. 0-486-65363-3

THREE PEARLS OF NUMBER THEORY, A. Y. Khinchin. Three compelling puzzles require proof of a basic law governing the world of numbers. Challenges concern van der Waerden's theorem, the Landau-Schnirelmann hypothesis and Mann's theorem, and a solution to Waring's problem. Solutions included. 64pp. 5⅜ x 8½.
0-486-40026-3

THE PHILOSOPHY OF MATHEMATICS: AN INTRODUCTORY ESSAY, Stephan Körner. Surveys the views of Plato, Aristotle, Leibniz & Kant concerning propositions and theories of applied and pure mathematics. Introduction. Two appendices. Index. 198pp. 5⅜ x 8½. 0-486-25048-2

TENSOR CALCULUS, J.L. Synge and A. Schild. Widely used introductory text covers spaces and tensors, basic operations in Riemannian space, non-Riemannian spaces, etc. 324pp. 5⅜ x 8¼.   0-486-63612-7

ORDINARY DIFFERENTIAL EQUATIONS, Morris Tenenbaum and Harry Pollard. Exhaustive survey of ordinary differential equations for undergraduates in mathematics, engineering, science. Thorough analysis of theorems. Diagrams. Bibliography. Index. 818pp. 5⅜ x 8½.   0-486-64940-7

INTEGRAL EQUATIONS, F. G. Tricomi. Authoritative, well-written treatment of extremely useful mathematical tool with wide applications. Volterra Equations, Fredholm Equations, much more. Advanced undergraduate to graduate level. Exercises. Bibliography. 238pp. 5⅜ x 8½.   0-486-64828-1

FOURIER SERIES, Georgi P. Tolstov. Translated by Richard A. Silverman. A valuable addition to the literature on the subject, moving clearly from subject to subject and theorem to theorem. 107 problems, answers. 336pp. 5⅜ x 8½.   0-486-63317-9

INTRODUCTION TO MATHEMATICAL THINKING, Friedrich Waismann. Examinations of arithmetic, geometry, and theory of integers; rational and natural numbers; complete induction; limit and point of accumulation; remarkable curves; complex and hypercomplex numbers, more. 1959 ed. 27 figures. xii+260pp. 5⅜ x 8½.
0-486-63317-9

POPULAR LECTURES ON MATHEMATICAL LOGIC, Hao Wang. Noted logician's lucid treatment of historical developments, set theory, model theory, recursion theory and constructivism, proof theory, more. 3 appendixes. Bibliography. 1981 edition. ix + 283pp. 5⅜ x 8½.   0-486-67632-3

CALCULUS OF VARIATIONS, Robert Weinstock. Basic introduction covering isoperimetric problems, theory of elasticity, quantum mechanics, electrostatics, etc. Exercises throughout. 326pp. 5⅜ x 8½.   0-486-63069-2

THE CONTINUUM: A CRITICAL EXAMINATION OF THE FOUNDATION OF ANALYSIS, Hermann Weyl. Classic of 20th-century foundational research deals with the conceptual problem posed by the continuum. 156pp. 5⅜ x 8½.
0-486-67982-9

CHALLENGING MATHEMATICAL PROBLEMS WITH ELEMENTARY SOLUTIONS, A. M. Yaglom and I. M. Yaglom. Over 170 challenging problems on probability theory, combinatorial analysis, points and lines, topology, convex polygons, many other topics. Solutions. Total of 445pp. 5⅜ x 8½. Two-vol. set.
Vol. I: 0-486-65536-9   Vol. II: 0-486-65537-7

Paperbound unless otherwise indicated. Available at your book dealer, online at **www.doverpublications.com**, or by writing to Dept. GI, Dover Publications, Inc., 31 East 2nd Street, Mineola, NY 11501. For current price information or for free catalogues (please indicate field of interest), write to Dover Publications or log on to **www.doverpublications.com** and see every Dover book in print. Dover publishes more than 500 books each year on science, elementary and advanced mathematics, biology, music, art, literary history, social sciences, and other areas.